STANDARD
PLANT
OPERATORS'
MANUAL Second Edition

*A basic reference on the operation and maintenance
of energy systems equipment, complete with
license examination questions and answers*

STANDARD PLANT OPERATORS' MANUAL Second Edition

Stephen Michael Elonka

Contributing Editor, Power magazine. Licensed Chief Marine Steam Engineer, Oceans, Unlimited Horsepower.
Licensed as Regular Instructor of Vocational High School, New York State.
Member: Instrument Society of America, National Association of Power Engineers (life, honorary).
Author: The Marmaduke Surfaceblow Story. Co-author: Standard Plant Operator's Questions and Answers, Volumes I and II; Standard Refrigeration and Air Conditioning Questions and Answers; Standard Instrumentation Questions and Answers, Volumes I and II; Standard Electronics Questions and Answers, Volumes I and II; Handbook of Mechanical Packings; Standard Industrial Hydraulics Questions and Answers; Standard Boiler Operators' Questions and Answers; Electrical Systems and Equipment for Industry

McGRAW-HILL BOOK COMPANY

New York   St. Louis   San Francisco   Auckland   Düsseldorf
Johannesburg   Kuala Lumpur   London   Mexico   Montreal
New Delhi   Panama   Paris   São Paulo   Singapore
Sydney   Tokyo   Toronto

Library of Congress Cataloging in Publication Data

Elonka, Stephen Michael.
    Standard plant operators' manual.

    Published in 1956 and 1965 under title: Plant
operators' manual.
    1. Power-plants. 2. Plant maintenance. I. Title.
TJ164.E4   1974   621.4   73-14983
ISBN 0-07-019296-0

3 4 5 6 7 8 9 0   HDMA   7 9 8 7 6

PREFACE

Our energy crisis makes this book a *must* for everyone responsible for squeezing more Btu out of each ounce of fuel in his power plant. This second edition, greatly expanded and revised, provides more complete information on safe and efficient operation and maintenance of boilers, air conditioning systems, electrical and many other major pieces of energy systems equipment than did the revised edition. New pages of questions and answers have been added to round out chapters.

Never before has so much practical information on plant operation and maintenance been crammed into one book; the cream of my most popular *basic* articles, and also some I edited for several authorities in my many years with *Power* magazine.

The 1,968 illustrations given here will help more than volumes of words to make clear the many practical operations needed to keep power services shipshape. This book will help the reader to pass examinations for steam operating engineer's licenses, both stationary and marine, and for passing refrigeration and air conditioning examinations, as well as for operating diesel engine plants. Thus, it will round out any library used for training, operating, and maintaining energy systems equipment.

Plant and power engineers write me frequently, asking for study material for their operators and maintenance men. Now, here is a volume that fits well into training programs. Firms will profit by placing a copy of *Standard Plant Operators' Manual* in the hands of every employee who is entrusted with their expensive machinery. Mistakes cost money, especially in lost production and repair bills. Any one of these chapters should help technicians advance to higher pay.

Steve Elonka

Acknowledgement

In addition to the numerous manufacturers credited in the individual articles, I especially thank the following: F. J. Ardolino, J. H. Braverman, R. H. Brenner, L. J. Cormack, B. Emerson, F. E. Graves, K. J. Karassik, R. J. Kembel, A. L. Kohan, R. H. Marks, H. B. McDermid, J. R. McDonnell, W. E. Mooney, J. Murphy, J. J. O'Connor, E. J. Parker, J. R. Sears, J. Todd, and W. B. Warner for giving permission to reproduce a few pages of material I handled for them in *Power* magazine.

Steve Elonka

CONTENTS

1
Boilers, steam equipment

1 Pipes out in the open and easily seen can be traced without trouble, but don't overlook those hidden in trenches just because they're hard to get at Lift the cover plates so you can follow the pipes. Traps are often hidden in trenches

Training Yourself to Be a Good Operating Engineer

It takes time and effort to get thoroughly acquainted with a plant, but such know-how pays off when things go haywire. Learn what makes things tick by following a plan like this

▶ AT A National Power Exposition held recently, Jenkins Bros exhibited the model of a power plant to show where and how their valves can serve the power engineer. Let's see if we can't make the model serve equally well to demonstrate for young engineers how major items connect together in a modern power plant. Here, in the photographs, is a complete hookup without obstructing floor beams, building columns, or floor trench plates. Everything is out in the open where pipe runs can be traced without confusion, Fig. 1.

Sketch Each Line. Now you might ask, why spend time tracing pipe runs?

The answer is simple: To be a good operator you must know every valve and switch in the plant, be able to locate each one instantly, and understand how its manipulation affects individual pieces of equipment as well as the plant as a whole. Piping diagrams and blueprints, easily understood by an experienced engineer, only confuse a beginner.

The easiest, quickest and soundest way to get to know a pipe run is to arm yourself with a writing board, paper and pencil, and sketch each line as you trace it out physically. Follow each branch under floor plates and behind

beams so every tap-off and valve can be put on your sketch.

One-line diagrams are sufficient: use an X to represent valves. Completeness is more important than a fancy drawing. Get the details down on paper and save finishing touches until later.

Drain and Vent Lines. Don't overlook drain and vent lines. In one plant, it was common practice to drain an outside fire line before cold weather set in. After being asked to drain this line, the workman reported the job completed. When tested a month later the pipe was found frozen, and every section and branch valve broken. The man

2 Learn the purpose of every valve connected in a pipe line. There is a reason for its being there. When opportunity arises, study its internal construction

Condensate Line. Going back to the fuel-oil heaters *A* and *B*, Fig. 2, trace condensate line to traps *L* and *M* and main discharge line *N*. Below *M* we find another trap *O*, which drains the heating coil fed by valve *H*. The circuit inside vault *I* should be investigated at the first opportunity. Now surprise yourself by seeing how easy it is to sit down and draw the layout without referring to your original sketch of the actual piping.

Having in mind a clear picture of the system, what shall we look for if the furnace flame begins to show sparks and smoke streamers? A quick look at the oil thermometer shows a temperature drop. What has happened? If you can't eliminate the trouble instantly, best move is to cut in the spare oil heater first thing to save time. Then start looking for the trouble.

Fig. 4, next page, shows a close-up view of Fig. 2. Starting with main stop valve *A*, we find it open. Next check pressure gage *B*. If reading is correct we know reducing valve *C* is working and the associated manually-operated valve properly set.

Suppose the oil temperature now shows signs of returning to normal. This points an accusing finger at the condensate trap on the main oil heater or tank coil unless someone has accidentally closed one or more valves at these units. Regardless of which trap is giving trouble, cutting in the spare heater provides sufficient capacity to carry the load.

But suppose the oil temperature still remains low. Next place to look for the trouble is temperature-regulating valve *D*, Fig. 4. Check it by cracking open its bypass valve and listening for steam hissing through. Another place to look for trouble is the condensate line. Has the discharge been obstructed?

Trouble Shooting. To become a competent trouble shooter, you must know your equipment. Having gained this knowledge, you can spot trouble without wasting valuable time by starting methodically at one end of the equipment and working to the other end.

For instance, suppose you know definitely that no one tampered with any valves and no cold oil was added to the fuel tank in the vault when oil temperature began to drop. This confines possible trouble spots to the pressure-reducing valve, temperature-regulating valve, the two traps then in service, and the condensate discharge line. A glance at the pressure gage either indicates or eliminates the pressure-reducing valve, and opening the bypass checks the temperature-regulating

had shut the main supply valve, but had forgotten to open the drain and vent valves. The line did not empty. Fortunately, no fires occurred in the meantime.

Depending on plant location, floor drain lines can introduce operating hazards if the operator is not acquainted with this built-in protection. Where there is danger of flood water backing up into the plant, valves are, or should be, installed where possible or plates provided so floor openings can be closed.

Such protection is valueless if the person responsible for applying it has not concerned himself with learning and practicing the job of preparing for high water. Overlooking one opening can prove extremely embarrassing to say the least.

Small Troublemakers. As is usually true, small things cause trouble oftener than big ones. Seldom is a turbine shut down by accidental closing of the main steam-header stop valve. Connection from header to turbine is simple, with-

out confusing branch lines, therefore easy to remember.

On the other hand, closing the stop valve in a cooling-water line that feeds several pieces of equipment can prove disastrous if such machines as running air compressors or turbine-bearing oil coolers are overlooked. Damaged bearings and scored compressor cylinders are hard to explain away if they result from carelessness.

Tracing a Pipe Run. Tracing out a pipe run is easy. Consider the steam supply for fuel-oil heaters *A* and *B* in Fig. 2. We know this steam comes from the boiler, so we follow both steam lines *C* and *D* back and find they come together at *E*. Just beyond junction *E* is the temperature-regulating valve *F* with its shutoff and bypass valve, and a branch line *G* with its temperature-regulating valve *H*, which feeds a set of oil-tank heating coils inside vault *I*. Entire system is fed by pressure-reducing valve *J* supplied by stop valve *K*.

3 Become familiar with boiler valves. Main steam valve usually operates automatically to stop flow into a burst header or backflow because of a ruptured boiler tube. Feedwater line always contains a check valve near the boiler

4 The study of automatic regulating valves is a subject in itself. Learn their purpose in a system and what care they need to keep them working smoothly

comes from, Fig 3. Feedwater from heater A comes through line B and interconnected valves, and enters the boiler drum at C. Saturated steam leaves the drum and travels through line D to the far side of the boiler where it enters the superheater. Leaving this unit through valve E, steam enters the ring header connecting the two boilers. Oil-heater supply comes off at F and turbine takeoff is at G.

A ring header such as this can get you into trouble if the various valves are manipulated improperly. Best precaution is to check and double-check your actions if, upon closing a certain valve, steam is heard hissing through as the valve begins to throttle near closed position. This can mean that steam to some piece of running equipment is being shut off because you have left a valve closed that should be open.

Another Caution: Never get into the habit of spinning a valve open or closed. Take it easy. On large manually operated valves fitted with a bypass, open the bypass and let downstream pipe get up to pressure before opening the main valve. On the reverse operation, have the bypass open before closing main valve; then close bypass.

Remember that tracing pipe runs is only one phase of learning power-plant operation, but it is just as important as any other. Another important step is to study every piece of equipment in the piping system. Learn what makes a pressure-reducing or temperature-regulating valve work, and what trouble signals to look for.

valve. Thus within a minute everything up to the trap and condensate line can be checked.

Effect of cutting in the auxiliary heater should soon appear on the oil temperature to indicate whether condensate is getting away from the main heater. Accurate thinking and quick action in a case of this kind mean the difference between a plant shutdown and a minor incident.

Quick action, without taking time to think, or without knowing what to think about, invites disaster. Suppose upon noticing the flame action you immediately attempt to cure the trouble by throttling oil flow to the burner. This appears to help, but you have failed to check all the angles, and meanwhile the oil is getting cooler. Eventually the flame cannot be maintained. Result is that steam pressure falls, load is dropped, and the plant has to be shut down completely.

More delay occurs before the trouble is finally located and the oil brought back to atomizing temperature so the fire can be relighted. Such avoidable interruptions make a black mark on an engineer's record.

Tracing Steam Travel. While on the subject of steam, let's see where it

1 The tubes should be cleaned by blowing the soot with hand lance or automatic soot blowers. For thorough job, use both automatic soot blowers and lance

2 The boiler should be emptied by opening the bottom blowoff valves only after the pressure reaches zero on the gage. Then your boiler is safe to dump

Steps In Preparing Boilers For Inspection

Water-tube boilers need a lot of attention before inspection time. Tubes must be blown, boiler drained and washed. Clean the fireside thoroughly also, so the inspector can find flaws and defects easily. A good engineer also cleans his boiler room before the inspector arrives. Things to look for in boiler are oil, scale, and corrosion. Follow these steps carefully for best results

► MOST STATES AND CITIES require that all steam boilers of a certain size and operated above 15 psi be inspected internally at least once a year. This inspection is made either by a state or city boiler inspector, or by an insurance company inspector authorized for that territory. In some cities, the municipality requires their own inspection, entirely independent of any insurance firm's inspection.

These "double inspections" don't happen often. When they do, both inspectors usually arrange to be present on the same day.

When the inspection date has been set it is the engineer's duty to have the boiler opened and ready at the appointed time. Preparation is important, so follow these steps to save time.

Blowing Tubes. Before taking the boiler out of service, clean the soot from the fireside of the boiler. This is usually done by blowing air or steam against the tubes, either with a hand lance or automatic soot blowers, Fig. 1. Most old boilers are cleaned with a hand lance while newer ones have automatic soot blowers.

Blowing tubes with wet steam is bad for the tubes. The moisture cakes the soot, plugging the spaces between tubes. This not only stops up the gas passages and cuts down the tube heat absorption surface, but forms sulphuric acid, which corrodes the tubes.

To avoid this, use superheated steam if possible. In any case, make sure all condensate is drained before using the soot blowers by opening the drain valve on the header.

Draining Boiler. After taking the boiler off the line and blocking the fuel supply, close all fire and ashpit doors, dampers, and other air inlets. Allow the unit to cool slowly.

When the pressure drops to zero on the gage, and a reasonable time has passed to allow the boiler and furnace walls to cool, open the blowoff valves so the water drains to the blowoff tank, Fig. 2.

Caution: Never empty a boiler through the blowoff connections while it shows steam pressure. That's because the water protects the shell and tubes against heat stored in the furnace brickwork. Otherwise, unequal strains might set in, causing riveted joints and tube ends to leak.

3 Wash drums and tubes with a hose before sediment has chance to harden. That makes job easy

4 Be sure to remove all handhole plates on header boilers for tube inspection for the inspector

5 The steam-drum water-level line is the best place to check for traces of oil. Rub your hand along shell

Washing Out Boiler. After opening the water-side, wash out the shells and tubes with a spray of high-pressure water. Be sure all tubes are washed thoroughly, Fig. 3.

If it is a header-type boiler, either sectional or solid, remove all handhole caps for washing and examining the tubes for scale, Fig. 4.

Caution: Wash immediately after draining the water from the boiler. This prevents any heat contained in the brick setting or furnace walls baking the soft muddy scale to the shell and tubes.

Look for Oil. Check along the steam-drum water level for signs of oil. It's usually hard to detect, especially if only a very small amount is present. One sure way to tell is to run the back of your finger along the water line, Fig. 5. If your fingers stain and the stain can't be washed off with soap and water, oil is getting into the boiler and that's a dangerous condition in any boiler.

Oil enters with feedwater. Then it's distributed throughout the boiler by circulation. Because it's lighter than water, it gradually rises and forms a scum along the water level. The real danger comes from tiny solid particles sticking to the oil before it adheres to the drum sides. Gradually this weighted oil settles to the heating surfaces. That's when tubes blister, or fail completely.

Removing Scale. Examine the inner shell and tubes for scale. If it's heavier

6 Boilers of this type have access doors for fireside inspection, cleaning, and maintenance. Check baffles and dampers

7 Turbine the boiler tubes for inspection. If tubes show no trace of scale, turbine them at least once in five years

8 If the mud drum is insulated, as in this boiler, be sure to remove the insulation from drum ends so metal is visible. Then inspector can do a better job for you

9 Open gage-glass connections and renew fusible plugs before the inspector arrives. He wants them in top-notch shape so you do not get into trouble

than "eggshell' thickness, remove it. This is done by hand or mechanical means, or both.

If you use the right feedwater treatment, the thin scale should peel off easily by running a hand scraper lightly over the metal. The tubes should be turbined, Fig. 7. Even if they are free of scale, it's good practice to turbine tubes at least once every five years.

If the boiler has been neglected and is badly scaled, the drums will have to be hand-chipped. Chipping is a long and tough job. It requires patience. If you're pressed for time, several concerns are equipped to remove all scale completely by the acidizing process.

Cleaning Fireside. Enter the combustion chamber through the access doors and clean the soot from tubes, mud drum, blowoff lines and combustion-chamber floor, Fig. 6.

Next enter the furnace and wire-brush the lower bank of tubes. Be sure to examine the fire arch and furnace side walls for needed repairs and leaks. Then open all access doors in the side-wall settings, and remove the soot from tubes and steam drum. While you're in there, examine the baffles for breaks. Make sure they haven't shifted.

In bent-tube boilers, remove the insulation from the outside surfaces of both mud-drum heads and between the brick setting and the heads, Fig. 8. The ends are now ready for a complete inspection.

That's important because these drums are subject to severe corrosion. Of course, their condition depends on the upkeep they get and the boiler's age.

Fusible Plugs. Remove the cross tee plugs from the water-column connecting piping so the lines can be examined, Fig. 9. They often get plugged and cause a false gage-glass reading. Running a brush through them every time the boiler is cleaned internally is good practice.

If the steam drums have fusible-plugs, renew them. This should be done once a year, regardless of their condition.

Your boiler is now in good shape and it's ready for the boiler inspector. It is your duty to give the inspector any information that will help him do the job right.

Don't Block Safety Devices

► CERTAIN TYPES of automatic oil-fired boilers have a push-button bypass switch connected in the burner flame-failure circuit. Making a habit of blocking the switch closed when lighting off, instead of holding it by hand, can lead to serious trouble if the attendant walks away and forgets to release the switch.

The danger lies in the fact that rising steam pressure may reduce the oil feed until the flame goes out. But the bypass switch is closed and flame failure does not shut the burner down. What happens? Steam pressure falls and oil flow is restored to a hot fire-box without a lighting-off torch being inserted. Result—a bad furnace explosion and serious injury.

THINGS TO DO

1 Remove soot from tubes
2 Drain water from boiler
3 Remove handhole manhole plates
4 Wash waterside
5 Remove scale from waterside
6 Remove soot from fireside
7 Remove oil from waterside
8 Check boiler accessories
9 Remove grates and check setting

1 Automatic soot blowers remove soot from inside tubes while boiler's hot

2 This combination scale remover and brush does double-duty inside tubes

3 Brush tubes in cold boiler by hand to help inspector find any trouble

9 Scotch-marine-type boiler's corrugated furnace has tough scale above the fire

10 Look for scale in upright boiler on tubesheet and tube ends above fire

11 Rub back of your clean fingers along water line to test for oily deposit

Firetube Boilers . .

. . . need plenty of preparing before the boiler inspector shows up.

▶ REMOVE SOOT FIRST. Whatever type fire-tube boiler you have, blow tubes before removing boiler from service.

Use steam or air, but don't blow moisture on tubes. Hand lance, automatic soot blower, Fig. 1, or both are needed. If boiler is "dead" Fig. 3, blow tubes, brush, scrape, or do all three. Brush and scrape if soot is caked. Combination tube scraper and brush, Fig. 2, cuts hardened soot loose and brushes tube at same time.

Always drain condensate from steam before blowing tubes. Soot caked on tubes causes corrosion.

Draining Boiler. Caution! Don't drain water from boiler before boiler setting has cooled enough to prevent damage, Fig. 4.

Fire doors and dampers open while cooling cause unequal expansion, mak-ing riveted seams leak. Keep them closed.

Remove Plates. Remove all manhole and handhole plates to wash waterside.

ASME Code specifies all boilers must have manholes, handholes, or other openings for examining and cleaning boiler's inside. Insert high-pressure spray hose if it's impossible to enter waterside.

Start washing as soon as plates are off. Then soft scale is mushy and easy to wash out. If allowed to dry, it hardens and makes removal tough.

Removing Scale. Run scraper lightly over softer deposits in hrt boiler. Remove hard scale with hand chipping hammer or mechanically. In either case, don't injure boiler material.

Look for scale on crown sheet in firebox boiler, Fig. 5. In hrt boiler,

4 Never empty hot boiler until furnace settings have cooled. Empty boiler heats up unevenly and riveted seams will leak

5 Scale forms in this firebox boiler on crown sheet above fire. Remove scale thoroughly with scraper or wirebrush

6 Wash small upright boiler through these handholes   **7** Remove all manhole plates and clean **waterside well**

8 Scale in the boiler is heaviest directly above the fire. Remove all scale so boiler inspector knows metal's condition

12 Check water column **a**, internal feed line **b**, fusible plug **c**, safety valve **d**, dry pipe **e**, and blow-off pipe, **f**

13 Remove grates **a**, and check bridge wall **b**, setting **e**, blow-off piping **d**, and clean combustion chamber **c** thoroughly

---

Fig. 8, look on lower shell. In scotch marine, Fig. 9, look above corrugated furnace. In vertical boiler, Fig. 10, look on crown sheets. Trouble-spot is directly above fire in these boilers. Remove *all* scale.

Removing scale from fire-tube boilers is a tedious job. Tubes may be rattled with pneumatic "rattler" from fireside. Or flat metal pushed back and forth between tube banks. If scale is heavy, it may be worth while to cut out and remove all tubes, then scale shell by hand.

Oil In Boiler. Oil is a non-conductor of heat so remove it from waterside. In boiler, oil eventually causes overheating and plate bulges when it gets in with feed water. Exception is new boilers not cleaned properly. Usually oil is in tiny particles over

heating surface, or an emulsion along water level, Fig. 11.

Check water gage glass for oil while boiler's steaming. If boiler "foams" open and check for oil. Then remedy the cause at once.

Boiler Accessories. (1) Remove plug *a*, on water column across tees, Fig. 12. Check for plugged connecting lines by poking rod through. (2) Check internal feed line, *b*, for holes. See that support is okay and line's end is not plugged with scale. (3) Examine fusible plug, *c*, and renew annually. (4) Test safety valve, *d*, for working condition. See that blow-off opening isn't plugged. (5) Examine dry pipe, *e*, for holes or slots being plugged.

Furnace and Fire Side. (1) If coalfired boiler, Fig. 13, remove ashes from grates, *a*, and from combustion cham-

ber. If internally-fired boiler, such as vertical-tubular, scotch-marine, and locomotive-firebox, remove grates. (2) Check furnace walls and bridge-wall condition, *b*, boiler setting, *e*, brick pier of hrt boiler for protecting blow-off line *d*. (3) Remove blowoff-pipe insulation, *d*, if exposed to combustion, whether rope asbestos, cast-iron sleeve, or fire brick. (4) Brush soot from both tube sheets and any portion of shell exposed to combustion. (5) Remove all soot in combustion chamber, *c*. (6) If small "package-type" scotchmarine boiler, where furnace cannot be entered from front, remove rear plate for access to furnace and rear tube sheet.

By doing these things before the boiler inspector arrives, you will get the inspection you should have.

Declare War Now On

WHY TUBES ARE ROLLED

BOILER TUBES are expanded into tube sheets to seal opening around tube and to stay tube firmly in its position

NEW WATERTUBE BOILER, has about 1/32-in. clearance between tube and tube-sheet hole at both ends of the tube

FIRETUBE BOILER tubes become scaled on outside. To help remove tube, clearance is 1/16 in. at end shown

HOW TO REMOVE AND ROLL

TO REMOVE TUBE, chisel off flare, flush with tube sheet. Then split tube with diamond point chisel like this

CUT BOTTOM-ROW TUBES inside tube sheet with cutter so they drop into furnace. Collapse and remove ends

SUPPORT LONG TUBE with pipe. If tube is badly scaled, drill hole in collapsed end and pull with chainfall

DON'T OVERROLL

OUTSIDE FERRULE fills loose space around enlarged holes. Buy, or make from thin copper sheet; use as shown

UNDER-ROLL, never over-roll tubes. Roll slowly and STOP when tiny flakes start leaving tube sheet around the tube

OVER-ROLLING enlarges and deforms tube sheet. Tube is thinned and weakened inside tube sheet to a dangerous point

Sloppy Tube Rolling!

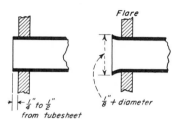

ASME CODE requires ends of watertube boiler tubes extend ¼ to ½ in. and flared ⅛-in. dia larger than tube

FLARING is usually done with tube expander that has belling or flaring rolls, with skirt to control flare depth

EXPANDERS are hand or motor operated. Hand rolling is harder work but there's less chance of damaging costly tubes

BEFORE INSTALLING new tube, check tube-sheet hole for damage. Stick in new tube and extend ¼ to ½ in. on ends

CUT TUBE at other end with inside tube cutter. If no cutter, mark tube in place (remove before wedging), and saw

TO HOLD TUBE in place while rolling one end, drive thin tapered brass wedge between tube and tube-sheet hole

OVER-ROLLING forces tube sheets apart. Long tubes of small diameter bow, then collect sediment and bag-fail

OVER-ROLLING 'work hardens' inside of tube at tube sheet. Hardness makes metal brittle and causes early failure

OVER-ROLLING causes surface tears inside rolled part. Tears weaken tube in time and again cause early failure

Now You Can Slash

1 Save expensive labor by doing one thing at a time. Start at top row and loosen all of the nuts with special wrench

2 Next operation on each plate is to remove nut, take off dog and screw nut on a few turns, to protect threads

3 Start at top row again and knock in each plate with one good smack near top of nut. Work fast, no lost motion

7 Brush graphite and cylinder oil mixture over threads. Run nut all the way down and out to within a few threads

8 Fill bucket with water and dump soft handhole gaskets in it. Water swells gaskets and helps the graphite stick

9 Work powdered graphite over wet gasket. DON'T USE OIL—IT HARMS BOILER'S WATER SIDE. Place on gasket

13 Start from top and smack each nut from four sides to help plate find its right position. Then tighten nut

14 Fill boiler slowly with feed pump. As boiler fills, take up on each leaky plate just enough to stop leak

15 If plate still leaks, loosen nut and smack four sides, then tighten. If leak won't stop, pump water below plate

Handhole Plate Time

4 Now grab each plate's stud with one hand, turn plate slightly, remove from hole and throw into one pile on floor

5 Grab plate by its stud and slide on piece of floor plate, turning plate so gasket is loosened all around outside

6 Scrape pieces of joint and scale from sealing face of plate. Do only this operation and stack plates in neat pile

10 Next run fingers around on inside of joint surface. This is important because foreign matter can cause leak

11 Scrape broken gasket pieces and other obstructions off with bent scraper. Don't overlook any dirty joints

12 Replace plate, dog and nut. Hold plate so joint is centered and against header. Screw nut to hold plate

16 Remove leaky plate and examine joint for obstruction—also plate might be warped or corroded. Replace it

17 With boiler full and feed pump holding pressure on boiler, go over all handhole plates and tighten nuts

18 While raising steam, paint headers, plates, dogs, nuts and studs with Bunker C fuel oil to prevent rusting

Removing, Rolling, Beading

TAKING OUT OLD TUBE ▶

1 Chisel bead from smoke box end so tube can be removed from tube sheet

2 Collapse end of the tube with "oyster knife" until the expanded part is free

6 Remove the welded tube end and then drive tube out so it can be removed

7 Rolled end may still not be free to clear tube sheet hole. Collapse it

8 Hammer lightly along tube while withdrawing if scale gives trouble

11 With tube marked to extend 3/8 in. from both ends, pull out and wedge

12 With tube wedged to keep it from turning, cut at mark with tube cutter

13 Position the tube in tube sheets and hold it in place with a thin wedge

17 When tube is rolled just enough to make it tight, peen end over lightly

18 Ductile tube is peened over quickly with hammer but finish with this tool

19 Move beading tool about half a tool width after each light hammer blow

And Welding Your Boiler Tubes

3 Burn bead off welded end of same tube with torch. Be sure it's same tube

4 Cut tube all around with torch about two inches inside of the tube sheet

5 Split the collapsed end of tube with acetylene torch if it bulges inside

9 Sand tube-sheet bearing-surface so new tube bears against smooth surface

PUTTING IN NEW TUBE ▶

10 Slide new tube in. You might have to guide other end with a long rod

14 This end extends ⅜ in., is ready for rolling. Wedged tube won't turn

15 Set tube roller so the small flaring rollers are at right place from stop

16 Oil rollers, insert the tool and keep steady push against tool while rolling

20 After rolling tube's other end, this power impact hammer beads it fast

21 Finish rolling after beading to make sure tube didn't loosen from beading

22 Here's finished job. Furnace end of tube is welded to prevent leakage

Keep Round Handhole

Face Seat and Plate

1 With boiler drained, remove nut from handhole plate assembly, so tubes can be turbined, inspected or even replaced

2 Loosen nut only few turns, then remove assembly by tilting as shown. Gaskets are not now used on these plates

5 Check the ground joint for smoothness so the steaming boiler will not leak

6 Remove sludge and scale from inside of the cap with scraper ground to fit

7 Grind cap joint against disk bench grinder that also has flexible shaft

Use One-Piece Tapered Plug

TAPERED PLUGS ARE SIMPLE, HANDY

- One-piece tapered plugs can be used on boilers having master handholes on the headers as shown

- These large holes are needed so plugs can be placed inside, between the header and tube sheet

- Boiler's pressure holds plugs firmly in place. Then header can be insulated to keep heat from escaping

Tube
Tube sheet
Tapered plug
Header

1 Tapered plug fits into hole opposite tube, with room to clear tube sheet

2 Remove plug by tapping lightly with brass maul. Hit caps in center only

6 Before replacing plugs, remove grease by carefully wiping joint with a rag

Graphite paste

7 Coat tapered hole contact-surfaces with an even coat of graphite paste

Plates From Leaking

3 Disk with grinding cloth held on flexible shaft now grinds header joints smooth. Flexible shaft is important here

4 Hold disk firmly against joint while grinding. Spring prevents uneven joint if electric drill isn't held perfectly square

8 Finished job is smooth, level joint that will mate perfectly with header

9 Coat ground joint lightly with graphite paste so it seals joints tightly

10 Corrugated-metal gasket not needed now that both joints are ground

3 Plugs fall between header and tube sheet. Remove from master handhole

4 Grease contact-surfaces of tapered holes to prevent joint from rusting

5 Clean tapered walls of plugs thoroughly by wire-brushing like this

8 Insert plug immediately after removing grease. Give it a slight twist so it seats well. Then tap firmly in place

9 Fill boiler and raise steam to 100 psi. If plug weeps, tap header around plug. Pull up on plug with tool shown

Watch boiler inspector

1 How do you go inside the steam drum? That depends on many things. Here the inspector goes in head first. Take only low-voltage electrical equipment inside

2 Check shell carefully for corrosion, such as pits, by tapping with hammer

6 Small holes in dry-pipe top and tiny drain-holes at bottom must be free

7 Head seams in drum often groove or corrode. Check all suspected spots

8 Openings in drum for water column must be free. Watch lower hole

▶ REGISTERED POWER BOILERS must be inspected at regular intervals, says the National Board Inspection Code. Owner or user has to prepare boiler for inspection when notified by inspector.

Beforehand, boiler must be opened and prepared for internal and external inspection. Caution! In cooling down a boiler, don't dump water until unit has cooled slowly, or you will damage settings and the tubes. Let it cool down naturally, if possible. Remove all manhole plates and other openings.

If deterioration of inaccessible parts is suspected, you may have to remove insulation material, masonry or fixed parts of boiler. Where moisture or vapor shows through covering, remove covering for inspection.

Inspector must get as close to boiler parts as possible with strong light. If attachments or apparatus need testing, plant operator does it in inspector's presence if he orders it.

How to prepare boiler for inspector. 1: Dump water and wash waterside with hose. 2: Remove manhole and handhole plates, wash out plugs and water-column connection. 3: Clean furnace and combustion chambers. blow soot from tubes and all surfaces to be inspected. 4: Remove fuel grates. 5: Remove section of brickwork if ordered by inspector. 6: Remove steam gage for testing. 7: Disconnect pipe or valve if steam or water leaks into boiler.

Certificate of Inspection can be withheld by inspector until boiler is properly prepared. You might have to remove enough jacketing or setting along longitudinal seams of shells, drums or domes so he can inspect rivets, pitch of rivets for data needed in determining safety of boiler. Or you might have to drill suspected thin spots at his request so he

check your boiler's waterside: I

3 Look into all downcomers at drum's rear for scale, also condition of tubes

4 Upcomers under baffles may corrode or have scale deposits. Hammer test

5 Inspect longitudinal seams inside the shell carefully for corrosion as here

9 Check manhole plate for corrosion on seat, also cracks from bellows effect

10 Tubes' inside headers may be heavily scaled. Check lower tubes well

11 Mud-drum header may be badly scaled. Check condition of scale

can gage thickness for possibly reducing steam pressure in older boilers.

Here are inspector's internal checks: Scale, oil, etc. Inspector examines all surfaces of exposed metal inside to check action caused by water treatment, scale solvents, oil or other substances that might have entered with feedwater. Oil inside a boiler is dangerous; you must prevent more entering at once. Oil or scale in tubes of watertube boilers or on plates over fire of any boiler weakens metal, causing bagging or rupture.

Corrosion along or next to a seam is more serious than in solid plate away from seams. Strength of longitudinal joint is less than solid sheet, so thinning here is dangerous. Severe corrosion is likely where water circulation is poor. Inspector checks these places.

Grooving and cracks along longitudinal seams are bad because they usually occur when material is highly stressed. Internal grooving, in fillets of unstayed heads, and external grooving, in outer surfaces of heads concave to pressure, are common because of slight movement in heads.

Stays are inspected for even tension, fastened ends are examined for cracks

where stays are punched or drilled for rivets or bolts.

Manholes and other openings are examined internally and externally for deformity or cracks. Opening to water-column connections, dry pipes and safety valves must be free of obstructions, such as mud, scale, etc.

Ligaments between tube holes in heads of all boiler types often crack, then leak and weaken boiler. Also check beading on tube ends for erosion and corrosion, especially in type of boiler here. Remember, the cleaner the boiler, the better job the inspector does.

Inspect the fire side

1 Check the cross box and down-comers in rear pass for leakage—corrosion

2 Trouble often starts on generating tubes near soot blowers. Look for signs of thinning and erosion from steam impinging on tubes. This is an important check

7 Easiest way to find bowed tubes in many boilers is to use straightedge

8 Inspect overhead baffles and bridge-wall for leakage, tubes for corrosion

6 Look along the tubes for signs of overheating, bulging and also sagging

9 Leakage between tube and wall here can cause a blasting effect on tube

10 Check brickwork for cracks, spalling, and for its general condition

# of your boiler carefully: 2

3 Check position of soot-blower holes for their even spacing between tubes

4 Look into superheater header to see condition of tubes, also the headers

5 Check refractory around burner throat for spalling—for general condition

Here are important trouble spots for you to check

► WE INSPECTED water side of boiler on pages 18 and 19. The article points out that the National Board Inspection Code requires registered power boilers to be inspected at regular intervals. It's up to you to prepare boilers.

After boiler has been emptied, cooled and dumped, open for internal and external inspection. You can't find all defects if boiler is not clean.

Note: Insurance men tell me some of the largest firms won't cooperate in cleaning boiler thoroughly for inspector. No marine engineer thinks of having a marine inspector come aboard until each pass of fire side is blown with hand lance, tubes inside furnace wire-brushed, water side of tubes turbined and boiler thoroughly washed out. Unless boiler is prepared properly, inspection is not made. Why don't insurance firms stick together and refuse to insure a boiler unless it is properly prepared for inspection?

After inspecting water side, inspect fire side thoroughly as shown here.

Bulging, blistering, leaks. Be sure to inspect carefully plate and tube surfaces that are exposed to fire. Look for places that might become deformed by bulging or blistering during operation. Solids in water side of lower generating tubes cause blisters when sludge settles in tubes and water cannot carry away heat. Inspect water side of such de-formed areas, then **remove** whatever solids caused the bulges or blisters.

Bulges and blisters can seriously weaken the plates or tubes. Boiler must be taken out of service until defective part or parts have been properly repaired. Blistered tubes must be cut out and replaced with new, and not hammered into shape.

In vicinity of seams and tube ends, observe boiler structure carefully for leakage. Firetubes sometimes blister, but rarely collapse.

Lap joints, fire cracks. Lap-joint boilers are apt to crack where plates lap in longitudinal or straight seam. If there is evidence of leakage or trouble at this point, you might have to remove rivets or slot the plate to determine whether cracks exist in the seam. Cracks in shell plates are usually dangerous, except fire cracks that run from edge of plate into the rivet holes of girth seams. Usually, a limited number of such fire cracks are not very serious.

Test staybolts by tapping one end of each bolt with a hammer. You get better results by holding hammer or heavy tool at opposite end while tapping.

Tube defects. Tubes in horizontal firetube boilers deteriorate faster at ends toward the fire. Tapping outer surface with light hammer shows if there's serious thinness. Tubes of vertical tubular boilers usually thin at upper ends when exposed to products of combustion. Lack of water cooling is the cause here.

Tubes subject to strong draft often thin from erosion caused by impingement of fuel and ash particles. Soot blowers, improperly used will also thin the tubes. Leaky tube often causes tubes in its vicinity to corrode seriously from water striking soot, causing acid condition. You can lose many tubes in this way.

Short tubes or nipples joining drums or headers lodge fuel and ash, then cause corrosion if moisture is present. Clean, then examine all such places thoroughly.

Baffling in watertube boiler. Check to see if baffling is in place. Combustion gas, short circuiting through baffles, raises temperatures on portions of boiler, causing trouble. Check location of combustion arches for right position with relation to tube surfaces.

Heat localization from improper or defective burner or stoker, or operation causing a blowpipe effect, must be corrected to prevent overheating.

When repairs are made, inspect workmanship. Excessive tube rolling is common. Then tube ends are thinned at tube sheets, tube-sheet openings enlarged—both harmful to boiler. Under-rolling is also common, especially if tubes are hard to get to. Hydrostatic test will show such leaks and leaky tubes can be rerolled (see how to roll and reroll tubes, pp 10, 11, 18, 19).

Now we'll inspect external

1 Safety valve is most important fitting on boiler. Examine spring to see that no foreign material is wedged between coils. Check the general outside condition

2 Valve lever must be in working order; but don't lift valve on dead boiler

6 If stop valve is in water column, install the outside screw-and-yoke type

7 Globe valve must be between drum and check valve, pressure under seat

8 Main stop must work freely. No leak around packing gland or insulation

11 Asbestos rope between the sinuous headers must be tight; no air leaks

12 Blowdown valve and fittings must be shipshape: can shut down boiler

Inspection Code is

▶ WE INSPECTED boiler waterside and fireside on pages 18 to 21 and explained that National Board Inspection Code requires registered power boiler to be inspected at regular intervals. Now let's wind up this job by inspecting the external fittings with Inspector Ed Zeman, of Hartford Boiler Inspection and Insurance Co.

Safety valves are the most important attachments on boiler; so they must always be in perfect condition to work. There should be no rust, scale or foreign matter in casings to hinder free operation. Best way to test setting and freedom of safety valves is by pop-

fittings of your boiler: 3

3 Escape pipe has to be well-bracketed; drain hole clear to carry condensate

4 Superheater safety must blow first to keep flow through the superheater

5 Gage-glass shut-off valves should not bind or leak. Check condition

9 U bolts through channel-iron supports help hold weight of steam drum. Examine bolt threads and everything connected with support to insure first-class order

10 Soot blowers may not work or they may be ready to fail. Inspect well

very firm on these twelve important points—you must know them

ping valves with steam. If this can't be done, test by try-levers. Inspect discharge pipe to make sure it's secure, won't blow away when valve pops. Operators have been killed as valve discharging into fire room fills space with steam in few seconds. Drain-opening in discharge line must not be plugged.

Steam gages have to be removed to test by comparing with a standard test gage. Steam gage must not be located so it's exposed to externally high temperature or to internal heat. Blow out steam pipe leading to steam gage.

Water column. Steam pipe should drain toward column, water-pipe connection should drain toward boiler so water cannot accumulate in steam connection, giving false reading.

Observe relative position of water column to fire surfaces of boiler for conforming to Code requirements for your boiler. Examine condition of attachments.

Suspended boilers. Examine supports where boiler structure comes near setting walls or floor. Make sure ash and soot won't bind boiler structure to produce excessive strains from expansion under operating conditions.

Check-valve and stop-valve in feed-water inlet line must be placed so stop valve is next to drum, with pressure under seat of globe valve.

Main stop must be in working condition, check externally.

Soot blowers need checking for doing their job right. They can cut tubes.

Sinuous headers must be calked or air leaks into the boiler between headers and cools it, raising excess air.

Blowdown valves must work freely, be packed, and all external piping and fittings be inspected to make sure they are not corroded or otherwise damaged.

Study National Board Inspection Code and avoid outage and accidents by keeping your boiler safe—and efficient.

Laying Up Boilers?

DRY METHOD

CORROSION works fast in idle boilers. Water side is attacked if no precautions are taken. Idle boilers are laid up either dry or wet to prevent this damage

DRY METHOD is best if boiler won't be needed for a month or more. First step after boiler is cool is to drain boiler completely dry by opening blowdown valve

CLEAN BOILER'S fire side thoroughly by blowing tubes, removing soot and inspecting thoroughly. Soot on idle boilers collects moisture and corrodes metal

QUICK LIME in wooden tray placed on top of tubes in locomotive-type boiler is next step. Use about 30 pounds of lime for each 100-hp boiler capacity. Place lime inside as soon as boiler has dried out. Then close up boiler tightly

MULTI-DRUM BOILER like this must have wooden tray of quick lime in each drum. Reason is most moisture collects on large drum areas. Lime is very hygroscopic, and absorbs moisture rapidly, before it collects on internal surfaces

WET METHOD

USE WET METHOD if standby boiler is idle only a week. Drain, clean, inspect. Fill to normal level, raise two to five psi steam to vent dissolved gases

MAKE WATER ALKALINE in excess of 400 ppm by three pounds of caustic soda for each 1000 gallons of water in boiler. Then add about 1.5 pounds of sodium sulfite per 1000 gallons of water. Test to see if water has minimum residual sulfite content of 100 ppm. Refer to page 54, Apr. 1, 1948, **OE** for test method

Wet or Dry?

INSPECT WATERSIDE carefully. Enter the steam drum or shell with low-voltage electric light. Remove any corrosion found, with a scraper. Scrape down to bare metal. Make sure valves are tight so no steam or water enters boiler

DRY BOILER next. This is important. One way is to make light wood fire in empty boiler's furnace with thin packing-box boards. Because there is no water in boiler, burn only a few boards at a time. Another way is to use a hot-air stove

HRT BOILERS like this usually need only one wooden tray placed on shell's bottom through manhole. Wooden tray is easy to build, holds enough lime and fits through manhole. Lime is inexpensive, can be bought at most building-supply firms

CLOSE UP BOILER after trays are in place. That means all manhole, handhole and other connections must be tight. If valves leak, blank them off. If boiler is idle for long, open it every three months and renew damp lime with new lime

FILL BOILER with water until it comes out of vent (after boiler cools enough so pressure returns to zero, but before vacuum forms). Close all connections

TEST WATER in idle boiler every week. Keep akalinity and sulfite concentration where it belongs. The colder water is in idle boiler the better for the boiler

BEFORE CUTTING boiler in on line, give blowdown to reduce alkalinity. This prevents carryover. Test water and give blowdowns until water is right

This Spring, Lay Up

HOW BOILER IS KNOCKED OUT

CAST-IRON BOILER corrodes very badly if left dirty in summer. Soot absorbs moisture, forms corrosive sulfuric acid

STEEL BOILER corrodes quickly in the summer if boiler is dirty and is not laid up properly for the idle period

UPTAKES and breechings coated with soot corrode in summer months when there's not enough heat to keep moisture out

HOW TO ADD YEARS OF LIFE

CLEAN boiler by removing ashes and punching tubes with wire brush until all heating surfaces are perfectly clean

REMOVE GRATES if coal burner. Clean grates with wire brush, renew burned or warped grates and swab with lube oil

WIRE BRUSH all heating surfaces of every type boiler. Remove thoroughly all soot and scale from between sections

SWAB all tubes with used lube oil by placing a rag over brush. Then swab the tubesheet thoroughly with used lube oil

REMOVE SMOKE PIPES on smaller boilers. Clean soot, then oil swab inside. Stack in dry place until heating season

GREASE all machined surfaces on both steel and cast-iron boilers. This prevents air leakage past corroded parts

Heating Boilers Right

DAMP BOILER ROOM causes sweating in idle boiler that's filled with water. Then corrosion knocks out dirty boiler fast

CAST-IRON BOILER is attacked by corrosion between sooty sections, at sooty tie-rods, and at machined surfaces of doors

BURNING TRASH in idle boiler is bad. Many boilers are badly damaged from firing with low water or no water at all

REMOVE MANHOLE and handhole plates on steel boilers (cast iron boilers don't have them) after draining off all water

WASH INSIDE with highest pressure water you have soon as boiler is dumped. Then sediment is soft and washes out easily

IF AFTER OPENING and washing large boiler, lime and mineral scale are on tubes outside, remove with this tube cleaner

REMOVE gage glass and wash in household ammonia. Renew washers and make sure the gage glass lines are not plugged

HANG SIGN on boiler saying whether filled or dry. Operators Notebook 35 told how to lay up waterside WET or DRY

CLEANED, oiled and left open for air circulation, boiler rates long life. Cast-iron boiler full; steel boiler usually dry

Look inside your jet-flow

How valve works___

▶ Three types of high-pressure steam-drum safety valves in general use are: (1) jet flow (2) huddling chamber (3) nozzle reaction.

Jet flow makes use of both reaction and velocity of escaping steam. Static pressure on disk overcomes spring tension, causing steam flow. Escaping steam strikes against piston and is deflected downward against nozzle-ring, discharging into body. This reactive force lifts disk higher, thereby increasing area of flow, therefore velocity.

As velocity increases, partial steam discharge takes place through the controlled orifices in guide assembly. Because of its confined velocity, the steam creates an upward force, lifting the valve still higher. Position of disk at "pop" lift is such that area of discharge is greater than area of nozzle; excess accumulation of pressure needed for full lift in huddling chamber and nozzle-reaction types is unnecessary for full capacity in jet-flow valve.

As pressure in boiler falls, velocity through orifices is reduced and valve's disk drops to intermediate lift momentarily. After that, diminishing reactive force controls further reseating, the same as in a reactive valve.

Blowdown adjustment is by an entirely different method from that in other type valves. In this jet-flow-type relief valve, the blowdown is controlled by the location of the exhaust belt reaction-lip, which is varied by adjusting the guide vertically.

Adjusting. If valve opens cleanly but does not seat sharply, nozzle ring is up too high. Lower by removing adjusting ring pin and turning nozzle adjusting ring to left one notch at a time between trials until valve seats sharply with a metallic thud. Adjust ring with screwdriver.

If blowdown is too long, nozzle adjusting ring is too high and must be lowered. If valve warns before popping, nozzle adjusting ring is down too low and must be raised. Turn ring to right one notch at a time until the warning stops completely.

Caution: When raising nozzle-adjusting ring, keep boiler pressure well below popping pressure or gag the valve while making adjustments. To protect personnel, always gag safety valves when adjusting. This is important.

Photo, courtesy Foster Engineering Co

Valve stem nut
Adjusting screw
Lever
Yoke
Spring step
Valve stem
Spring saddle
Deflector
Guide
Guide ring nut
Piston
Adjusting ring pin
Nozzle adjusting ring
Nozzle
Disk
Metering valve

safety valve, keep it working

How to overhaul

1 Remove lifting lever and screw down valve-stem nut. Spring assembly can be removed without changing the setting

2 Now lift out the valve assembly in one piece and turn upside down on work bench so you can lap in the valve disk

3 Lap disk with lapping ring and paste to remove slight depressions. Give it oil rub; clean by dipping in solution

4 Unscrew adjusting ring from nozzle, examine nozzle seat that contacts valve disk. Only perfect seat is steam tight

5 Lap in nozzle seat with lapping ring and paste. Again, give oil rub after lapping and rinse in cleaning solution

6 Replace adjusting ring and measure specified distance for that size from straightedge to lip of the adjusting ring

7 With adjusting ring locked in place and all the internal parts thoroughly cleaned, replace guide for disk piston

8 Valve spring assembly is now ready to go into place. Spring adjustment has not been disturbed and valve is set

9 With valve assembled, check the piston by moving with fingers. Piston must be free for valve to work properly

Here's information on

Castle nut
Lifting disk
Compression screw
Stem
Guide
Disk holder
Disk
Warn ring
Warn ring pin
Nozzle

CLOSED. Pressure is contained in throat chamber, no pressure on conical baffle

How valve works

▶ THREE TYPES of high-pressure steam-drum safety valves in general use are: (1) jet flow (2) nozzle reaction (3) huddling chamber. Jet-flow type was covered in detail in other pages in this chapter.

Nozzle-reaction type. In the opening phase of any safety valve, spring force keeping valve closed must be counter-acted by a greater force on pressure side. But as disk rises, spring load increases from compression. So total force needed to obtain high lift must also increase. In nozzle reaction valve

How to Overhaul

1 Remove seal, yoke pin, cap, yoke, castle nut and lifting disk from valve

2 Measure compression screw height to save time when reassembling valve

7 Measure distance from top of disk-holder guide to top of the valve body

3 Loosen the locknut, screw the com-pression screw out of the valve bonnet

8 Remove warn-ring pin and also the disk-holder guide pin from valve boc'y

nozzle-reaction safety valves

THREE-QUARTERS OPEN while popping. Partial pressure is on the conical baffle

FULL OPEN. Full pressure against enlarged baffle area, full reactive forces

HALF CLOSED. Pressure is against conical baffle on a rapidly diminishing area

shown, energy is produced by a conical baffle skirting the disk. This baffle causes lifting energy to be greatest at or near end of lift, thus giving full-bore lifts or greater at low accumulations.

Opening. As valve opens, steam escaping across seats impinges against conical baffle—then steam also acts against additional in-between areas. This large increase in head-pressure area gives sudden upward force, creating "pop" action. Initial force is increased by dynamic force of jet from nozzle—and impinging against disk face. This force increases rapidly with the flow.

Disk rises quickly from these forces, which constantly increase from increased velocity and with help of enlarging baffle area. Thus port orifice closes almost instantly.

At the same time, another lifting energy source comes into play. Conical baffle turns fluid jet downward, giving thrust-effect as in steam-turbine blades. This reactive force takes up rising compression load of spring. At usual overpressures of 3% for steam, 5% for gases, 25% for liquids, reactive force quickly lifts disk full-open.

Closing. System pressure-drop is usually gradual, with disk falling slowly at first. Jet tends to hold disk open much longer than needed. In this valve,

a new control principle comes into play.

As disk falls in response to system pressure-drop to about 75% of full-open position, ports begin to vent pressure under the baffle. This cancels part of reactive force holding valve open. At same time, decreasing pressure under baffle works against a constantly reducing baffle area. Finally, total upstream forces become less than spring force—and disk closes sharply from about 50% of its rated lift.

When this or any safety valve needs overhauling, it's best to return it to manufacturer. If you must repair valve yourself, follow instructions carefully.

Photos, Courtesy J E Lonergan Co

---

4 Mark warning ring with chalk so you can establish same setting later on

5 Bonnet is not under spring tension now; remove body bolts, lift bonnet

6 Remove stem with disk and holder Disk floats in holder, no misalignment

9 Screw out disk-holder guide. With special tool, remove the steam nozzle

10 If light imperfections on seat or disk, tap in both with lapping block, shown. Remove disk from holder for this job. Never lap disk against seat; use the block

Get acquainted with your

Lifting gear

Flanged yoke

Steel spring

Compression retaining nut

Closing control

Grooved guide

Single ring blowdown control

Thermodisk seat

A

C

D

B

How valve works and other facts you should know

▶ THREE TYPES of high-pressure steam-drum safety valves in general use are: (1) jet flow (2) nozzle reaction (3) huddling chamber.

▶ TWO OF THESE types of safety valves are covered in detail in this chapter. Make sure you know what types you have and keep them in working condition.

Huddling-chamber type. In this valve the static pressure acting on the disk area causes initial opening. As valve pops, steam space within huddling chamber between seat and blowdown ring fills with steam and builds up more pressure on roof of disk holder. This temporary pressure increases the upward thrust against spring, causing disk and its holder to lift to full pop opening. After a predetermined pressure drop (blowdown), valve closes with a positive action by trapping steam on top of disk holder.

Blowdown is adjusted by raising or lowering the blowdown adjusting ring. Raising this ring increases blowdown; lowering decreases blowdown. Capacity figure as stamped on name plate and as required by the ASME Code for steam is 90% of valve's actual capacity when flowing at 3% overpressure.

Huddling-chamber-type valve is intended primarily for steam service but will work with gases. However, it is rarely used for such services as it doesn't have a closed bonnet. It's never used for liquid service.

Precision-closing control. With valve in open position and discharging, steam is bled into chamber A through three bleed holes B in roof of disk holder. Likewise, spindle overlap C rises to a predetermined position above the floating washer D. Area between floating washer and spindle is thereby increased by difference in two spindle diameters.

Under this condition, steam in chamber A escapes to the atmosphere through the secondary area formed by floating washer D and the spindle.

At instant of closing, spindle overlap C is so adjusted that it moves down into floating washer D, thereby effectively reducing escape of steam from chamber A.

Resulting momentary pressure buildup in chamber A produces a downward thrust in direction of spring loading. Combined thrust of pressure and spring loading results in tight, positive precision closing without wire drawing or dragging of seat surfaces.

—*Courtesy: Manning, Maxwell & Moore, Inc*

huddling-chamber safety valve

HOW TO OVERHAUL ⟶

1 Remove cap and lever assembly, screw compression retainer against spring

2 Remove bonnet, then lift the spindle guide and feather assembly from valve

3 Unscrew the feather (valve disk) from spindle. Hold spindle; turn the disk

4 Check blowdown's ring's position by screwing the ring up against disk

5 Insert power-driven abrasive-cloth recutting tool through feather guide

6 Regrind seat if necessary, using an electric drill, as shown, for speed

7 After grinding disk on seat, adjust the spindle overlap to keep the right back pressure on the disk for positive closing. Use the scale as illustrated here

Install Gage Glasses

1 Glass thins mostly above water line; also breaks with fittings out of line

2 First shut off top and bottom cocks with chains when the glass breaks

3 Then open drain valve on column that is used for blowing down the glass

4 Remove broken glass, then open cocks from floor to blow out broken pieces

5 Cut glass to right length by revolving between cutters until it breaks

6 Place top washer, and then nuts (not shown); bottom washer; then install

7 Add graphite to washer so nut does not twist glass; hand tighten only

8 Heat slowly by blowing steam at top. Open cocks from floor; close drain

Round Glass

LIKE OTHER CERAMICS, glass is brittle, does not deform before failure. It fractures only from tensile stresses, never from shear or compression. Stress-strain curve is straight line to the breaking point.

Glass has high degree of "notch-sensitivity." Tiny surface scratches, bruises or imperfections cause failure. So be careful when handling gage glasses. Keep them in wrapper until ready to install. If you must put them down, place on soft, clean surfaces.

Glass can be heat-treated to raise its endurance limit. This is known as "tempering." First, glass temperature is raised high, then it's chilled. That sets up residual stresses that are tensile in interior and compressive at surface. And it's at surface where tension usually starts. Tempering also tends to reduce harmful effects of surface bruises and scratches. But tubular gage glasses are not as "adaptable" to tempering as flat glasses.

Corrosion is tough on gage glasses—caused by alkalinity and temperature. Most corrosion (thinning) on tubular glasses is above water line at upper half. Here steam condenses and continually flows down glass, eroding it. Boiler water with pH of 11.5, for example, attacks glass 30 times faster than water with pH of 8.5. That shows importance of low pH if you have glass-thinning problem. Rate of attack from temperature increases 100 times from 265 to 500 F. Solution then is to use flat glass, protected by mica so water does not contact glass.

Of course, water in gage is cooler than water in boiler. Some condensate, formed in boiler connection above gage, flows into gage and lowers alkalinity. This condensate dissolves some silica in glass because of decomposition.

Silica content increases as condensate flows down over glass. Rate of corrosion below water level depends on silica remaining in gage water. Rate of water attack above water level is less as distance increases from top end of glass to bottom. Boiler water, not treated for silica removal, corrodes glass.

For over 300 psi, it's best to use flat glass, protected with mica. Pressure limits for tubular glass decrease with glass length. Buckling stress is the result of eccentric column loading.

Right for Long Service

1 Mica, protecting glass from steam and water, turns opaque before failure

2 Remove clamps, glass gaskets, mica. Graphite the threads, run them down

3 Remove old gasket but don't scrape low spots in joint surface of gage

Polish surface perfectly smooth or the tiny high spots will cause failure

5 Clean both ends of gage so scrapings won't plug cocks when back on drum

6 Clean surface thoroughly, oil with molybdenum disulfide so glass slips

7 Place new gasket, mica, new clean glass, another gasket; place clamps

8 Glass mustn't touch metal. Use only molybdenum disulfide on high temp

Flat Glass

ON HIGH-PRESSURE boilers above 1000 psi, there have been many costly gage-glass failures running into thousands of dollars. But fault has been from gage design, as well as from method of installing glass, rough handling and glass imperfections.

At Corning Glass Works, I saw a nail driven into wood with a flat glass. This is possible because high residual stresses push out against the outside high compressive-stress, which was set by tempering. Here are failure causes, ways to prevent:

(1) Never reuse gage glasses. (2) Use special glasses only for high-temperature service. (3) Under extreme pressure, gage metal tends to bulge out, so gage bends if designed poorly—starts a crack because glass won't bend. (4) Scraping gasket off metal raises burr —causing high unit loading. Always polish gage surfaces perfectly smooth. (5) Numerous scrapings to remove gaskets create high and low spots. Level old surfaces. (6) Pulling up unevenly on bolts loads glass unevenly. (7) Asbestos is "lumpy" in gasket, causing high spots. Use only best gasket material.

(8) Many failures occur on high pressures when starting up because glass was not heated gradually. Don't open gage cocks until new glass has chance to heat by conduction. Then crack top-cock only. Also, cool down slowly. (9) Don't blow down oftener than needed. (10) Use two valves for blowing down. Keep one throttled for slow flow through glass, blow with other valve only.

(11) Lubricate contact surfaces of glass with molybdenum disulfide only. This does not crystallize, allows travel between glass and fitting. (12) Don't use cover plate if it has permanent deflection, if it's of poor design, or if clamping-device design is poor. (13) Mica discoloration means that mica failed and water contacts glass. Don't wait for glass to fail—change it.

High pressures and temperatures in newer boilers (around 2000 psi) bring new problems. Flat glasses are strengthened by tempering, but residual stresses or strain set up may release at high temperatures. Over long periods, glass loses much strength that was added by tempering. So when ordering glass, consider pressure and temperature.

Handle Blowoff Valves

ASME CODE REQUIREMENTS

Slow-opening
Extra-heavy pipe
Slow-opening

Slow-opening
Quick-opening

Slow-opening
Plug-cock

PARAGRAPH 308, ASME Code for Power Boilers, requires all boilers carrying over 100-psi working pressure, except traction or portable boilers, to have two blowoff valves on each blowoff pipe. These may be two slow-opening valves, or one slow-open-ing and one quick-opening valve, or one slow-opening valve and one plug cock. Traction and portable boilers must have one slow- or one quick-opening blowoff valve. On all types boilers, connect the blowoff valves with only extra-heavy pipe

DETAILS AND OPERATION

Lip
Hard seat valve
Seatless sealing valve

Back seat
Cooling chamber

COMBINATION seatless sealing valve in same steel block with hard-seat valve, 400-2500 psi. Lip protects seat from wire-drawing. Ported valve needs no seat

QUICK-OPENING valve, up to 600 psi. Spring holds disk to sealing surface of port. That keeps grit off sealing faces. Disk rotates in use, keeps joints tight

STRAIGHTWAY flow, good to 1800 psi. For back-seating to pack under pressure with valve open. Straight passage of sedi-ment prevents scale clogging or eroding

Scale

BE CAREFUL when blowing boiler. First open quick-opening valve slowly. Then open slow-opening valve slowly enough to prevent shock, but fast enough so seat won't wiredraw. To stop blowing, close slow-opening quickly, then fast-opening

NEVER jam valve if it won't close. Open few turns fast to clear, close again slow. Jamming on scale wire-draws

BLOW when boiler has banked fire or low loads — sediment settles more then. Watch glass, don't leave open valve

With Kid Gloves

USES

EMPTY cold boiler through blowdown line for cleaning, inspection or for repairs. Don't empty until water is fairly cool, or boiler 'seams and joints warp, leak

BLOW OUT mud, scale or sediment while boiler is steaming. *Caution:* Make sure blowdown valves are closed on idle units or scalding water will blow into them

BEFORE CUTTING boiler in on line, give blowdown to reduce the alkalinity. This prevents carryover. Test the water and give blowdowns until water is just right

ANGLE blowoff valve, 200 psi. Seating surface is on *outside* of seat ring. That protects seat from cutting effect of the water, scale and sediment passing through

PLUGCOCK, 175 psi. Plug is pressure-lubricated through stem so tapered cock turns easily. Since valve opens, closes by turning, seating surfaces stay clean

DON'T use globe valve on blowdown line. Pocket in valve clogs with mud, sediment and scale. Then boiler cannot be blown, valve can't be cleared under pressure

HERE ARE BLOWDOWN TANK DETAILS

Tank prevents damage. Blowing directly to sewer instead of to tank blows hot water and steam in other sewer connections. This tank is always full of water. When hot water blows in, cooler water in bottom overflows to sewer as it's displaced. Large open vent prevents backpressure. Small siphon breaker keeps tank from emptying. Manhole is for internal inspections

Key questions and answers on boiler safety

Where pressure vessels are used, safe operation is as important as efficiency and economics. Code reflects current practice

The disastrous New York Telephone Co boiler explosion (POWER, January 1963, p 64) sparked many changes in ASME's Boiler and Pressure Vessel Code. As a result, more emphasis than ever before is placed on boiler safety. Here are key questions and answers about safety regulations from a book* that all boiler operators should know.

Q—Are state laws being passed to include furnace explosion protection in the existing Boiler Codes?

A—New York State has adopted a law for automatic heating boilers, including requirements on combustion safeguards.

EXAMPLE: Some of the following are now in the State Code:

1. Gas-fired boilers: (a) Pilot has to be proved, whether manual or automatic, before permitting the main gas valve to open, either manually or automatically, by completing an electrical circuit. (b) A timed trial for the ignition period is established on the basis of the input rating of the burner. For instance, for input rating of 400,-000 to 5,000,000 Btu/hr per combustion chamber, the trial for the ignition period for the pilot of automatically fired boilers cannot exceed 15 sec. And the main burner trial for ignition also cannot exceed 15 sec. (c) The burner flame-failure controls must shut off the fuel within a stipulated time, again depending upon the fuel input of the burner. For a burner rated with an input of 400,000 Btu/hr or higher, the electrical circuit to the main fuel valve must be deenergized automatically within 4 sec after flame failure. And the deenergized valve must close automatically within the next 5 sec.

2. Oil-fired boilers. Similar provisions have been adopted, with requirements on response time for controls to shut off the burner based on fuel input in gal/hr, instead of Btu/hr. The

*Elonka, S M and Kohan, A L "Standard boiler operators' questions and answers," McGraw-Hill Book Co, NY 1969.

flame must be supervised continuously by the controls.

No doubt, many states will incorporate requirements on safety combustion controls in the future as furnace explosions continue on their destructive path.

Q—Does the ASME Code require a low-water fuel cutoff on a cast-iron steam-heating boiler?

A—Yes. Each automatically fired steam or vapor system cast-iron boiler (or low-pressure-steam steel boiler) must be equipped with an automatic low-water fuel cutoff. It must be so located as to automatically cut off the fuel supply when the water level drops to the lowest safe waterline. The safe waterline cannot be lower than the lowest visible part of the water gage glass at the boiler.

Q—What two control parameters are usually considered in a flame failure system to prevent furnace explosions?

A—1. Control the input composition so that it cannot accumulate to an explosive batch or mixture, as graph on facing page describes. This is called input control.

2. Ignite in proper sequence all combustible combinations of fuel and air as they enter the furnace. This is called ignition control.

Input control depends on how much fuel can be put into a furnace before combustion must ensue, so that no dangerous mixture is present for a delayed light-off leading to an explosion. The time factor is called the grace period, which is the time needed to build up an explosive charge. It

varies with the design of the furnace: A small, fast-heating flash boiler might have a grace period of about 1 sec, while a large, commercial-type furnace might have 4 sec or more for ignition or input cutoff.

As a safety measure, ignition protection and shutdown of fuel and air into the furnace must be completed before the end of the grace period. As a rule of thumb, a fuel-air mixture greater in volume than 30% of the volume of the combustion chamber may be a damaging mixture.

Q—Describe a continuous, automatic check system for a flame-failure safety circuit.

A—Since it is possible to check the flame sensor by manually cutting off the flame, the same sort of check has been developed to do it automatically. Thus the flame safety circuit, including the sensor, is checked automatically and continuously by simulating flameout. As circuit diagram on opposite page shows, flame failure is simulated by means of a swinging shutter that interrupts the line of sight of an optical flame detector intermittently. Each time the sensor's view is blocked, a flameout is simulated long enough to prove out the sensor but not long enough to shut down the system. Should the sensor "see" flame while its view is blocked, the system immediately shuts down on malfunction. But if the sensor detects the intermittent flameout, its no-go signal of one-second duration is too brief to shut down the system, since the logic network has a time delay to prevent nuisance shutdowns. Thus, the system is constantly checking on its flame sensor, logic network and flame-failure safety circuit.

Q—Which inspectors make the legal inspections and reports (to a jurisdiction) that a boiler is safe or unsafe to operate, or that it requires repairs before it can be operated?

Methane-air mixture, showing flammability and explosive limits. Mixture is controlled to avoid explosive batch

Automatic continuous check system of flame safety circuit uses shutter that simulates flameout by blocking sensor

Tank for disposing of blowoff boiler water is necessary when no open space is available around boiler for blowoff

A—Three types of inspectors are:

1. State, province or city inspectors, who see that all provisions of the boiler and pressure vessel laws, and all the rules and regulations of the jurisdiction, are observed. Any orders these inspectors give must be complied with, unless the owner or operator petitions and is granted relief.

2. Insurance company inspectors qualified to make ASME Code inspections. If commissioned under the law of the jurisdiction where the unit is located, they can also make the required periodic reinspection. As commissioned inspectors, they require compliance with all the provisions of the law and rules and regulations of the authorities.

3. Owner-user inspectors, employed by a company to inspect unfired pressure vessels for direct use and not for resale by such a company. They must also be qualified under the rules of any state or municipality which has adopted the Code. Most states do not permit this group of inspectors to serve in lieu of state or insurance company inspectors.

Q—Why and when is a blowoff tank necessary for boilers?

A—Blowoff tank is necessary when there is no open space available into which blowoff from the boilers can discharge without danger of accident or damage to personnel or property. For example, discharging to a sewer would probably damage the sewer by blowing hot water under high pressure directly into it. As the schematic at bottom shows, a good blowoff tank installation is always full of water.

Q—Do not the integral chevron driers, separators and other hardware restrict the flow of steam outward from any opening in the drum of a steam generator, and thus restrict the opening to a safety valve?

A—The opening to the main steam line will usually be adequate to handle the design flow capacity. The ASME Boiler Code has definite rules on the opening to a safety valve. For example, internal collecting steam pipes, splash plates, or pans are permitted to be used near safety valve openings, provided "the total area for inlet of steam hereto is not less than twice the aggregate areas of the inlet con-

nections of the attached safety valves. The holes in such collecting pipes must be at least ¼ in. in diameter."

In the case of steam scrubbers or driers, the ¼-in. diam opening does not apply, provided the "net steam inlet area of the scrubber or drier is at least 10 times the total area of the boiler outlets for the saftey valves." When inspecting drum internals, therefore, check the condition of the ¼-in. diam holes in the collecting pipes. They must be free and clear. The same applies to the openings on driers, because plugged driers on collecting pipes could lead to restrictions in the safety-valve openings. That would reduce relieving capacity flow, which could be dangerous.

Q—What type of safety valve should be installed on a boiler?

A—An ASME- or NB-approved and registered direct spring-loaded pop type, properly marked as to pressure and capacity and equipped with a testing lever. The pressure setting must match either the maximum allowable pressure for which the boiler is designed or, on older boilers, the maximum pressure allowed by state or city laws and codes.

Q—How can one recognize an ASME-approved safety valve, and how does one secure permission from ASME to make approved safety valves?

A—All ASME-approved safety valves are stamped with an ASME symbol, as shown below. Permission to use the symbol is granted by the ASME to any manufacturer complying with the provisions of the Boiler and Pressure Vessel Code. The manufacturer must agree, by filling out forms issued by the Society, that any safety valve to which the symbol is applied will be constructed in accordance with the Code and that it has the capacity stamped upon the valve under the stated conditions. The manufacturer must also agree that he will not misuse, or allow others to use, the stamp with which the symbol is applied. ●

Official symbol denoting ASME stamp of approval for safety valves built in accordance with Code requirements and limits

Stop C-I Hot-

WHY BOILERS CRACK

SEDIMENT PILES UP month after month, until hottest part of casting next to fire is cut off from circulation. Casting is distorted and finally cracks, causing costly repair and shutdown

RUNNING COLD WATER directly into boiler is another reason why cast-iron boilers crack. With 60 F water striking hot cast iron, expansion cracks boiler even without sediment in water

HOW TO PREVENT CRACKS

EXTENSION keeps most sediment from reaching boiler. Run cold water into tank as shown so 200-F water near top takes chill off 60-F water before it reaches boiler near hottest part

BEWARE—just because your hot-water tank has extension, don't neglect tank or boiler. Extension might be badly corroded or broken off. Examine once a year to check condition

FLUSH TANK (if it has no manhole) before installing extension. First close discharge to building supply. Then drain boiler completely dry. Open cold water to tank and flush tank until discharge from hose shows that water is completely clear

FLUSH BOILER by hooking high-pressure hose to one side after draining boiler. Run water until its discharge clears up

Water-Boiler Cracks

SEDIMENT AGAINST HEATING SURFACE of cast-iron, sectional hot-water-heating boilers cracks many boilers yearly. Owners and operators suddenly find ashpit flooded like this

ALL WATER HAS SEDIMENT, whether from well, lake or other sources. Sediment is often minute particles. It travels through tank and settles in lowest part of the boiler to do much harm

ONE TENTH OF TANK DIAMETER is right extension length. Longer extension is bad because space under end is dead space. it cuts down circulation, cools hot water too much besides

CLEAN TANK AT SET PERIODS after installing extension. Good practice is once a year. With manhole on tank, it's easy to remove sediment and also examine tank's inside thoroughly

FLUSH FROM OTHER SIDE like this until water clears. Cast-iron boilers have no manhole or handhole plates to take off

SIMPLE EXTENSION is made by running pipe die over pipe until extension is right length. Then screw into the tank

SWEATED-FITTING EXTENSION is made by sweating pipe into bushing screwed into tank. Then sweat line into bushing

MAKE MINOR WALL REPAIRS

1 Here is a badly eroded firebrick lining that's backed by common brick outer wall. Because anchor bricks are broken, furnace wall has separated from lining. To find loose area of any furnace wall, tap suspected section with wooden mallet. Loose wall has a hollow sound. Unless repaired, it may cave in under load

2 Never trust a loose wall. Only way to be on safe side is to tear it out and rebuild right. Tear out loose wall up to first row of solid anchor bricks. This operation takes good judgment so good wall section isn't scrapped. Next clean out anchor brick grooves and brush all chips and dust from the bricks and grooves

3 Mix a good-quality high-temperature bonding mortar thoroughly with water until it's creamy. Dip bricks in mortar, letting excess mortar drip off so joints are thin. Fit anchor bricks in firmly and build wall as shown. If top bricks don't fit, hammer in plastic firebrick, Fig. 4. Brush new wall with bonding mortar to seal

4 Small irregular areas can be repaired quickly with plastic firebrick. Hammer into anchor brick areas well with wooden mallet. Trim finished wall with trowel's edge so finished surface is porous. This allows moisture to escape. Low- and high-temperature plastic firebrick is wrapped in 100-lb cartons

Courtesy—A P Green Fire Brick Co, Mexico, Mo.

USE SUPER REFRACTORIES

1 Hand-fired furnace walls last longer when side and back walls have three rows of silicon carbide bricks at fuel-bed area. At hotter area above, use four rows of mullite bricks on side walls, five rows on back wall. Front wall needs none

2 For single-retort stoker jobs, this pattern of silicon carbide bricks protects side and bridge wall. Here again the front wall isn't lined with super refractory. By following fuel-bed pattern, only trouble spots are lined. Cost is reduced

3 Multiple-retort stoker furnace walls are under higher temperature and need super refractory lining on all four walls. Silicon-carbide bricks are best for good grade coal. For some fuels, mullite bricks in upper wall portions are desirable

4 Spreader stoker walls need silicon carbide at grate line to fight wear and clinker adherence. Back walls sometimes lined higher to resist abrasion of chunks of coal during heavy feed rates. Mullite above the fuel bed resists slag erosion

▶ Do you want to build a furnace wall that stays up longer? OK, then use super refractories in the fuel bed area. An 85% silicon-carbide brick really has abrasion resistance. Being very dense and not subject to corrosion by most coal ash, it resists clinker adhesion and erosion far better than fireclay.

For coals with ashes that fuse easily, silicon carbide works best when placed to fuel-bed depth because it resists abuse and wear.

In higher temperature regions above the fuel bed, harmful chemical conditions caused by floating ash particles containing much iron may cause slagging of silicon carbide bricks. This doesn't happen in the cooler zone lower down.

For upper walls, the best refractories are made of mullite, which is a composition of aluminum and silicon oxides. This refractory is ideal because it forms inert crystals of mullite that prevent fluxing with ash. To insure the correct composition, mullite is produced artificially from electric-furnace melts.

Bonded mullite and bonded silicon-carbide brick both stand temperatures well above 3000 F. Mullite has abrasion resistance much greater than firebrick.

Its resistance to clinker attack exceeds that of silicon carbide a little, but firebrick by a wide margin.

Above are a few ways to install super refractories in furnace types shown. Some shapes are laid, hung, wedged or bolted in place. They should be laid in cements of their own composition with thin joints and carefully bonded into the firebrick backing. In most cases they are installed as standard brick shapes and laid with standard bricklaying methods. When chosen and applied right, their life is several times common refractories.

Courtesy, THE CARBORUNDUM CO

Why boilers explode

EXPLODED BOILER caused $228,000 in property damage and business interruption

We asked John Todd for the reasons, since he has been an authority on boiler explosions for many years. His experience points up one fundamental shortcoming—the assumption by owners that a boiler built to standards and furnished with automatic controls requires no expert attention.

Q With the many safety devices and automatic controls on present-day boilers, why are there so many boiler failures?

A The main reason, in my opinion, is the lowering or complete absence of operating standards. Too many owners have only one thought in mind when buying boilers provided with automatic controls—they hope to operate without the benefit of qualified operators. Please remember that until such time as all interested parties (these include governmental and insurance company inspectors, boiler manufacturers, code-making organizations and engineers) can convince boiler owners that they cannot shed all responsibilities toward safe operating practices by substituting automatic devices, boiler accidents will continue at the present alarming rate.

Q What is the most common type of boiler accident occurring today?

A The greatest accident producer today is low water which usually results in overheating and loosening of tubes, collapse of furnaces and, in some cases, complete destruction of the boiler. In certain classes of boilers a low-water condition can set the stage for a disastrous explosion that can cause serious loss.

Q Why is low water responsible for most boiler accidents in the past decade?

A Primarily it is due to a lack of boiler-operating and maintenance standards. This is true despite the fact that most boilers, especially heating ones, are provided with automatic feed devices, just one of the many automatic controls. But it is these devices that lull the owners into a false sense of security. Many of them feel that the boiler is fully automatic and completely protected from accidents. Not understanding fully how potentially dangerous a fired pressure vessel can be, the owner (or any one else) does not seem to take the slightest interest after installing one of these so-called automatic boilers.

The fact is that automatic feed devices, like all other automatic devices, will work perhaps a thousand times, perhaps many thousand times. But at some time they will probably fail, usually with disastrous results. That is why it's the duty of everyone concerned with boilers to realize that unless proper operating and maintenance standards are instituted, accidents are certain to occur.

Q You said that low-water accidents are caused primarily by failure of automatic feed devices. But isn't it true that, in practically all cases, low-water fuel cutouts are installed on all automatically-fired boilers to protect the boiler from overheating, in the event of failure of the feeding device? If such is the case, what causes the failure?

A That's correct. Practically all boilers today are provided with low-water fuel cutouts. What most people don't realize is that, as is true in most accidents regardless of the type of cutouts, a series of failures occurs. Thus, the basic fault may be failure of the automatic feed device to operate. Then we experience failure of the low-water fuel cutout. The net result is overheating and burning of the boiler metal. The failure of the automatic feed device and the subsequent failure of the low-water fuel cutout to operate stem from the same basic cause—lack of operatning and maintenance standards practiced by the owner.

Q How should a low-water cutout be tested?

A The only positive method of testing a LWCO is by duplicating an actual low-water condition. This is done by draining the boiler slowly while under pressure. Draining is done through the boiler blowdown line. We find that many heating boilers are not provided with facilities for proper draining—an important consideration.

Many operators mistakenly feel that draining the float chamber of the cutout is the proper test. But this particular drain line is only provided for blowing out sediment that may collect in the float chamber. In most cases the float will drop when this drain is opened due to the sudden rush of water from the float chamber. Every boiler inspector can tell you of numerous experiences of draining the float chamber and having the cutout perform satisfactorily. But when proper testing was done by draining the boiler, the cutout failed to function.

Q What percentage of boiler losses are caused by low water?

A Approximately 75 per cent.

Q Various articles have been published on the subject of safety valves, particularly the low-pressure type with a setting of only 15 psi. If it is true that all boilers furnished today are provided with ASME-approved valves, why should anyone take exception to them and why shouldn't they work?

A In the first place, most people have a misconception of the term "ASME-approved." To set the record straight, the ASME itself does not approve a type of safety valve. By referring to the ASME Code, Section I—Power Boilers, and Section IV—Low Pressure Heating Boilers, you will see that both Codes contain limited-design criteria. The Codes also require a manufacturer to submit valves for testing. Such tests are solely for pressure-setting and relieving capacity.

In brief, the ASME symbol on a safety valve attests to the fact that the limited design criteria and the materials outlined in the code have been supplied by the manufacturer and that the relieving capacity and set pressure stamped on the valve have been proved.

Getting back to the first part of your specific questions, in some cases experience indicates that a particular type of safety valve has an inherent design weakness. After a short period of operation, the disk may be subject to sticking due to close clearances. This condition will render the valve useless, and the boiler will be without the benefit of over-pressure protection.

In regard to the second part of your question, the failure of safety valves to work is usually due to build-up of foreign deposits that result in "freezing." This is an indication that the valve has not been regularly tested or examined. One of the greatest causes of foreign deposit build-up is due to a "weeping" or leaking condition. The only way to be sure that a valve is in proper operating condition is to set up and adhere to a regular program of testing the valves by hand while the boiler is under pressure. Also, any weeping or leaking valve should be immediately replaced or repaired. This is important.

Q How often and in what manner should boiler safety valves be tested to make certain they are in proper working order?

A Low-pressure (15 psi) safety valves should be lifted at least once a month while the boiler is under steam pressure. The valve should be opened fully and the try lever released, so the valve will snap closed. For boilers operating between 16 and 225 psi, the safety valves should be tested weekly by lifting the valves by hand. On these higher-pressure boilers it is good practice to test the safety valves by raising the pressure on the boiler. This can usually be done when the boiler is being taken off the line. Then, if the valve feathers from improper seating, it can be corrected when the boiler is cold.

Q What can be done to prevent boiler failures?

A Boiler failures can at least be greatly reduced if boilers are placed under the custody of properly trained operators. This means that boiler owners must use sound judgment when employing boiler operators. Everyone with an interest in boilers must be encouraged in educating boiler owners and operators in proper operating procedures.

A very important step is establishing a regular program for testing of controls and safety devices, then faithfully following through.

Further, a program must be established for periodic maintenance of controls and safety devices. First thing to realize is that providing boilers with the most modern proved controls and safety devices is no guarantee that you will not have boiler failures. Any control or safety device is only as good as the testing and maintenance it receives. You cannot ever relax on these two.

Q What can a boiler owner do to assure that he has taken all possible steps to prevent boiler failure?

A First, purchase the best equipment available for a given service. Second, make certain that the boiler is properly installed and equipped with all the necessary appurtenances and safety devices. Third, before taking final acceptance, specify that the installation be inspected by a commissioned inspector in the employ of an insurance company or the state or municipality. By doing so, the owner will be assured that the equipment and the installation meet the legal requirements of the particular state or municipality. Fourth, the owner should provide the operator with a log book and a set of preventive maintenance and testing procedures. He should insist that such procedures be followed religiously and that the results of the tests and maintenance be recorded and be made a permanent part of the boiler room log.

These Five Steps Help Prevent Furnace Explosion

1 Get competent firemen, instruct them properly on equipment and controls. They're your first line of defense

2 Follow lighting-off procedure, below, faithfully

3 Provide good-sized inspection window so lighting torch, pilot, main burner flame, flame rod, if used, can be easily seen

4 Set up periodic inspection and maintenance program as follows (Include checking fuel-air ratio, draft adjustments, pilot flame settings):
Test safety shutoff fuel valves for tight shutoff
See how combustion safeguard responds to loss of flame contact
Test fan interlocks
Keep test records for insurance inspector's review

5 Get the insurance company's approval in advance for new installations or proposed changes. Submit plans showing burners, fuel piping, draft fans

LIGHTING-OFF TIME for oil- or gas-fired boilers represents a danger period for possible furnace-explosion build-ups

▶ THERE ARE REALLY only two danger periods when oil- or gas-fired furnaces can explode. These periods cover a good share of the time when a boiler should be getting the most attention. They begin with lighting off, and carry over to light-load firing. But they can happen only when there has been a build-up to dangerous proportions of unburned fuel or flammable products of incomplete combustion.

According to one insurance company's* findings 87% of all boiler furnace losses for a 13-year period (1936-1948) took place in manual-lighted boilers. These are units where fuel to the main burner can be turned on only by hand. Once lighted, though, the firing rate can be automatically controlled.

What's even more interesting, the waterwall type of boiler furnace construction is the most likely to suffer from a furnace explosion. Look at figures, facing page. Waterwall types have only relatively weak wall heads for support. Refractory furnace, though, is of much more rugged construction.

Manual-lighted boilers begin, usually, at about 10,000-lb-per-hr steaming capacity and 100 psig, and go up from

there. Automatic-lighted units take in the smaller sizes.

What can you do to make sure your boiler furnace never builds up an explosive mixture? Five steps, above, give some basic precautions you can take. In addition, here are helpful operating tips.

Lighting Off. Let's establish this most important point first. A high-caliber well-trained fireman is the primary factor in safe lighting off. He instinctively goes through these steps in lighting off a boiler:

1. See that all drain valves are closed; water is in the boiler; access and observation doors are closed.

2. Check fuel valves at all burners for closure.

3. Thoroughly purge the furnace and passes for 5 minutes at ¼ maximum air-flow rate.

4. Insert a reliable lighting torch. (A dozen turns of ½-in. asbestos rope, secured by a hairpin bend, on a 5-ft length of ⅜-in.-dia steel rod is considered reliable.) Be sure torch has a substantial flame before putting it into the burner lighting part, and put it in place in such a way that main or pilot burners, if present, light off promptly.

5. Open the fuel valves at the burner, and fire the burner in keeping with the manufacturer's instructions.

6. If a main burner fails to ignite within 5 seconds after opening the fuel valve, or if the torch, pilot or main-burner flames blow out:

(a) Close the fuel valve at the main burner right away.

(b) Pull out the torch and turn off fuel to pilot, if present.

(c) Adjust dampers, burner registers and pass for no less than ¼ rated air flow through setting.

(d) Purge at this rate for 5 minutes or more before readjusting dampers, etc, and inserting the lighting torch a second time.

General Operation. A few operating conditions stand out in a study of gas- and oil-fired furnace explosions. Flame failures nearly always occur while firing with only one burner at low ratings —say below 30%—under low loads or while warming up shortly after lighting off. Under these conditions the boiler-furnace temperature drops off to a point below the prompt reignition temperature for the fuel.

Because of this, when boilers are fired below 30% rating, reduce the number of burners. Never try to cut back on burners once you get below the stable minimum firing rate. Instead, completely shut off individual burners so remaining ones have enough load to stay within stable operating range.

REFRACTORY-FURNACE design, above, will give the furnace a ruggedness that is helpful in the event of any explosion

WATERWALL TYPE of furnace suffers most from any explosion because the wall supports are relatively few and far apart →

How about flame failure or other combustion safeguards? Quick-acting electronic devices of the "flame-rod" type with gas, or "photoelectric" with oil, help if properly used. The flame-rod cannot sense unburned gas during lighting off, but it can see a pilot flame. While firing, both types respond to flame failure, but neither can do anything about dangerous build-ups of flammable products.

Since each and every boiler has its own characteristics that can bring on unforeseen operating troubles, let your combustion safeguards just sound an alarm to start off with. Then when you have reliable operation and you've made some necessary adjustments in firing and control equipment you can connect your combustion safeguards to shut off fuel.

As far as automatic combustion-control adjustments go there are some definite rules you ought to follow. For instance, make sure air-flow rate always follows steam flow and does not lead it. Then see that the fuel-flow rate is automatically adjusted to follow the air-flow rate. This cuts down the chance of over-rich combustion and gives safe response if fans fail.

Gas Hazards. Manual-lighted gas-fired boilers, particularly with large waterwall furnaces, have figured in some very severe explosions. The lighting-off explosion damage runs much higher than that during regular firing operation. In fact, there've been so many more lighting-off explosions with gas-firing compared to those with other fuels that previous practice has been closely checked. One thing stands out and that

is a failure in the past to put in reliable pilots for each main burner. These pilots prove their worth not only in lighting off but in low-load service and warm-up periods when the boiler is bringing up steam pressure.

Chief among the hazards in gas-fired operation is leaving gas-burner cocks open too long without purging, especially during unsuccessful attempts to light off. Next comes accidentally leaving burner cocks open.

High among any list of hazards is accidental flame failure while warming up after lighting off or while firing at low ratings. Biggest reason for this is unfavorable air-gas ratio adjustments.

As more or less of a follow-up of unfavorable air-gas adjustments is the danger of not enough air for complete combustion.

Oil Hazards. Similarly there are several outstanding hazards in oil-burning operation. Most important is accidental flame-failure while warming up shortly after lighting off or firing at low rating. Its causes are many. They include unfavorable oil-air ratio, poor atomization, interruption of oil flow by oil-pressure failure or a slug of water.

Next in order of hazard importance is incomplete combustion of oil. It may result from (1) too low an atomizing pressure—either from too low a steam pressure or too large a sprayer plate (2) too low a temperature, or (3) firing at too low a rate to maintain proper furnace temperature.

Then comes the hazard created by leaving the burner oil valves open too long without purging during unsuccessful attempts to light off.

The last two, in order of importance, are (1) attempts to light burners from hot furnace refractory and (2) not enough air for complete combustion either from fan failure or failure to start draft fans or open dampers.

Recommendations. To avoid furnace explosion, there are a number of practical steps you can take in addition to the five steps already given. For gas-firing they involve: (1) providing a continuous, fixed, premixed pilot for each main burner (2) installing, where a boiler has more than one burner or burner unit, an approved cock and gas safety-control system (3) protecting against low gas pressure on single-burner boilers (4) interlocking the main safety shutoff gas valve with electrically driven draft fan (5) adding approved flame-rod combustion safeguards on 8,000-20,000 lb per hr boilers where practical.

For oil firing, these recommendations hold: (1) thermometers in the oil piping close to burners using heavy oil (2) an oil-pressure gage on the burner oil-header supplying pressure-atomizing or steam-atomizing burners (3) photoelectric combustion safeguards and interlocks with furnaces having more than two of the six walls as waterwalls or any other boiler where damage would badly cripple plant production no matter what its construction.

Further, in oil firing operate soot blowers only when the burners are firing at high ratings. Then open wide the off-take damper or induced draft to vent steam. Meanwhile throttle the air flow by means of the forced-air damper to prevent blowing out the fires.

To POP...
or not
to POP?

Once a day? Once a month?
During the full moon?
Just on ground-hog day?
Here is the straight dope

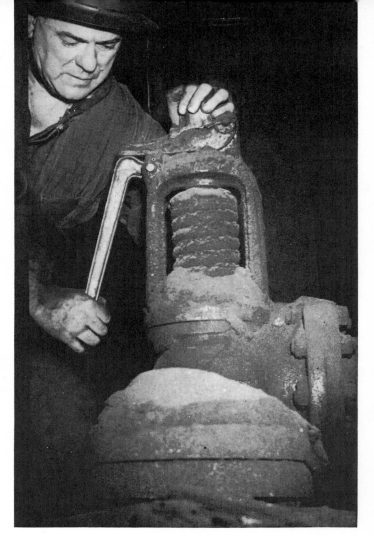

▶ *Power* magazine printed a reader's question asking HOW and HOW OFTEN he should test his boiler safety valves.

We got a flock of answers. All agreed that safety valves must be tested. All agreed pretty much on the HOW. Big difference of opinion was the HOW OFTEN.

We think the matter is important. Since it turned out to be so controversial, we did some research. We talked with engineers: consultants, and chief engineers in large plants and small. With valve manufacturers and with boiler inspectors. We got a cross section of opinion. Here's what they say:

Boiler inspector. Years ago, when weighted-lever safety valves were used, firemen moved weights, added weights, and even wedged timbers between the levers and the boiler-room roof. Held the valves shut. Easier to keep an even head of steam that way. But selfish chief engineers, thinking more of preserving their boilers than the feelings of the firemen, ordered daily testing of safety valves. This became established practice, more or less. With the advent of springloaded valves, daily testing continued from habit rather than from reason.

ASME Code recommends all boilers operating above 15 psi receive at least one annual internal and external inspection. Since it isn't practical to pop safety valves on high-pressure boilers, as in large utility and industrial plants, annual inspection should include removal of safety valves for complete overhaul and inspection.

Below 50 psi, pop valves once a month if plant is in poor condition. If in good shape, every six months is OK up to 600 psi. If you use the hand testing lever, have steam pressure at least 75% of safety-valve setting.

Chief engineer. Safety valves should be dismantled annually for cleaning and inspection. If plant operation leaves much dirt on valves, 6-month intervals are better. If plant is clean and valves in good condition, popping is not necessary oftener than every two months or so, up to about 650 psi. At lower pressures, increase the time between tests in proportion, but inspect often for clogged drains and dirt build-up in bodies.

Testing is a necessary evil, but should not be done oftener than absolutely necessary. Always a chance a piece of scale will be caught between disk and seat when the valve slams shut. Play it smart and have plenty of steam, at least 75% of set pressure, if you lift valves by hand. Best bet is to run pressure up to popping point. You know if the valves are working right—and you clear the seating surfaces. Another good angle: valves should be popped before a boiler outage. This checks need for repair or adjustment.

Valve manufacturer. Popping safety valves every day may keep the log right on the ball, but it's unnecessary—and costly. If valves are clean, and this depends on good housekeeping, including feedwater treatment, don't induce valve leakage by frequent testing.

Tendency to leak involves two factors: (1) possibility of dirt or scale being trapped when valve closes (2) more important, the differential between setting and operating pressures. The nearer the operating pressure to setting pressure, the greater the possibility of potential leakage.

Setting up a testing schedule to cover all types of plants wouldn't work any better than a schedule for rainfall. Tailor testing to suit your own conditions. If you're in a chemical plant, chances are your valves take a beating from "gunk" in the atmosphere, and should be tested every two months or so. If you're in a spick-and-span plant, testing every six months should be enough. Best guide to a test schedule is experience. Try a set interval. Then a longer interval. So long as popping pressure remains constant, and valves show no inclination to stick, you're doing all right. Somewhere along the line you can establish a pattern. Since safety is most important, satisfy yourself that selected frequency meets this need.

Consulting engineer. Your testing schedule is dictated by your individual operating conditions. ASME *Suggested Rules for Care of Power Boilers* means

just this by saying, "Test the valves in accordance with the instructions given for the particular plant."

It's no secret that frequent testing is an invitation to leakage or damage. Particularly if the valve setting is little higher than operating pressure. During normal operation, loading of the valve disk, being proportional to differential in operating pressure, is relatively small. The less valves open, the less chance of difficulty.

Safety valves, particularly those in dirty atmospheres, should be tested at intervals founded upon experience. In clean plants, at pressures up to 650 psi, testing at 6-month intervals is practical. Once a year, valves should be removed for inspection and testing. Plants in the high-pressure range, using complex controls, can't permit haphazard opening of safety valves. Such plants have rigid outage and inspection schedules. Prior to an outage, pressure is usually run up to popping point. During outage, valves are removed, inspected and overhauled.

In speaking of valve damage, we usually think of wiredrawing, or entrapment of foreign matter. However, we can't ignore the effect of thrust. Where discharge is perpendicular to stem axis, developed thrust can be high, depending on design and capacity. Mountings, particularly the flange bolts, are subjected to severe loading.

Temperature cycling, plus shear stress could add up to bolt failure. That's another reason for less popping—and it points up the importance of inspection.

Testing tells we have protection against overpressure. But the act of testing is a form of valve maintenance and should be thought of as such. Proper function within design limits is worthy of more attention.

If slight leakage occurs, tendency too often is to increase spring compression just enough to stop leakage. Fact that this increases the setting is dismissed as negligible. Cause of leakage isn't eliminated. Though increase doesn't conflict with designed working pressure of boiler, or exceed spring-adjustment range, the end does not justify the means. Smart upkeep calls for valve overhaul, not fiddling with the spring.

More common disregard for designed function: continued use of higher-pressure springs following a permanent reduction of boiler operating pressure. ASME Code permits spring-adjustment range of 10% up to 250 psi; 5% on higher pressures. Example: reduction from 160 to 140 psi requires a change to a 140-psi spring, not merely reducing compression of the 160-psi spring to a 140-psi setting.

Popping safety valves

Here's the story . . . in a nutshell

- **Frequent popping is harmful, costly**

- **Weekly testing rarely necessary; daily—never**

- **Dirty or clean plant, high or low pressure determine how often**

- **With hand lever, have steam pressure not less than 75% of valve setting**

- **Best way: Run pressure up to popping point**

- **Remove, test and inspect valves yearly**

1—FIRST FIND THE EXPANSION FACTOR

To protect the low side of a reducing valve . .

Find your high-side pressure at bottom of chart, draw line straight up until you meet your low-side steam-pressure line, then read expansion factor on left-hand scale

... How Big a Safety Valve?

The old rule of "safety valve same size as low-pressure piping" is all wet according to Mutual Boiler and Machinery Insurance Co engineers. Their tests show you need more relieving capacity than you can get with a single safety valve same size as piping*

HERE'S HOW TO INSTALL THE SAFETY VALVE

Insert tee or manifold same size as safety-valve inlet

Valves same size as line. Total capacity equal to that of single valve determined from chart

This hookup is **NO GOOD** Tee fitting restricts flow

2—PICK OFF THE RIGHT VALVE SIZE

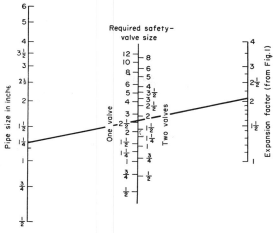

Draw line between expansion factor (from Chart 1) and pipe size of low-side piping to find required safety-valve size

Here's How to Pick Right Valve Size

PROBLEM. What size valve would you put on a 1¼-inch, 20-psi steam line fed from a 160-psi source?

SOLUTION. Follow lines drawn on charts. On Chart 1 enter with high-side pressure of 160 psi, run up to 20-psi low-side line, then left to read expansion factor of 2.1.

Now go to Chart 2. Draw line between 2.1 expansion factor and 1¼-in. pipe size. You can use either one 2½-in. safety valve or two 2-in. valves.

Finally check valve's capacity in Chart 3. Valve should be able to relieve 1830 lb of steam per hour. Any combination of smaller valves 1¼ in. or larger is okay, too, if total capacity is 1830 and they're installed in approved manner.

3—THEN CHECK VALVE'S CAPACITY (LB STEAM PER HOUR)

| | Safety-Valve Setting (Pounds per Square Inch) | | | | | | | | | | | | | | |
	5	10	15	20	25	30	35	40	45	50	60	80	100	120	140
½ in.	60	68	75	80	85	90	94	99	102	105	113	126	140	150	155
¾ in.	112	125	136	147	156	165	174	181	189	195	209	234	258	279	292
1 in.	198	219	237	255	271	287	301	315	328	341	366	410	452	486	507
1¼ in.	389	431	470	505	536	566	594	621	647	672	721	809	889	956	997
1½ in.	540	600	650	700	750	800	845	885	920	955	1030	1150	1260	1350	1430
2 in.	980	1065	1150	1240	1310	1395	1455	1530	1595	1655	1780	2000	2195	2365	2470
2½ in.	1435	1560	1705	1830	1945	2055	2160	2255	2355	2450	2640	2955	3250	3500	3650
3 in.	2375	2625	2855	3055	3255	3445	3620	3790	3940	4105	4395	4945	5440	5840	6080
4 in.	4350	4840	5250	5625	5985	6310	6625	6935	7230	7525	8070	9060	9850	10,500	11,025
6 in.	10,900	12,000	13,050	14,000	14,950	15,800	16,550	17,400	18,050	18,800	20,050	22,500	24,500	26,400	28,000
8 in.	18,900	21,000	23,000	24,750	26,500	28,050	29,050	31,000	32,400	33,550	36,100	40,500	44,350	47,700	50,950

Safety-Valve Size (left axis label)

▶ SUPPOSE THE DIAPHRAGM of a pressure-reducing valve ruptured, leaving the valve wide open! What would happen on the low-pressure side? Would a conventionally installed safety valve keep the low-side pressure down?

A near explosion in a policyholder's plant led Mutual Boiler and Machinery Insurance Co engineers to have some doubt. They started looking into the problem and came up with some startling facts.

For years it's been common practice to install safety valves, on the low-pressure side of a reducing valve, of the same size as the low-pressure piping, set at the allowable pressure of the equipment to be protected. To find out if this was safe enough, Mutual's engineers ran the following test.

The Test. They took a 2-in. reducing valve with 2-in. piping on the low-pressure side, protected by a conventional 2-in. safety valve set at 30 psi.

Pressure on the high side was 143 psi. Then the reducing valve was held open (to simulate a ruptured diaphragm). The safety valve let go okay, but the low-side pressure rose to 105 psi, even with the valve blowing.

The engineers repeated the test with *two* 2-in. safety valves, and the low-side pressure rose to 70 psi with both valves blowing.

So they tested safety valves, same size, of another reputable make. Same story. With one safety valve the pressure shot to 120 psi; with two valves blowing the low-side pressure hit 87 psi.

Conclusion was that safety valves same size as piping did not have sufficient relieving capacity for safe use on the low side of pressure-reducing valves.

Charts. To find out just what capacity was necessary, they continued their tests and came up with the above charts. Box (above right) tells how to use these charts to find the proper-size valve for most installations and the required relieving capacity. You need to know

both size and capacity because a valve the right size may not have the required relieving capacity. In this case you need either a larger safety valve or two or more valves whose total capacity equals the required.

Hooking Up Valve. Box (facing page) shows two approved methods for hooking up safety valves and one method that's NO GOOD. The bottom hookup is n.g. because the tee fitting slows down flow and sometimes lets pressure build up in the up-stream piping faster than safety valve can relieve.

Exceptions. However it's not always necessary to go to a safety valve larger than the piping. One manufacturer has developed a special set of orifice nipples to go on the down-stream side of the reducing valve to limit the maximum possible flow through the reducing valve to a rate that can be handled by the safety valve. Or you can install a rupture diaphragm same size as piping that will let go at a pressure not greater than 20% above safety-valve setting.

Know your basic controls

What you must know

► YOUR HOT-WATER BOILER must have a safety relief valve to keep the pressure at or below the maximum allowable working pressure. But, until recently, ways of doing this were not clearly understood. Today, the ASME Boiler Code defines the correct procedure in detail. But, in spite of this information, many hot-water jobs are still fouled up, Fig. 1. Is yours fouled? Insurance magazines are filled with reports of life and property loss from needless explosions.

What causes a relief valve to open? The average hot-water boiler is built for maximum working pressure of 30 psi. Relief valve is set for same pressure. So anything that ups boiler pressure to 30 psi causes valve to open, Fig. 3.

Critical demand is placed on the safety relief valve during emergency when it must discharge both high-temperature water and steam, Fig. 4. When relief valve discharges 212 F, sudden drop causes higher temperature water to flash into steam—that's why ASME relief valve is tested and rated on *steam*. Usually valve lifts when heating system can't dissipate heat as fast as it's developed by boiler.

How about low-water conditions? Many hot-water boilers are knocked out by this condition, Fig. 5. Erroneous belief that the pressure-reducing valve used to fill the hot-water system initially will keep boiler and system full causes trouble. Pressure-reducing valve is usually set at 12 to 18 psi while safety relief opens at 30 psi and closes at 26 psi —so pressure reducing valve is ineffective during time relief valve is working. If a hand-fill valve is used, any leak in system can quickly cause a low-water condition.

Minimum basic control you should install on a hot-water boiler is a float-operated low-water fuel cutoff, Fig. 6. Because of higher working pressures of hot-water boiler, low-water cutoff is for working pressures over 30 psi.

Fig. 7 shows low-water cutoff action in emergency. Lowering boiler water and water in float chamber at same time causes float to drop. That opens electrical circuit and stops automatic burner.

Low-water control, Fig. 8, is about foolproof. But you can't rely 100% on low-water cutoff to stop automatic burner each time there's low water. Reason is that under some conditions low-water cutoff cannot work. Then you must use combination boiler water-feeder and low-water cutoff shown in Figs. 8, 9 and 10.

Several kinds of overfiring can occur to make low-water cutoff useless. In such emergencies only way your boiler remains safe is to feed water to boiler as fast as water discharges through the relief valve, Fig. 9.

Large leak or drained boiler causes low-water cutoff switch to stop burner, Fig. 10. How about firing a dry boiler? Fired continuously by a runaway burner and carrying no heating load, this boiler may not explode because it has a modern ASME approved relief valve to protect against over pressure. But the boiler may be ruined because it goes dry, or because it tries to operate as a steam boiler though it isn't hooked up for this service. Check your boiler today.

—*Courtesy, McDonnell & Miller, Inc, Chicago, Ill.*

1 Non-code relief valve—capacity isn't marked, valve's in wrong place, scale in feed line isolates valve from boiler. Don't use relief valve as a reducing valve

5 Causes of low water: Carelessness in firing dry boiler, drawing hot water; piping, boiler, pump, etc leaks; relief-valve discharge caused by overfiring

9 Low-water cutoff's useless if boiler's hand fired, fuel feed sticks open, burner's put in 'on' position, zone controls isolate boiler from hot-fuel bed

for hot-water heating boilers

2 Code-approved relief valve has Btu rated capacity, also try-lever for testing. Valve is atop boiler, rated capacity is matched to gross output of boiler

3 Relief valve opens from city-water pressure, hydrostatic test, water-logged or small compression tanks, high-static head or from high pump pressure

4 Causes of overfiring. Limit control doesn't stop burner, fuel valve fails, burner on manual, residual heat from coal firing, burner is greatly oversized

6 Low-water cutoff. Locate control above lowest permissible water level as there's no normal water line in hot-water boiler. Sketch shows safe hookup

7 Remember: attach low-water cutoff that cuts off fuel before water level hits danger point, or attach feeding device with cutoff to steam-water boilers

8 Here, water's fed mechanically as fast as relief valve discharges it, burner stops electrically for low water. Setup is best for steam-water boilers

10 Safe hot-water boiler has ASME relief valve to prevent over-pressure, low water cutoff or preferably feeder cutoff combination for maximum safety

Right way...and common piping errors

Right   Wrong

FIRST SKETCH is correct, all others wrong. Check your water feeders and low-water cutouts for these common mistakes. Piping errors shown cause everything from high water in gage glass to low water—regardless of where water is in your boiler. Unless corrected, your boiler will eventually fail. Don't be surprised to find one of these errors in your piping. Install an ASME approved safety valve.

Have you examined your boiler lately? Regular inspection is best way to prove effectiveness of feedwater treatment. Check your water program with help of these . . .

22 Q & A's on today's feedwater treatment for packaged boilers

1 How do packaged boilers differ from designs?
A Most packaged boilers are merely modifications of conventional designs. But, since they are completely fabricated *before* shipment, installation is quick and easy. Large amounts of steam are produced with equipment of comparatively small physical size. This calls for relatively high heat-transfer rates and high steam-generating area per unit of packaged-boiler volume. Water holding capacity is much smaller than conventional boilers.

2 Do packaged boilers need high-quality feedwater?
A Packaged boilers don't require higher quality feedwater than conventional boilers run at equivalent ratings. But clean boiler surfaces are more critical when operating at high evaporation rates. Scale builds up at a more rapid pace. Smaller steam space promotes priming and foaming. Therefore treatment must be carefully selected and controlled to produce high-quality feedwater. Don't compare packaged boilers to older steam-generators run at much lower ratings. These units sometimes do get by with minimum attention to feedwater treatment.

3 What is the first step in any water-treatment problem?
A Analyze the water. Chemically pure water is rare. Few water supplies are suitable for domestic or industrial use without treatment. And chemical composition of different water supplies vary greatly. So it is impractical to prescribe any one ideal treatment. The water must be completely analyzed before treatment can be sensibly chosen.

4 What are the basic aims of feedwater treatment?
A Basic aims are continuity of boiler service, maximum boiler and thermal efficiency and minimum maintenance charges. Chemical composition of makeup, boiler water and condensate returns must be carefully controlled.

5 How many treatment methods are available?
A There are many methods available. All come under three broad classifications—mechanical, heat or chemical treatment. Mechanical treatment includes filtration and boiler blowdown. Heat treatment includes makeup distillation and steam purification. Distillation is limited to boilers with small amounts of makeup. Steam purifiers are used where the process demands very dry steam. Chemical treatment, both internal and external, is most widely used. External treatment adjusts raw-water analysis before entering the boiler. Internal treatment adjusts boiler-water analysis by feeding chemicals directly to the boiler.

6 What is major treatment problem with heating boilers?
A Corrosion or pitting is the main problem. Scale is not a problem because the same feedwater is used continuously. Initial treatment is usually good for the entire year.

7 Why does boiler feedwater with high percent of makeup magnify the treatment problem?
A Often live steam cannot be returned to the boiler as condensate. Makeup water is added in its place. When steam is formed, any solids present in the feedwater remain behind in the boiler. Every gallon of makeup water gradually increases the concentration of the boiler water.

8 Is a small amount of scale formation harmful?
A Scale causes localized overheating that can seriously damage your boiler. It definitely impairs heat transfer. Here's what scale does to heating efficiency.

Thickness of scale inches	Percent loss of heat		
	Soft carbonate	Hard carbonate	Hard sulfate
1/50	3.5	5.2	3.0
1/32	7.0	8.3	6.0
1/25	8.0	9.9	9.0
1/20	10.0	11.2	11.0
1/16	12.5	12.6	12.6
1/11	15.0	14.3	14.3
1/9	—	16.0	16.0

9 How is scale prevented?
A Scale-forming ingredients are calcium and magnesium which form carbonate and sulfate scale when heated in the boiler. It can be prevented by external zeolite softening of makeup plus supplementary phosphate treatment in the boiler. Internal treatment can carry the full load if ex-

ternal softener is too costly. Phosphate precipitates calcium and magnesium phosphate, a nonadherent sludge, removed by blowing down the boiler. Soda ash, sodium aluminate, sodium silicate and various organic chemicals are also used to prevent scale.

10 Why are dissolved gases important?
A Dissolved oxygen is the chief cause of piping and boiler corrosion. Carbon dioxide, in combination with oxygen, greatly increases rate of corrosion. Oxygen and carbon dioxide carried over with the steam are dissolved in the condensate to cause return-line corrosion.

11 What is the difference between fireside and waterside corrosion?
A Fireside corrosion usually occurs in summer when the boiler is not in use. Look for concentrated pitting along bottom of firetubes or a general thinning of metal at the entrance of direct tubes. Damage is caused by sulfuric acid, formed by reaction of moisture and sulfur on carbon or soot-covered surfaces.

Waterside corrosion oxidizes the boiler metal forming pit marks or holes in the tubes. Examine for small mounds of powdered black iron oxide along the top and sides of tubes. Concentrated pitting takes place there.

12 How do you combat corrosion?
A Prevent fireside corrosion by keeping fire surfaces clean and dry during any boiler shutdown. Waterside corrosion is generally treated chemically. Sodium sulfite is very effective. It combines readily with oxygen to form sodium sulfate. Hydrazine and organic chemicals are also used. Filming amines and ammonia combat return-line corrosion.

Oxygen and carbon dioxide can also be removed by deaeration. But most small installations cannot justify the cost of equipment. A small packaged deaerator for this service is not yet available. Larger installations, or those where percent makeup is high, can often justify deaeration.

13 What is priming and foaming?
A Priming is the discharge of water slugs with steam caused by violent bursting of steam bubbles. Causes include high water level, boiler overload, highly concentrated boiler water and dirt or oil in water.

Foaming takes place on the surface of the boiler water. Causes include finely divided suspended matter and high boiler-water concentrations.

14 Foaming and priming—How is it prevented?
A Assuming that boiler is properly designed, check for possible overload. Next see if steam separator is working. Then analyze boiler salines for alkalinity and total dissolved solids. Use blowdown if solids are too high. If alkalinity is too high, reduced feed of caustic may be the answer. Antifoam chemicals will help prevent foaming.

15 Is embrittlement a problem?
A Embrittlement or intercrystalline cracking of boiler metal causes damage that's expensive to repair. It may mean replacing whole sections of the boiler. Many feel that excess boiler-water caustic is the cause. To prevent this, maintain recommended ratios of sodium sulfate to alkalinity in the boiler. Compounds containing nitrate, lignin or tannin are often helpful.

16 What are boiler compounds?
A Compounds are chemicals used for internal treatment. Organic and/or inorganic materials are compounded into liquid, paste, solid or powder form. Common ingredients include: bark and wood extracts, starch, agar, flaxseed, gelatine, sodium silicate, sodium hydroxide, sodium phosphate, soda ash and sodium chromate. Plant operators find them a real asset in feedwater treatment. Standard compounds are available in easy to handle form for almost every feedwater problem. Operators' job then boils down to feeding recommended amounts according to routine control tests.

17 How are compounds fed to the boiler?
A Method used depends on feedwater cycle, amount of makeup, and boiler pressure. Cost of feeding equipment is also a factor. Chemical solution is often pumped into the boiler drum distribution line. When makeup is low, pump can be operated manually at regular intervals or automatically on a timed cycle. Where the makeup is high and control more critical, use a pump arranged to automatically feed chemicals in proportion to boiler makeup flow.

Pressure solution feeders or pot type feeds are often used where accuracy of feed is not critical. Chemical can also be added in batches to heater or hot well if care is taken to avoid deposits in feed lines to the boiler.

18 What external feedwater treatment is commonly used?
A External treatment usually has a place where the percent makeup is high. Some installations use a zeolite softener supplemented by internal treatment. Other boilers require more elaborate treatment. Newest method is salt regenerated anion dealkalizer following zeolite softener. Alkalinity is reduced by exchanging bicarbonates and sulfates for chlorides. Effluent has about 15 to 30 ppm alkalinity.

If water supply is turbid and contains bacteria, a packaged type coagulator is the answer. Follow this unit with filters and zeolite softeners.

Demineralizing process may be called for if the boiler pressure is unusually high. High makeup may also call for its use. But this equipment is expensive compared to other treatments and regeneration requires acid and caustic handling. So, although its use for the packaged-boiler field is limited, keep it in mind when you *do* need high-quality solids-free water.

19 What maximum total dissolved solids is allowed in the concentrated boiler water?
A Most packaged boilers allow from 1000 to 3500 ppm. Actual figure for a given job depends on the type of boiler, operating pressure, percent makeup and makeup analysis.

20 How do you estimate amount of blowdown?
A Blowdown is about equal to total-dissolved feedwater solids divided by total-dissolved boiler solids, usually expressed as a percentage of the total feedwater flow.

21 Blowdown—How often?
A Blowdown low-pressure heating boilers very intermittently to remove sediment from bottom of the boiler and lower boiler solids during startup. Since there is no makeup, the blowdown valve should seldom be used.

Regular blowdown schedule is needed for high-pressure boilers and those with high makeup. Frequency depends on rate of boiler operation, amount and analysis of makeup.

22 Who supervises the feedwater treatment program?
A Boiler-water treatment involves more than just scale prevention. Kind and control of chemical treatment, regulating preheating and pretreating of makeup, regulation and control of boiler blowdown to keep right boiler-water solids are all just as important. Supervision and control of this program calls for a qualified operator or a firm specializing in boiler-chemical service. Prevention of feedwater trouble usually costs far less than the repair of neglected equipment. Final proof of any feedwater program is to observe its effect on the boiler waterside. Make internal boiler inspections a regular part of your routine maintenance program.

Here's your pressure-pilot

Applications are many and varied

Oil burner. Atomizing steam pressure is regulated to oil burner at preset amount, above the oil pressure. Steam pressure under diaphragm is balanced by oil pressure and adjusting spring on top. Once set, adjusting spring force is constant. As combustion-control valve raises oil pressure, regulator raises steam pressure until pilot diaphragm is again balanced. Improves combustion.

Boiler feedwater. Valve controls pressure drop across boiler level-regulator when boiler is fed by motor-driven centrifugal pump. As demand for feed decreases and level-regulator throttles, main regulator also throttles, offsetting increased pump-discharge pressure. It maintains inlet pressure to level-regulator at constant differential over boiler pressure. Self-operated, no air needed.

Makeup air. Here we provide control of makeup-air temperature with varying outdoor-air temperatures. A maximum coil pressure is set on pressure control of pilot to heat incoming air to needed temperature at lowest outdoor temperatures. Thermostat element then throttles the pilot (and therefore main valve) to maintain constant outlet-air temperature needed. Only valve needed.

Gas-pressure control. Here main pressure is constant from propane-gas generator; valve limits steam pressure in the generator. Usually pilot double spring controls main valve, admitting just enough steam to maintain set gas pressure. When gas pressure drops and calls for more steam, the pilot (small) limits steam pressure on exchanger to the pressure for which pilot is set.

Remote regulator. Hookup regulates pressure at a distant point from regulator itself. Slight increase in controlled pressure throttles the pilot, which in turn throttles the main valve. One big advantage is there's no limitation on distance so long as control pipe is solidly filled with water. Use hookup where noise or appearance isn't too important at accessible locations.

Hot water. Valve controls converters for hot-water heating systems. Pressure element of pilot limits heater pressure on start-up with cold water. As return temperatures increase, thermostat throttles the pilot and main valve. That maintains a constant converter outlet temperature. Operation of this hookup does not depend on outside power source. Modulating control saves steam.

Now let's look

One big difference in this regulator from those featured in the following pages is that its pilot valve is on outside, not in housing. Here, pilot can be removed from valve in a few minutes, replaced if needed. Maintenance men can thus have several cleaned pilots on a shelf, ready to service a large number of pressure regulators throughout

single-seated packless regulator

Electric shutoff. Remote shutoff is by electric control where frequent interruption is needed. When electric power is supplied to the pilot, it opens. That admits steam to the pilot and allows it to control the main valve as a simple pressure regulator. Here interrupting current flow allows pilot to close. That closes main valve. Costs less than separate electrically operated shutoff.

Pressure shutoff. Hook-up provides remote shutoff by fluid pressure where frequent interruption is needed. When pressure is applied to diaphragm (double spring), pilot closes. That denies pilot (single spring) steam needed to open main valve. Pilot, when unloaded, opens, which allows it to control main valve as a simple pressure regulator. Can operate a distance from regulator.

Makeup steam. Valve supplies steam to turbine exhaust main. When turbine load decreases and turbine exhaust is not enough for steam load, very slight drop in exhaust pressure causes pressure regulator to feed correct amount of steam to meet demand. Here pilot-operated accuracy avoids undue pressure drop before makeup starts, which makes steam flow to plant very constant.

Temperature control. This is economical hookup for warehouses, shops, kilns and other spaces where appearance is not too important. Pilot controls steam flow to the installed radiation that's indicated by the temperature at thermostat bulb. Pilot controls the pressure on the heating surface when the temperature pilot opens wide, as at start-up, for example. No air or electricity needed.

Oil preheater. Here's way to control fuel-oil temperature from an oil preheater. When the oil temperature is near the setting of thermostat element, the element begins to remove pressure loading (imposed by the pressure-loading spring) from the steam-pressure control diaphragm. That allows the valve to throttle and control. Limiting pressure minimizes carbonization here.

Domestic hot water. Valve controls hot-water temperature from a tankless heater as shown. Rising water temperature resets pressure loading. This reduces the heater steam pressure to that needed to maintain the load. This hookup eliminates need for a second regulator of the pressure-reducing type. Thermal element is built to withstand accidental over-temperatures.

inside this valve and see how it works

plant. Then pilot can be cleaned in leisure time, placed back on shelf.

Operating cycle. Main valve is normally closed. Placed in service, pressure acts on under side of small pilot valve. Control line then connects pilot diaphragm chamber to delivery piping. Now compressed adjusting-spring forces pilot valve open, see details in drawing.

Connecting tubing conducts fluid from pilot to main valve diaphragm and bleed port. When pilot opens, fluid flows through pilot faster than it can escape at bleed port, creating a loading pressure that forces main valve open.

Delivery pipe and control line are now being filled with fluid flowing

through main valve. As delivery pressure rises, it overcomes force exerted by adjusting spring and pilot throttles. This, in turn, allows main valve to throttle just enough to maintain the set delivery pressure. When demand stops, pilot closes, allowing main valve to close. *Courtesy: Spence Engineering Co, Inc, Walden, New York.*

For best results know

How valve works

▶ PRESSURE-REDUCING VALVE is an automatic regulating valve for controlling the downstream pressure of a fluid from a higher-pressure source. Three general types are: (1) self-contained internal-pilot piston-operated (2) self-contained external-pilot operated (3) spring- or weight-loaded direct-operated, with diaphragm, bellows or piston. A self-contained or self-operated valve uses fluid being controlled to operate its main valve.

Valve shown is for steam, air or gas service, with initial pressures from 25 to 1500 psi. This type is built for temperature up to 1000 F, for reduced pressures from 2 to 600 psi.

Principle is simple. Screw down on handwheel to compress the adjusting spring against the metallic diaphragm. That opens controlling valve, which admits high pressure from inlet-body port to top of piston. Piston opens main valve and admits reduced pressure to outlet piping. This reduced pressure acts through the outlet body port (or through an external control pipe on some valves) on the underside of diaphragm to balance compressed spring.

Action caused will throttle the controlling valve and control the reduced pressure at the set value. Any load change immediately reacts with a pressure change on diaphragm, instantly repositioning the main valve. That maintains the constant reduced-pressure setting. Change in reduced pressure affects the valve action.

Don't guess at upstream pressure. Use gage at control point to indicate line losses due to friction, which may result in undersizing a valve. Be sure that valve you select is suitable for minimum and maximum load and for superheat conditions.

Never buy a reducing valve to match the pipeline size without checking flow requirement as the line may not be right size to start with. Get information from manufacturer's capacity tables to show size for a given flow and valve design. When reducing to 25% of inlet pressure or less, steam expands in volume two and one half to five times. If you don't know the flow, find it with a flowmeter. Or estimate on basis of equipment performance that valve will service.

Photo, courtesy, Leslie Company

*self-contained internal-pilot piston-operated

Controlling valve seat

Controlling valve

Cylinder liner

Seat ring

High pressure

Low pressure

Main valve

Piston

Main line stop-valve

Strainer and blowoff valve

Typical installation of pressure-reducing valve has these elements

Bypass valve

Bypass

Pressure reducing valve

Expander (where required)

your s-c i-p p-o* reducing valve

How to keep it working

1 Close stop valves on each side of reducing valve, release compression on adjusting spring by unscrewing handwheel

2 Unscrew adjusting-spring case with wrench. Remove case, take out spring. Remove diaphragm with a suction cup

3 Unscrew controlling valve-seat with socket wrench provided for purpose. Remove valve spring and the tiny valve

4 Examine seat and valve. If scored or cut, regrind with fine grinding compound. Clean by dipping in kerosene

5 Check clearance between control valve and diaphragm with straightedge and feelers, against maker's recommendations

6 After assembling spring with valve and screwing controlling valve-seat into top cap, remove the cap and gasket

7 Use bolt to remove piston. Examine rings for wear and fit. Remove cylinder liner, inspect the inside for wear

8 Unscrew bottom cap and take out with main valve and spring. Examine for cuts or scored seat, for free movement

9 If needed, grind in valve with liner and piston in place. Clean parts, check for free movement, then reassemble

Here's your air-operated

How the control valve works . . .

AIR PUSHING DOWN on the diaphragm provides power for working this control valve. Valve shown has direct-spring return; also made with reverse-spring return, where air pushes up under diaphragm. Increase in operating air pressure above the initial spring setting, of 3 psi for 3- to 15-psi range, pushes motor stem down. This motion is opposed by the spring compression.

Resulting valve action is known as air-to-close when valve plug is moved toward valve seat; air-to-open when plug is moved away from seat. Either action is from direct motor where design permits inverting valve body and plug. Where design doesn't permit air-to-open action with direct motor, use the reverse-motor type (not shown).

Stop in upper case holds initial diaphragm position. Spring is normally set so motor stem starts to move when air pressure to diaphragm is equal to the minimum spring range. Compression can be changed with external adjusting screw.

Spring load is carried by a ball-thrust bearing in motor having more than 100 sq in. effective diaphragm area. Ball reduces force needed to adjust. It also minimizes torsional stress imposed on spring under compression. Travel indicator shows exact plug position between full open and full closed.

Plug design is important. It directly governs flow through valve. Two most common types are double-seated top-and-bottom guided V-port (shown) and parabolic plug (not shown).

Some applications . . .

HIGH-PRESSURE GAS to transmission line is usually reduced. The angle valve shown is recommended for this service

DISCHARGE PRESSURE on steam-driven reciprocating pump may be held closely by controlling the steam to the pump

STEAM MAKEUP to low-pressure process or heating. A second controller can relieve the excess on the light loads

HEATING-STEAM PRESSURE to reboilers and heat exchangers is controlled to meet the exact process requirements

diaphragm-controlled valve

Pressure controller is this valve's brains . . .

In previous pages of this book we learned that a control valve has a pilot valve. Small pilot is either built inside the larger regulator body or is placed on outside. Pilot of this control valve is part of the pressure controller placed on outside of valve; it is connected to valve's diaphragm.

Supply air passes to under side of pilot plug. A little air passes through metering orifice and serves as nozzle supply. When nozzle is covered by flapper, these things happen: (1) Nozzle pressure builds up on top of upper diaphragm. (2) Diaphragm block moves downward. (3) Pilot plug seals exhaust seat in diaphragm block and opens inlet seat. (4) Output pressure builds up under lower diaphragm. (5) Diaphragm block moves upward.

Inlet seat remains open until equilibrium is reached between nozzle pressure and output. When thus balanced, there's no flow of air into or out of pilot—theoretically. But small bleed between supply and outlet increases

pilot responsiveness when at equilibrium.

Decrease in nozzle pressure opens exhaust seat, allowing output air to escape between two diaphragms— through horizontal ports in intermediate plate. Ratio of effective areas between upper and lower diaphragm is such that a change of about 3.5-psi nozzle pressure causes output pressure change of 3 to 15 psi.

Flapper pivots on pin at movable end of proportional bellows. Bellows is connected to output pressure from pilot. Travel of this bellows is proportional to output pressure.

Increase in controlled pressure causes free end of bourdon tube to move flapper clockwise, tending to cover nozzle and increase output pressure. Proportional (feed-back) bellows tends to lower flapper bearing, causing flapper to cover nozzle. In normal operation, reaction of pressure in proportional bellows keeps flapper in position to throttle a steady output pressure.

Movement of feed-back bellows is proportional to: (1) movement of free end of bourdon tube (2) output pressure of pilot. So output pressure must be exactly proportional to movement of free end of bourdon tube.

Function of proportional band-adjustment knob is to vary effective length of proportional flexure rod.

Control-setting knob is for setting controlled pressure. Index setting scale directly above knob has graduated reference marks between Low and HIGH. Direction knob is moved along scale; indicates effect of adjustment.

Range of pressure is for 0 to 10.000 psi, but all bourdons furnished will withstand 25% overpressure without damage, according to manufacturer. Standard controller takes clean dry air or noncorrosive gas at 20 psi. Air set of reducing-relief valve with filter-drip-well is good insurance in air-supply line as shown in application sketches. *Courtesy: Mason-Neilan Regulator Co, Boston, Mass.*

Water Hammer...

RUNAWAY WATER

CONDENSATE in steam lines cuts down steam's efficiency besides being harmful to system. Water collects from boiler carryover or from condensing in poorly insulated lines. If lines sag like these or are not drained properly, steam blows small pools of water together, collecting them into waves, then slugs. Because steam travels fast, slug is carried at that speed until it reaches turn in line. Then incompressible water slug hits piping with loud hammer. This condition is dangerous

DAMAGES

THERMOSTAT BELLOWS is knocked out quickly by badly pounding steam lines. With water slug slammed against delicate bellows, it's deformed so valve won't seat, then steam blows through trap

GASKET JOINTS are usually weakest part, especially if piping is old and gasket material is other than metallic. Water hammering at joints soon forces water through or blows out gasket

DAMAGED PIPING can be costly, dangerous. A high-speed slug of water often shoots through elbow or tee at end of long line, like a cannon ball. Or fitting may crack, making it necessary to replace

CAUSES — CURES

LOOPS AND DEAD ENDS trap water. Then water's picked up by steam for knockout wallop. Go over old system and eliminate all useless fittings or hookups that condense steam — collect water

VERTICAL LOOP around horizontal line is bad. This 4-in.-dia loop has flow resistance of 32 ft straight piping. Try to run vertical lines against wall, with horizontal lines a foot away from wall

RUNNING DRAIN into large pipe does not do good job. Water shoots over small opening. Right way is to place fitting in line that has large "pot" for collecting water from pipe's lower half

...Damage, Causes & Cures

TRAP WATER BEFORE IT DOES DAMAGE

STOP WATER HAMMER completely if from boiler carryover, water condensing in long lines, lines not sloped downward, or if poorly insulated. First step is to place tees in line as shown, with drain line to traps from flanged bottom of tees. Make sure lines have slight drop toward discharge end if carryover problem is tough. Next, insulate lines completely and of right material for steam temperature. Piping layout shown here will cure old system, save money and prevent outage in long run

STEAM-ENGINE CYLINDER HEADS crack or let go completely from a large slug of water. Water can't leave cylinder fast enough in a high rpm engine. Steam separator above engine gets flooded

BENT PISTON RODS are common when water carries over into steam engines or into steam reciprocating pumps. This is more common when pump looses prime and races, then gets slug of water

DAMAGED TRAPS are common when a heavy shot of water is forced against working parts. And it's no fault of trap's construction. Steady pounding by water hammer distorts or deforms vital parts

DRAINING AT ELBOW of vertical line is important. Tee with drain line before elbow is good. But simpler and most effective drain is to replace the el with a tee and hook to steam trap

DEAD END of headers should be drained. But drain must be at lowest part so water won't collect. Right way is to use tee and tap into side of bottom leg. Then dirt is kept out of the trap

PLACE TRAP well below steam line. At 24 ft shown, pressure of water leg above trap is 10 psi, in addition to steam pressure above water. The lower trap is the higher head pressure is on trap

'Here's *My* Idea, Boss'

THIS PHOTO MAKES ME THINK of an old engineer friend who worked so successfully with people that he eventually ended up as the firm's president. When asked why people always tried to break their necks doing things for him, he summed up his philosophy as follows:

1. Always ask a subordinate how *he* thinks a job should be done before assigning the work.
2. Pay the full price for work, whether in money, materials or manpower.
3. Face issues squarely. Putting them off only stores up trouble, and is a bad example, besides.
4. Use new methods. Don't be afraid to pioneer just because no one else has done it before.
5. Hire consultants who have solved problems similar to yours. Cash in on their experience without wasting your time to find out the answers they already know.
6. Put all instructions in writing to make them definite and specific.
7. Start and finish one big job at a time.
8. Snap decisions are usually bad. If a "yes" or "no" is demanded, usually answer "no."
9. Share honors and bonuses for department with all in department, right down to last fireman. You can't have a "team" by treating only the bosses as prima donnas, or your team will let you carry the ball by yourself. This last is the master secret.

I've seen a lot of success formulas, but this one strikes me as having the slightest margin of error. If you are a youngster just starting, take my advice and paste this in your hat.

2
Measuring combustion

Save fuel and your product

STEAM BOILERS. Boilers that generate power or process steam usually have CO_2 recorders for controlling combustion. But power engineers know that a high CO_2 reading alone isn't proof of efficient combustion. To get the full story, you must also check for presence of unburned gas. Instrument shown takes readings in last boiler-pass to check accuracy of your CO_2 recorder. It also shows completeness of combustion. Charts and tables supplied with instrument find exact combustion efficiency. These charts and tables also show exact CO_2 and excess air in combustion products. They are easy to use.

STEEL FURNACES. Open hearth furnace, steel industries' largest fuel user, has many operating variables. Furnace deterioration causes air to leak in. Variations in fuels and other combustion components also give more trouble. Waste-gas analysis measures over-all effect of these variables. This instrument gives information for adjusting furnace controls. It analyzes gases, disregards fuel variations and isn't affected by gases from the charge. Continuous indication of oxygen and combustible content allows operator to make correction at all ranges of fuel flow needed for top efficiency.

CERAMIC, REFRACTORY KILNS. Steady flow of high-quality product through kilns is important in making brick, refractory and abrasive products, pottery and porcelain. Kiln's atmosphere has a direct bearing on product's quality. So for more accurate control, nature of atmosphere must be correct in all portions of kiln that work passes through. Here instrument serves as accurate, time-saving combustion tool to check atmospheres in vicinity of work, or in all parts of kiln, for that matter. For this service, instrument helps bring each burner into right combustion alinement. Readings can be taken by furnace operators.

Wheatstone-bridge

▶ GONE ARE THE DAYS of guessing at combustion efficiency by glancing at character and color of flame in furnace, measuring temperature and then looking at smoke coming from stack. Today, we depend on instruments to control combustion components: fuel, air, temperature and pressure. But you must make chemical gas-analysis to check efficiency of control process.

Heat prover shown is portable, weighs 25 lb. It has sampling tip and hose, with thermocouple for measuring temperature. Instrument operates from an ac power source, 110 v, 25-60 cycles. It analyzes gas sample for oxygen and

dollars with better combustion

GLASS TANKS. Total fuel cost to operate modern. glass tanks is very large. And, of course, economical generation and application of heat calls for efficient combustion practice. Here, again, production rates and product quality benefit from precise control of glass-tank atmosphere. Problems before combustion engineers in glass industry are like those of open-hearth fuel engineer we tell you about in second item here. Correct control settings are found from analysis of tank gases. So here, again, this instrument accurately analyzes combustion products in the presence of contaminating gases of the bath.

INDUSTRIAL FURNACES. Heating and heat-treating furnaces in metals and metals-processing industries need right combustion. Reason: Work is in direct contact with combustion products in many industrial furnaces. So burner controls must be set to produce oxidizing or reducing atmosphere in line with metallurgy of furnace product. These furnaces cover wide range in size, construction. Material is usually put in one side and comes out other, ready to process. Air-fuel ratio is generally regulated with temperature regulator. Quality of oxidizing or reducing atmosphere is important, making instrument invaluable.

CEMENT KILNS. Cement is highly competitive and cost per barrel depends largely on fuel used for process. As in steel and glass production, combustion products are diluted with the gases evolved from the chemical reactions within the kiln. So direct analysis for oxygen is needed to evaluate excess air in waste gases. In general, burners are adjusted to low oxygen and for zero combustible content. Oxygen analysis and check on combustion efficiency guide the kiln operator. It helps him set his controls so he saves fuel, yet operation is efficient, troublefree. Again, charts and tables show exact efficiency.

circuit measures resistance change

combustible-gas content. Percents by volume of these sample gases are shown continuously and simultaneously on instrument meters.

Steady flow of sample is drawn into instrument by rotary pump. After passing through filter, pump delivers sample simultaneously to both oxygen and combustible analyzing sections.

Sample flowing into combustible section is mixed with air and pumped through the active cell of combustible analyzer. Hot platinum-wire catalyst, suspended inside cell, forms one arm of a wheatstone-bridge circuit. Air-sample mixture burns at surface of wire, in-

creasing its temperature and electrical resistance in proportion to volume of combustible gas in sample. This change in electrical resistance is measured by wheatstone bridge. Bridge meter is calibrated to read directly in *Percent Combustibles*.

Sample flowing into oxygen section is mixed with hydrogen generated in electrolytic cell. Hydrogen-sample mixture passes into active cell of oxygen analyzer. Any oxygen in sample burns with hydrogen at surface of cell's platinum wire. Wheatstone-bridge circuit again measures change in electrical resistance of cell wire. Bridge meter here is cali-

brated to read in *Percent Oxygen*.

Oxygen and combustible meters have both 4% and 20% scale ranges. Dual-range feature affords greater instrument accuracy and sensitivity. Oxygen meter has additional scale for measuring output of instrument thermocouple in degrees F.

Operation is entirely automatic—all parts housed in unit. Scale ranges are selected by operating range switches below each meter. Instruments are checked periodically against calibrated cylinders of laboratory-analyzed gases by firm lending the instrument.
—*Courtesy, Cities Service Oil Company*

MECHANICAL instrument has two temperature-sensitive bulbs which deflect Bourdon gages to work linkage, meters

Here are 3 modern instruments to measure Btu's

1 Mechanical Btu meter . . .

Central heating and cooling in large shopping centers, airports, giant industrial, commercial or residential developments, as well as product quality in the process industries, all involve accurate measurements of heat energy used. The data is needed for accounting and quality control, too.

Today, these measurements are made with the *Btu meter*, an instrumentation system that measures the rate at which a flowing fluid gains or releases heat.

One Btu is the quantity of heat needed to raise the temperature of one lb of water through one deg F. Water is the frame of reference—6/10 Btu is enough to heat one lb of alcohol one deg F; 1/11 Btu is enough for copper. Industrial Btu meters obtain the information they need for calculating Btu by measuring the temperature difference (degree change) across a specific portion of the process. Mass flowrate of a liquid in lb per hr is proportional to square root of the pressure differential.

Specific heat, which is the thermal capacity of the fluid measured, also comes into the picture. Specific heat is simply the amount of Btu's needed to raise one lb of the particular substance one deg F. By the definition of the Btu, specific heat of water at 59 F is 1.

Multiplying specific heat by temperature differential by rate of flow, you get the rate of Btu exchange:

Specific heat (Btu per lb, F) × *temperature difference* (F) × *mass flowrate* (lb per hr) = *Btu per hr*

Thus to measure amount of heat change in a fluid, we must know characteristics of the fluid, then measure temperature change and flow rate. To obtain total Btu exchanged in the process, Btu rate is integrated with respect to time. Either mechanical or electrical integrators are used to do the job.

All the integrator does is register instantaneous Btu

measures rate of fluid flow with a positive-displacement meter which turns a meter shaft. Total flow is indicated on a flow integrator driven by a gear connected to the shaft. Temperature differential is measured by two mercury-filled bulb thermometers. They operate two spiral elements, linked to a cam hookup and a temperature-difference indicator.

A ratchet pawl connected to end of the meter shaft turns with the shaft. The pawl is also connected to a cam follower, which determines what percentage of rotation of the meter shaft it takes for the ratchet pawl to engage the ratchet wheel. Ratchet wheel is coupled to a gear which drives a Btu totalizer. That records total Btu exchanged in measured flow.

As temperature difference increases, pawl engages the ratchet wheel for a greater portion of each revolution of flowmeter shaft. A bigger Btu reading is added to the totalizer indicator for each unit of flow. Cam and linkage fix the proper relationship between flow, differential temperature and specific heat. Output: direct indication of heat exchanged in system.

rate against a time-reference base. For example, suppose in the course of one hr the Btu rate is 10,000 Btu per hr for ½ hr, 40,000 Btu per hr for ¼ hr and 20,000 Btu per hr for another ¼ hr. At the end of that hr the totalizer would have recorded a total of 20,000 Btu: 5000 for the

ELECTROMECHANICAL instrument has flowmeter, temperature-measuring elements, servomechanism, units shown

ELECTRONIC instrument is similar to electromechanical, but this one multiplies flow and temperature electronically

2 Electromechanical Btu meter...

may utilize any type of flowmetering system, although the differential flowmeter (sketch above) is most common. This flowmeter operates in the conventional way. As sketch indicates, flowmeter is coupled to the differential-temperature receiver through mechanical linkage.

Measurement of differential temperature is made either by two resistance temperature detectors or two thermocouples installed in inlet and outlet lines of the system. Output of the temperature-measuring elements operates a servo-mechanism, which positions members of a mechanical multiplying linkage.

Multiplying linkage indicates rate of heat exchange and, with suitable integration, amount of total heat exchange.

This type of Btu-metering system has one advantage over the completely mechanical meter at left. A continuous record of rate of Btu exchange is available: data often needed in the process industries.

With suitable retransmitting accessories incorporated in the system, this meter is used for control.

3 Electronic Btu meter . . .

has same basic components as the electromechanical type. But the multiplication of flow and temperature differential is done through electronic circuitry. This meter has a differential flowmeter to measure rate of fluid flow in the process.

Linear flow signal from transmitter to flow receiver is sent over electrical lines via a differential-transformer-transmission system (sketch above).

Inlet and outlet temperatures are measured with resistance temperature detectors. Flow and differential temperature are multiplied in the Btu-receiver system, using retransmitting potentiometers on the differential-temperature and flow receivers as the source of input signal to Btu receiver.

Btu receiver is a slidewire type. It rebalances output from the multiplication network, indicates and records rate of heat exchange in the metered system. This slidewire receiver also operates an integrator which continuously integrates rate of Btu exchanged to maintain a record of total Btu transfer. This electronic system can also do control jobs.

---

first ½ hr, 10,000 for the next ¼ hr and 5000 Btu for the final 15 minutes of that given hr.

Three basic types of Btu meters are mechanical, electromechanical and electronic (above). But whichever of the three you use for measuring heat energy consumed

in any system, it must do the following: (1) measure flow-rate of the fluid (2) measure temperature difference between incoming and outgoing fluid (presuming no change of state between these two points) (3) compute and integrate product of (1) and (2) to give total Btu used.

Here's an Easy Rule For CO₂

Percent CO_2	Percent Excess Air	Percent loss for room temperature of 70 F and flue temperature of					
		400F	450F	500F	550F	600F	650F
5	314	28	32	36	40	44	48
6	245	23	27	30	34	37	41
7	196	20	23	26	29	32	35
8	159	18	20	23	25	28	31
9	130	16	18	20	23	25	27
10	107	14	16	18	20	23	25
11	88	13	15	17	19	21	23
12	72	12	14	15	17	19	21
13	59	11	13	14	16	18	19
14	48	10	12	13	15	16	18
15	38	10	11	12	14	15	17
16	29	9	10	12	13	14	16
17	22	8	9	11	12	14	15
21	0	7	8	9	10	11	12

The table above is a handy reference, but you won't really need it if you follow these easily remembered instructions

▶ COMBUSTION ARITHMETIC is one of those things you can make as hard or easy as you wish. Its largely a matter of how fine you want to figure. The minute you begin splitting hairs, the simplest combustion problem gets complicated.

On the other hand, if you view combustion from a practical operating angle, and are willing to accept close approximations, some of the important calculations will get about as simple as buying a sack of potatoes.

Rule for Coal and Coke. For this I shall take only one page to explain a simple rule for estimating the stack loss from the CO_2 (carbon dioxide) percentage without referring to any table or formula. This rule is for coal and coke only. Don't use it for wood, oil, gas, bagasse, etc.

It holds because the burnable part of coal and coke is mostly carbon, with only a small percentage of the burnable hydrogen which would throw the rule off in the case of fuel oil.

The exact amount of air required to burn one pound of carbon is 150 cu ft, consisting of 31 cu ft of oxygen and 119 cu ft of nitrogen. The nitrogen doesn't burn. The oxygen burns to the same volume of CO_2, so the products of the combustion are still 150 cu ft (measured cold), consisting of 31 cu ft of CO_2 and 119 cu ft of nitrogen.

The Unavoidable Loss. Now there isn't any way to avoid the waste of heating up the necessary 150 cu ft of flue gas from room temperature to the temperature at which it is thrown out of the stack.

To heat 150 cu ft of flue gas 100 deg takes 27 Btu, which is just under 2 percent of the 14,500 Btu released by the burning of one pound of carbon.

So now we see that if the room temperature is 70 F, for example, and the stack temperature 570 F, then the temperature rise is 5.0 hundred degrees and the stack loss is $5.0 \times 2 = 10$ percent, if the coal is burned without any excess air at all.

Now suppose, with these same temperatures, that 50% excess air is used. Then the total air is 150 percent of the theoretical or 1.5 times the theoretical. So the percent loss up the stack will increase to $1.5 \times 10 = 15\%$ of all the heat in the coal fired.

Total Air from CO_2. The only thing needed to complete the rule is some way of figuring total air from the CO_2 percentage reading of an orsat or CO_2 meter.

Whatever part of the oxygen in the air actually burns with carbon produces the same volume of CO_2. The remaining volume of oxygen goes through unchanged.

The percentage of oxygen in the air by volume is 21. If this number can represent the total air supplied, the CO_2 percentage will represent the part really needed for combustion.

To make clear, take a case where the excess air is 50 percent, meaning that *two thirds* of the air *burns* and *one third is excess*. Then, of the 21 percent oxygen, 14 shows up as CO_2. Working this backward, we divide 21 by 14 to get 1.5, which means that total air is 1.5 times the theoretical requirement.

Now we can set down the rule:

A. Divide 21 by the CO_2 percentage to get the *total-air factor*.

B. Subtract room temperature from flue temperature, and divide by 100 to get the number of hundreds of degrees temperature rise.

C. For the theoretical air volume allow a flue loss of 2 percent for every 100 deg temperature rise.

D. Then actual flue loss is $A \times B \times C$

Example: Room temperature is 80F. Flue temperature is 520F. CO_2 in flue gas is 12%. What is flue loss?

A. $21 \div 12 = 1.75$, total air factor
B. $520 - 80 = 440$, or 4.40 hundred degrees rise
C. Loss $= 1.75 \times 4.40 \times 2 = 15.4\%$

Don't try to carry these computations out to a lot of decimal places. The method isn't that accurate. But check it against several values in the attached table to convince yourself that it does give a result close enough for most practical purposes.

With a little practice you can work such problems through from CO_2 percentage of flue loss in less than one minute.

3
Fuel: oil, gas; burners

You must know <u>how</u>

Oil must be warm ————————————

Fuel oils must be stored and pumped through piping while warm. Too little thought is usually given to this important fact, often causing trouble after system is installed. Today, plant operators have a bigger headache since heavy fuel oils are becoming heavier with higher pour points and viscosities. This means your present fuel system won't do its job, although it once might have worked.

Three points must be considered. (1) You must store oil warm enough so it can flow toward your suction line for removing it from tank. (2) Oil must be easily handled in piping system from fuel oil storage tank by fuel pump and in quantities demanded by burner. (3) Oil must be delivered to the burner in a state ready to burn. So you can see that these are three separate steps.

Pumpability. Experience has shown that 5000 ssu is the limit of easy pumpability. Also that a No. 6 oil of 100 ssf at 122 F will reach the 5000 ssu when its temperature is about 83 F. Today, No. 6 oils have a commercial range of between 100 and 200 ssf at 122 F. Often this reaches 300 ssf at 122 F. For heavier grades, storage temp must be close to 110 F instead of 83 F.

Some heavy fuel oils may have a maximum pour point as high as 90 F although standard ASTM pour test may call for a much lower pour point. This is because pour point of residual fuels is not certain; it depends on initial temperature before cooling and time held at this initial temperature. So anything less than 110 F suction temperature might give pour trouble even if chart viscosity temperature characteristics show pumpability at lower temperature. Then multiple-tank installations, tanks exposed or in wet soil and tanks buried below ground in cold climates must have some means of heating oil at suction inlet to 110 F. And heating must be at rate you withdraw oil. Return unburned hot oil to tank bell at suction. But this may not solve problem on cold starts. *Courtesy, Preferred Utilities Mfg Corp, New York, N. Y.*

Heating in the storage

Suction → ← Hot return

Suction bell pockets warm oil

Suction bell

Most Codes require oil to be withdrawn from top of buried or vaulted tanks through suction stub reaching 6 to 8 in. from tank's bottom. Where heated oil is frequently pumped into smaller tanks, you may have no trouble if suction is near return, top sketch. But second sketch shows a better method. Here inverted bell traps the hot return oil that suction and return lines run into. This simple change of installing a bell may solve a serious pumping problem.

Ribbon heat ————————————

Thermostat
Lead-in wire —→
Capillary tube
Sensative bulb
Ribbon heat

440-v supply
← Step-down transformer
Magnetic switch
Thermostat
Heating cable

Now, electric line heaters keep your heavy oil hot on startup

Ribbon heat is a low-temperature lineal heater in form of flexible tape ½-in. wide and 1/10-in. thick. Heat is from electricity passing through four special resistance wires. All are insulated and waterproofed.

The end of this special heating tape is attached to waterproof connection of 3-ft non-metallic sheathed cable. Entire assembly is sealed at each end, not harmed by mild acids or alkalies. Use where maximum fuel temperature flowing through line is not above 170 F. Plants have been shut down where hot-storage oil flashed in the pump. All of ribbon-heat units are single phase of 110-220 or 440 volts. And they must be installed across one phase of the polyphase supply. The 440-v units in this line have a single-pole pilot thermostat operating the coil of a two-pole relay.

to heat residual fuel oil

tank is important

Electric heater

Suction-stub electric heater shown keeps oil at needed temperature inside suction-stub pipe section during shutdowns. This is a low-watt density, thermostatically - controlled immersion heater. Wattage available in this suction heater is low. It is not for heating oil other than to keep it pumpable during shutdowns. Thermostat is usually set for 100 to 110 F. Place a drain in concrete enclosure around manhole if water collects. This prevents flooding.

Coil and bell

Coil heater combined with suction bell is very efficient. Coil in bell heats oil as it is withdrawn by pump. Here very little heat is needed as heated oil does not mix with large quantities in the tank. Besides, hot return from burner is also to bell. Hot water may be used instead of steam in coil, depending on local ordinance and hot water available. If high-pressure steam is used, control with temperature control valve. To play safe, send exhaust to waste.

Field-storage tank heater

Large field storage tanks should have a suction- or tank-outlet heater. Open end of heater shell extends into tank. Fuel suction is on closed end, outside of shell. If you have oil return line, connect outer part of shell so hot return oil does not dissipate with stored oil. To repair coils, close large clapper valve so coils can be removed without emptying tank. Equalizing valve fills the heater with oil, helps you open clapper valve when heater is empty.

All electric heating

Electric suction-stub heater is in vertical suction line in storage tank. Ribbon heat around long suction piping keeps oil in pumpable state. Electric heaters in tank bring oil temperature to fire point needed for burning. Data for selecting proper unit or ribbon heat sizes are based on piping covered with at least 1-in. thick sponge-felted asbestos or its equivalent. Such insulation fits firmly over flat coils.

Electricity plus steam or water

Suction-type steam or hot-water heater is added here. Ribbon heat is for holding oil temperatures in insulated pipe lines during shutdown or out-of-service periods. Their purpose is to maintain the pipe line at a suitable temperature by making up for radiation and convection or conduction losses from piping. So don't expect ribbon heat alone to maintain oil temperature when flowing through fuel line.

Pipe Your Fuel Tank...

THE OIL SYSTEM

1 Here's a good hookup for industrial system burning heavy oil. Oil is preheated in storage tank for easy pumping. Suction and discharge strainers keep burner tips from plugging. Both steam and electric service pumps safeguard against most plant emergencies. Air chamber prevents joints leaking with burners closed, also dampens pulsations. Oil heaters keep oil at fire point. Return lines purge air into storage tank with unused oil. System is preferable to one with dead ends

ABOVEGROUND

7 Horizontal aboveground tank must have a fireproof support under one third of circumference. Here oil enters, leaves tank near bottom from both ends

8 Safer way to hook up same tank is to run fill line above tank and down inside. High and low suction lines with valve at the top assure against leakage

9 Inclose indoor tank in concrete box and cover with sand. Protect suction line from damage. Connect piping to tank with welded couplings or bushings

...For Best Results

PIPING DETAILS

2 Buried-tank capacity is usually less than 30,000 gal. It must be at least 5 ft. from building foundation, should be below all outlets attached to tank. Fill line should be below suction-line level, or it must have a trap so it won't act as vent. Pitch suction line toward tank; install check and shutoff valves near tank

3 Have the suction 2 in. from tank bottom for light oil, 4 in. for heavy oil. That keeps sludge, water, sediment out of the system. Remove impurities yearly.

4 Trap on return line prevents vapors escaping when oil is low. Tie vent in top of tank, have flame arrester. Size vent to prevent pressure when filling

5 For aboveground tank, place check valve in return oil line as shown. That prevents siphoning oil out of the tank. Also have trap on end of the line

6 Extend fill line below suction line or place a trap on fill line's end so it cannot vent. Don't cross-connect fill line with any other lines, or to vent

BURIED

10 Safest oil storage is underground tank. Line for pumping out water and sludge is needed in large tank. Piping hookups shown are based on good practice

11 Heavy underground tanks sink, cause trouble unless buried in good spot. To prevent floating in wet ground, lay on a concrete slab, anchored with rods

Stop Gas Leaks Now...

WRECKAGE AND ITS CAUSES

SERIOUS ACCIDENTS like this mount yearly as gas leaking into basement from the high-pressure main explodes. Cost to industry runs into millions now that gas is being widely used

FURNACE EXPLODES under all types of boilers when gas leaking into gas-fired boiler is not purged. Accumulated gas from leaking burner was ignited by hot shutdown furnace here

PREVENT WITH RIGHT PIPING

BYPASS gas around crawl space from outside so leaking gas won't follow the piping inside. Insulate against freezing

IF YOU MUST run long piping under crawl space, do as shown. Put meter in occupied space or in vented building

CASING around gas line coming up from crawl should have removable seal for inspection. Make space for checking

SAFEGUARD METER . . .

PLACE vacuum relief valve between gas compressor, or booster pump, and meter if meter is for less than 5 psig, *left*. Check-valve guards against backpressure where air or oxygen exceeds gas pressure or where the gas from the h-p system may back up

PIPE atmospheric chamber of pressure regulator to outdoors or to a constant burning pilot. *Never leave vent unpiped*

...From Wrecking Your Plant

SMALL ARC from switch wrecked this compressor house when operator flipped on light switch. Slow-leaking gas valves let in enough gas to damage many plants like this one yearly

CORRODED PIPE *(above)* at ground level is main hazard. Gas leaks into unused part of building. Cracked main *(right)* under building lets in gas, often is ignited by a pilot light

BURIED piping coming up through floor must be encased, sealed with removable seal at one end. Other end is as shown

VENTILATE crawl spaces or unused spaces if piped for gas. Seal walls, floors around piping to localize escaped gas

WET FLOORS corrode piping. Use floor sleeve high enough above floor to keep water from line. Paint all gas piping

... AND FURNACE

GAS-BURNING furnace needs this hookup for purging. Air-pressure switch in ventilating duct prevents ignition of accumulated gas before furnace is properly purged before relighting. Flame failure device shuts off gas if flame fails unexpectedly

How modern burners match their jobs

Characteristic new features of industrial oil and gas burners

Burner type	Fuel	Principal new features	Type of flame	Excess air	Applications	Remarks
Mechanical atomization	Oil	Higher fan and pump speeds for compacter units. Better pump cut-off. Safer flame control. Smaller capacities	Constant size	Low to medium	Steam and hot water boilers. Air heaters. Conventional industrial processes	Air heating, recirculation of gases cut smoke, save fuel. Europe setting pace
Steam or air atomization (high-pressure)	Oil	Wide turndown	Strong spray pattern	Low	"	"
Steam or air atomization (low-pressure)	Oil	Even wider turndown	"	Low, added air may be entrained	"	"
Air atomization (medium pressure)	Oil	Widest turndown of all	"	High, with designs to maintain stable flame	"	Recently introduced into the U.S.
Ultrasonic	Oil	Several types using ultrasonic or resonance principle, 0.3 to 25 gal/hr	Constant	Low to medium	"	Vibration obtained electrically or by steam or air
Vertical rotary (central or wall-flame)	Oil	Improvements over old burner types. For example, preheating of wall rings	Constant, blue or luminous	Low	Mostly space heating	Many of these, or combinations, used in Europe
Vaporization	Oil	Improvements over old types. Residence time of oil in burner reduced; excess oil drained out	Variable within narrow limits	Low to medium	Space heating where noise is strongly objectionable	In the U.S., API and oil companies contribute to R&D
Infrared	Gas	Protection against burning of product obtained by quick-cooling radiant refractory, or retractable burner elements	Radiant	Low	Drying, curing, baking, finishing—in textile, paper, plastics, veneers and ceramics industries	Must be applied judiciously to specific industrial processes
Torch, premix or nozzle mix	Gas	Variable heat input. Turndown of 40 to 1 possible. Good combustion, with little CO and NO_x	Tailored at will	Low to high	Direct heating of air for space heating or industrial processes	
High intensity	Gas or Oil	Good heat penetration even at 15 ft from tip. Advantageous with recuperative generators	Long, very stable, may develop 4×10^6 Btu/hr per cu ft	Very high, can be low with secondary air	Ceramic industry. Heating of castings, bars and billets. Rotary dryers, calcining and roasting	Used in current revival of batch ceramics kiln
Short-flame or disc-type	Gas or Oil	Simplifies construction of utilization equipment. Strong radiation or convection heating	Flat flame; no forward travel	Low to medium	Melting pots, crucibles, forge furnaces; also batch kilns with side-burners	
Variable luminosity	Gas and Oil	Radiant or convection heating. Small amount of oil cracked	10% of oil in gas gives luminosity	High	Application to processes when load or pattern changes	

4
Steam turbines

STEAM TURBINE OVERHAUL is a major job requiring skilled technicians, proper tools and a check-off list of repairs and checks

Let's Inspect and Overhaul Your Steam

▶ TURBINES ARE STURDY, need little attention. But you must open them for a general check-up every so often, depending on service and operating conditions.

Check-Up. A machine running at constant load with few shutdowns need not be opened as often as one on start-and-stop operation. That kind of service causes high-temperature stresses from quick-starting, sudden load changes.

If you're not prepared to do the job yourself, be sure to have turbine firm supervise this important work. An experienced man will know what telltale signs to look for and how to correct them so machine won't have to be opened again until next scheduled shutdown.

Bearings. Turbine shown has ball-bearing thrust. Most operators find it good insurance to remove thrust and replace with new every time turbine opening is scheduled, regardless of wear. Reason is that thrust bearing's failure can wreck internal parts by rubbing them against each other.

Flexible Coupling. If you have this type of coupling, dismantle and check for wear, shoulders and galled surfaces. If coupling does not work right, it may transmit thrust from connected machine to thrust bearing. Then overloading

may cause failure. Replace worn parts, clean and assemble coupling.

Right way to check clearance between split bearing and journal is with inside and outside micrometers. Mike shaft with outside micrometers, then mike assembled bearing with inside mikes at various points. Take the difference, and that's the clearance.

Rotor, Stationary Elements. Turbine's efficiency depends on high-velocity steam. So slightest internal defect can be costly. First, keep steam clean so there is no deposit on blades. Soluble deposits usually form at earlier turbine stages, insoluble deposits on middle and later stages. Whatever kind, deposit restricts steam flow, cuts efficiency.

Remove deposits, inspect moving and stationary blades carefully for erosion, corrosion and cracks. Cracks form around blades and spread out from

OPEN TURBINE CASING, remove rotor. In photo, hardwood spacer slab prevents sling's wire cables damaging the blades

to be made under supervision

Turbine Now

the jacking-holes in the turbine rotors.

Corrosion can be serious if valves don't seat properly in sealing, drain, throttle and header valves during idle periods. Then steam condenses in turbine, doing damage. Answer is to regrind valves and keep them tight.

It's good practice to indicate rims of turbine wheels while machine is open. Reason is that years of operation and temperature changes cause growth and distort turbine casing, rotor and stationary elements. Turn rotor slowly while indicating wheel rims. Also, take clearances between each wheel at rim —check with previous figures.

Safety. Don't take chances. Use right lifting gear and tools. Have only responsible men on job. Clean all parts thoroughly and remove dirt with vacuum hose. Check everything yourself to make sure no slipup increases outage.

INSPECT wheel jacking-holes for cracks; all moving and stationary blades and nozzles, for cracks, erosion, corrosion

QUICK clearance check of main bearings without taking leads is to use feelers as shown before taking bearing off shaft

DRIVE new thrust bearings on shaft if old are worn. It's good insurance to install new ball-thrust at each overhaul

RETAP all threaded holes so screws enter easily when parts are assembled

CARBON RINGS are used here for both interstage and casing-end sealing. Rings are made in segments, held together with garter springs, and kept from turning by metal stop on spring. These rings work like piston rings—sealing against shaft and side of groove they are housed in.

Clearance between shaft and ring must be right. Keep to turbine maker's recommendations. Before checking clearance, polish shaft section that rings seal, so surface is right. If too much clearance, polish ends of segments with fine abrasive cloth. But this job must be done right or ring ends will leak steam.

If there is much refitting, best practice is to turn piece of short stock same diameter as shaft and use it on work bench to fit rings onto. Carbon dust or solids from boiler carryover often cock rings in groove, causing them to break. Or impurities work into segment ends, causing leakage.

Usually, you have only to clean carbon dust from sides and ends of ring, clean sides of grooves and reassemble rings around shaft. Rings are precision made, so give job to good mechanics.

POLISH shaft sections with a fine abrasive strip where the carbon rings seal

RUB the ends of carbon rings lightly against fine abrasive cloth to clean them

Carbon seals prevent leakage between the various stages

ASSEMBLE the ring segments on shaft and hold together with garter springs

CHECK clearance against manufacturer's recommendation. Sand ring ends if needed

Right clearances do better sealing job, save shaft

LABEL all segments so they go back into the right places, to avoid trouble

REPLACE ring segments in stationary diaphragms after cleaning with vacuum

CLEAN groove sides of carbon-ring housing as sides are also the sealing surfaces

SMEAR sealing compound in turbine housing to seal the lower half of gland box

Carbon rings seal against groove side and shaft, like piston ring

High - pressure steam leaks under carbon-ring housing if it's not sealed

Use vacuum cleaner to remove all dust and foreign particles. Machine's inside must be kept clean

PLACE three strips of light condenser cord on sealing compound as shown here

LOWER half of carbon-ring casing is in place here and sealed against h-p steam

REPLACE soft packing at housing end that keeps oil from traveling on shaft

LOWER rotor assembly into place after removing the wooden blocks and the tools

WITH rotor in place, insert the garter springs in groove, fasten around the shaft

METAL stops prevent carbon rings turning. Move rings so the stops fit in slots

WHEN RINGS are all assembled, move metal stops down into position before closing up the carbon-ring casing. Rub light coating of sealing compound on casing joint, but be careful not to smear on the carbon rings or on the garter springs

Careful work, with every man doing his job right until machine is "buttoned up," will keep this unit running many months until the next major overhaul

PLACE top half of carbon-ring casing carefully in place over coated joint

POSITION carbon-ring segments into the diaphragms against the spring as shown

COAT diaphragm's joint face with graphite and cylinder oil before lowering in place

DIAPHRAGMS are all in place between these moving blades

CLOSING the turbine means an anxious period because many things can go wrong to cause delay. From around casing, clear all tools and small parts that might fall in, making you lift cover again—or, worse yet, that might go unnoticed until turbine is turned over with steam. That's why you should use only responsible mechanics.

Before spreading sealing compound on casing joint, **again** vacuum the joint surface so tiny foreign particles dropping on joint will be removed, instead of shoved inside. Compound used here on joint is Copaltite. Apply only when casing is ready to be lowered after asbestos string is placed as shown. On very large turbines, compound is usually applied with a short, stiff paint brush. Here, it was spread thin by hand to cover surfaces evenly and without smearing on

Use cylinder oil and graphite sparingly or it will end up in the boilers

COAT outside of all diaphragms with graphite and cylinder oil as shown here

RUN this fine asbestos string, three turns, all around the coated casing

HEALTHY SMACK against wood block helps seat diaphragms

CASING joint is well-coated with sealing compound—asbestos string placed

"LOWER AWAY, and here's smooth sailing because this turbine is shipshape"

the other parts, such as studs, carbon rings, rotors, etc.

Asbestos twine gives body to compound, and makes a triple seal for steam to work through if there are low spots in the joint surface.

After casing is lowered, take up on all nuts evenly, only hand-tight. Then tighten nuts by staggering from one side of the casing to the other, instead of starting with one nut and going all around.

Scheduled maintenance, as shown in these photos, is the only answer to preventive maintenance. But, again, I should remind you that unless you are set up to do the work, it will pay to have one of the manufacturer's field engineers supervise the job. Before opening a turbine, always have on hand spare parts that might be needed—in good shape.

Fix Those Seals

HOW GLANDS WORK

TO START, open gland steam. Steam blowing: out around shaft, seals it; through turbine, is condensed by cooling water

WITH MAIN STEAM CRACKED, leakage past shaft blows back in sealing line to end. No sealing steam needed.

WITH FULL PRESSURE on turbine chest, more steam's at hp end than needed. Excess leakage helps l-p turbine stages

GLAND TROUBLES AND CURES

TURBINE VIBRATES and screeches when carbon sealing steam pressure's too low. Then rings ride shaft; hug, overheat it

GIVE FULL STEAM BLAST to carbon rings to 'lubricate' shaft and lift rings. If vibration doesn't stop, slow turbine fast

SHAFT EXPANDS from friction when sealing steam fails. Carbon rings wear grooves fast and shaft glows with heat

PLACE EACH SEGMENT in vise between wood strips. With 3-cornered scraper, carefully remove shiny inside surface

FILE from one end of each segment with fine mill file until assembled ring has 0.001 in. clearance, for each in. shaft dia

PLACE RINGS in box, exactly as they were in gland (on big job). Ring sides are also sealing joints; watch assembly

Before They Knock Out Turbine

CARBON RINGS in gland restrict steam leakage out and air leakage in. Rings are in segments, held by garter springs

LABYRINTH GLAND has soft white-metal segments in dovetailed grooves around shaft. These restrict leakage at shaft

COMBINATION labyrinth with leakoff, plus steam-sealed carbon-ring gland. Grooves at labyrinth break steam path.

OVERHAUL CARBON RINGS at set periods. First step is to mike turbine shaft and turn short casting to exact shaft dia

REMOVE RING SEGMENTS from gland. Segments are numbered at each end and each segment has a number. Watch this

LAY SQUARE END of casting on surface plate. Assemble one carbon ring against it, and check clearance with feeler gage

REMOVE LABYRINTH SEGMENTS for inspection by light taps against wood strips. If they do not slide out easily, stop

RUST in gland housing often holds labyrinth segments. Fill dovetail slot with penetrating oil and let set overnight

REST SEGMENT in vise on wooden block. Upset sealing edges with flat-tipped chisels so metal almost touches shaft

Solid-Wheel Turbine

How Turbine Works

"SOLID WHEEL" TURBINE is an impulse, helical-flow type, used mostly for noncondensing service. Steam issues from expanding nozzle, at high velocity and enters side of wheel bucket. There steam direction is reversed 180 degrees. Stationary reversing chamber returns steam to moving wheel, where process is repeated until most of steam's energy is spent.

These machines are sturdy, compact, and can take a lot of abuse. Blades and wheel are one piece, and blades have large clearances, with double-rim protection. End play won't damage turbine, and blades won't foul.

Close blade clearance isn't too important because power-producing action of steam acts on curved surfaces at back of buckets. Blade wear has little effect on performance. Steam enters and leaves at right angles to shaft, so there is almost no end thrust. Rotor can be removed from shaft because it's not shrunk on.

If not loaded, turbine can be started up cold in a few seconds. (But always warm up, if possible.) With large wheel clearance and little load on thrust, danger of blades rubbing is small. But because of all these features, this turbine is often neglected and abused. For best results, give it the attention shown here.

Look Inside

STURDY machine stands punishment, but don't be afraid to open for needed maintenance. Check shaft bearings for wear. Allow clearance recommended by maker for your machine. Main bearings also take thrust wear. If over 0.025-in. thrust clearance, renew bearings. Check shaft to see if it's bent, worn at bearings and at carbon rings. If shaft must be replaced, loosen shaft nut on side of wheel, and press wheel off shaft.

Don't wait until the wheel rides on the bottom reversing chamber. With shaft and wheel removed, inspect nozzle, then run wire through it. Also, inspect reversing chamber for wear.

When starting up turbine, first open exhaust so machine warms up gradually. Make sure drain is open so condensate is removed as fast as it forms. Drain should be left open when machine is not running. After machine is warm, crack throttle and turn over slowly until all water works out. Make sure oil rings in bearings turn before leaving machine.

Carbon Seals

FRICTION expands shaft if sealing steam fails, grooves

CLEARANCE between shaft, carbon rings must be right

SEAL against casing—shaft, same as with a piston ring

CARRYOVER from boiler deposits solids, then rings stick

Is Tough Work Horse

Watch These Details_____

Dia wheel	Lap
18 in.	1/16 in.
24 in.	1/16 in.
36 in.	3/32 in.

CORROSION is tough if drain is closed on idle turbine. Water floods casing

LAP changes with thrust wear. Check every time turbine is open—inspected

WEAR of 1/8 in. to 1/4 in. on bucket lips opposite the nozzle will down efficiency

GOVERNOR is constant-speed fly-ball type, spring opposes centrifugal force of governor weights. As speed rises, weights move out, closing governor valve. When speed drops, weights move in, allow valve to open. Look for knife edge on weight lever's ends—so that governor will not stick. Slide contact and slide shoe wear. Governor yoke wears into shaft.

Seat of governor valve wears—avoid this by adjusting steam with nozzle so valve does not throttle. That also prevents chatter. If governor hunts, look for improper valve setting, sticking or lost motion in mechanism. Find lost motion: move governor lever by hand, observe all parts.

Common fault is tight valve-stem packing that doesn't allow valve to work freely. Loosen packing or repack. Sticking is from bent or bruised valve stem—or valve wings rusted or scaled. Molded sleeve metallic-packing used needs no lubrication. Just keep packing tight enough so it doesn't leak.

PARTICLES of wear—carry-over wedge ring ends open

RINGS COCK from sediment, then they break very easily

FILE ends until clearance is right between shaft—ring

SCRAPE uneven surfaces to fit shaft for a perfect fit

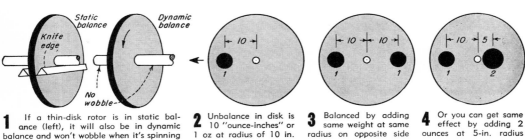

1 If a thin-disk rotor is in static balance (left), it will also be in dynamic balance and won't wobble when it's spinning

2 Unbalance in disk is 10 "ounce-inches" or 1 oz at radius of 10 in.

3 Balanced by adding same weight at same radius on opposite side

4 Or you can get same effect by adding 2 ounces at 5-in. radius

5 If you stack balanced disks to make a rotor, it will be in perfect dynamic balance

6 In this case, weight added opposite unbalance gives static and dynamic balance

7 But placing a weight this way leaves a bad wobble despite perfect static balance

8 Half turn later pressures on both bearings are reversed, meaning bad vibration

How to Balance

▶ IF A ROTOR WILL rest in any position on level knife edges (Fig. 1, left) it's in *static* balance. If it will also spin fast without shaft wobble, it's in *dynamic* balance.

Any rotor in dynamic balance is automatically in static balance, but the converse doesn't hold except for a thin disk. So don't assume a rotor is in running balance just because it balances perfectly on knife edges.

Fig. 2, 3, 4 show how to balance a disk. Find low point on knife edges. Add weight on light side, or drill metal away on heavy side until rotor balances in any position.

Stack of disks, each separately balanced in this way, makes a perfectly balanced rotor.

Fig. 7 and 8 shows that balancing a rotor statically won't balance it for running unless weights are at right point along length of shaft.

To balance a wobbling rotor, correct each end separately according to the wobble at that end as explained in Fig. 11.

A & B = unbalance
a & b = corresponding corrections

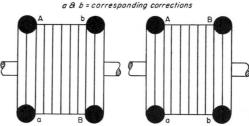

9 Here is perfect correction for two ends in unbalance on opposite sides of the rotor

10 This gives perfect correction where both ends of rotor unbalanced same side

Work in this disk to stop wobble of A

11 Spin the rotor in working bearings and pencil mark "High" side of shaft A. Add weight opposite this mark at left end of rotor until left-end wobble almost disappears. Then do same for right end. Repeat for both ends. Make a third round if it's necessary

NOW you can 'shoot' your bearings in line

At long last, here's a telescope that aligns your bearings within a thousandth of an inch or two, and does away with the ancient time-consuming tight-wire job we all dreaded

Turn page for details ▶

These targets with mikes are heart of scheme..

MICROMETER target holder is a steel plate shaped to hold three inside micrometers. With target usually placed in first and last bearing bore of machinery to be aligned, micrometers are adjusted to position the cross-wires in exact center of bearing bores. Telescope is set on zero, instrument adjusted horizontally and vertically so target cross-wires and telescope reticle cross-hairs are in same line of sight.

This line of sight then becomes a reference line for the entire length of machine. Targets are centered with micrometers in all other bores of turbine and generator stator. Then measure center-line displacement of these other bores from reference line on micrometer and record. From these readings, plot curve on cross-section paper showing "as found" alignment of the bearings.

INSIDE MICROMETERS center the targets, the telescope then establishes a reference line . . .

.here's your know-how

.you're ready to align other bearings

▶ NOT WIRE, but an alignment telescope is used to align turbine-generators at Consolidated Edison Company of New York. This is more accurate, easier to use, and faster than the old tight-wire method, which has been standard for years.

Alignment telescope has been used in the aircraft industry to position in space fuselage and airfoil wing sections without needing large steel jigs.

Problems. There are two basic problems in aligning a turbine-generator: (1) positioning the three to six bearing supports so turbine spindle or spindles are in line with generator rotor shaft at coupling or couplings. (2) allowing the stationary parts (generator stator, turbine cylinders and all fixed internals of turbine—diaphragms, blade rings, dummy rings, etc) proper clearance for moving parts of machine.

Tight-wire method is to stretch a small-diameter wire through turbine-generator with wire ends positioned on proposed center line of machine. Once tight wire is in place, the various parts are aligned to wire with inside micrometers. You measure from wire to round openings or bores in parts to be positioned. By comparing micrometer measurements and doing a lot of arithmetic, relative position of parts to wire are found and parts repositioned accordingly.

This may sound simple if you haven't tried to align with wire, but it's a time-consuming process. Wire is subject to vibration, and is a hazard and obstruction to work going on. Besides, the many micrometer measurements and involved arithmetic often lead to human errors.

Telescope Method. We selected a 60-power Farrand Alignment Telescope, capable of focusing from 18 inches to infinity. In the eye piece is a reticle, horizontal and vertical micrometer scales, and a precision level. Both the large range of displacement and level are desirable for turbine-generator alignment.

Telescope's optical center line replaces tight wire as center line is projected throughout machine's length. Optical center line is set level with the built-in precision level. This optical center line can be displaced vertically and horizontally, plus or minus 0.120 in. Telescope is sighted on cross-wire targets mounted in a target holder, fitted with standard inside micrometers at sides and bottom. Targets fit into bore diameters of 12 to 40 inches. Telescope may be mounted on an end-bearing pedestal or on a separate stand, apart from turbine, *photo.*

Results. Alignment-telescope method is more accurate, less costly, takes less time. Besides, it permits other work to go on in area, even installing or removing of internal turbine parts.

We developed telescope method for heavy machinery on two different turbine-generators. A third unit (53,000-kw topping job) was then completely realigned this new way. Existing alignment was found by centering optical center line through No. 2 and No. 4 bearings. For final realignment, we established a level optical center line through No. 6 bearing. We aligned other bearings, generator stator, and rotating and fixed turbine parts to this center line. As a result, we had to realign and regrout three bearing pedestals and turbine sole plates.

On installing the turbine spindle and generator rotor, all coupling checks showed satisfactory alignment.

Turn page for more details ▶

REFERENCE LINE is established with two targets at extreme ends of machine and telescope on stand. By placing the targets in other bearing bores to be aligned, as here, exact amount of misalignment in thousandths of an inch is quickly known

Plot curves of both plan and elevation alignment, *sketch*, as you line up the machine. To do this, set up telescope, as explained before. Center a target with inside micrometers in bearing bore at each end of machine for finding the reference line. With this line established by the telescope, you are now ready to center targets in all other bores to be aligned—or in bores you want information on, whether or not it is practical to align at this time.

From these readings, plot plan and elevation curves on cross-section paper showing the "as found" alignment. Indicate misalignment in thousandths (mils), using enough units to each square to have exaggerated lines far enough apart for easy reading. Also, mark off approximate positions of bearings being aligned with vertical lines.

After making needed shim changes under parts, such as generator stator and bearing pedestal sole-plates, use the telescope again to check and verify the adjustments. After all adjustments are complete, plot the "as left" curve for future reference. Result will be an easy-to-read chart similar to the one shown here.

While no time studies have been made either with tight wire or telescope, personnel using both feel that this method has saved much time and certainly has done a better job.

Use of telescope is naturally not confined to turbine-generator sets. No doubt you will find many uses as time goes on.

CURVES show reference line, "as found" and "as left" lines. Mark off approximate positions of the bearing bores being aligned. Cross-section paper is handy for this use, roughed out on job as each bearing is being aligned. Manufacturer's data furnish much of the information needed for this important work

If you can't make use of this new method,
you may want to align with tight wire

MEASURING distance from bore to wire calls for very fine sense of feel. If there are vibrations or kinks in wire, chances are that readings will not be very accurate

WEIGHTS hung on wire, at both ends of machine, and wire length affect wire sag

GOOD MECHANIC is needed to work with tight wire. Work is slow, holds up job

Tight-wire has been used for years to align bearings of heavy machinery. Wire acts as a datum (something actual or assumed) from which measurements are taken with inside micrometers. In this way parts, such as bearing pedestals, turbine cylinders, blade rings, dummy rings, diaphragms, etc, are aligned.

Wire is stretched tight through center of bearing spaces in machine, with each end of wire fixed in space, according to bench marks established by the surveyor. Micrometer measurements are taken from various bores against wire throughout length of machine.

By adding, subtracting and dividing the micrometer readings and allowing for tight-wire sag, you can find where the center of bore is in relation to where it should be. Then various parts found out of line are moved or adjusted into correct position.

Disadvantages. You can see that this tight-wire method has several disadvantages. One is that it's time-consuming. The job is tedious and must be done by a careful mechanic with the right feel for the job. Then, again, the wire sag varies on each job, depending on its length, diameter and tension. (Tension is obtained with either a spring or a weight at one end.) If any of these factors is not correct, wire sag is naturally inaccurate. Kinks in the wire also lead to inaccuracy.

Also, if the tight wire is displaced from center of bore, angularity of micrometer to wire to a true radius leads to inaccuracy. Errors are also made when manipulating the micrometer readings, together with sag of wire to find true position of a particular part of turbine with respect to datum line.

At times, the wire is bumped, and then it must be reset again. Or the wire vibrates, giving false micrometer measurements. Another disadvantage is that it's impossible to remove a cylinder, blade ring, diaphragm, bearing pedestal, etc, to make repairs or adjustments for alignment purposes while there is a tight wire through the machine's bore. These are some of the reasons that make optical tools a welcome newcomer to the modern maintenance man.

I Must'a Opened the Wrong Valve

O H NO, NOT *again*, BRUNO! *That's the second time you flooded that machinery room this week. And we just got through drying out all those motors and putting them back into service this morning.*

Does this picture bring back unhappy memories to you readers? Sure it does. This kind of thing is repeated almost daily—in some plant or another around the country. The wrong valve is opened or closed. The wrong switch is thrown or pulled. Equipment isn't lubricated, or it's lubricated with the wrong oil. So, slowing down our power output goes on and on, like "Ole Man River."

So what? Just this: all these expensive boners can be slashed way down by using check-off sheets for starting up and shutting down equipment. Label all your valves and controls to make them foolproof. Then employ well trained, competent help. And I've found the more modern and automatic the plant, the more need for well rounded men to keep it running. That also means training each new man before turning him loose to gum up your expensive works.

Careful screening will find men who intend to make plant operation a career, men who take pride in their profession. With no licensing required of operating engineers in most of our states, it just doesn't pay to hire the first jack-of-all-trades who comes along and is willing to work for less.

Mechanics of Bruno's caliber might cut down on the payroll but you'll surely pay through the nose before they are through with you.

5
Steam engines

Guess I Forgot To Oil It

Y OU SURE DID, Jocko, and there goes a week's top-priority space production shot to blazes. But we chiefs can't hold a guy like Jocko responsible—not if we have all our marbles.

That's why the trend today is to put the brightest mechanic in charge of lubrication and not trust it to a grease monkey. Why? One little slip, and *zingo*, the whole plant can be knocked out—at least temporarily.

So smart engineers in larger plants see that this key spot is filled by a man who keeps on his toes about lubrication practices. Instead of monkeying around, he keeps log sheets of all oil and grease changes, watches supplies so they're not contaminated, and so on. In smaller plants, engineers assign the oil can to a well paid mechanic who at least knows how to overhaul a bearing. A guy like that makes each drop of oil strike home—and he never misses squirting that drop to the right place, at the right time.

If one bearing failure can be so expensive, what cheaper insurance can a chief buy against breakdowns and costly wear than smart, well paid help?

Monkey business and lubrication don't mix. It's either one or the other, but never both.

SETTING ECCENTRIC "ON CENTER"

1 To find high point of eccentric throw: use template with pin through it; mark point where pin touches eccentric; swing template around, mark other side

2 Keep template firmly on hub or shaft when marking. One method of marking is back of eccentric where pin hits. Other is with scriber thru hole in template

3 Take pair of dividers and find point on eccentric edge halfway between two points of contact already marked. This is highest point of eccentric throw

4 Next set machinist's square to half shaft dia. Hold square level, mark **O**. With one end of trammel at **O**, mark points **A** and **B** anywhere on shaft side of eccentric strap

5 Use marks **A** and **B** to find center mark **Y** on eccentric strap with dividers. Mark **Y** is important for future reference. Make it permanent by using small chisel to stamp into strap

6 Next use mark **A** to make mark **C**. Then from **B** make mark **D**. Keep dividers near inside edge for accuracy

7 Use mark **C** and mark **D** to find center **Z**. **Y** and **Z** are at extreme opposites. Make **Z** permanent with chisel

8 When the eccentric sheave mark **X** is opposite mark **Z** or **Y** on eccentric strap, the engine's valve is known to be "on center". That means valve is at extreme position of valve travel, regardless of crank's position. These two reference marks are useful in case eccentric slips, and for finding valve's position without opening steam-valve chest

Bumping a Piston

1 "Bumping piston" means finding clearance between steam-engine cylinder head and piston. Clearances **X** should be equal. Wear in crank, main bearing and crosshead pin changes clearance. So center piston between ends before it cracks a head

4 Next, remove cylinder head and screw piston against crank-end cylinder head, **A**. On smaller engines, head doesn't have to be removed. Instead turn piston rod with pipe wrench over grinding cloth, **B**, or with open wrench on rod flats, **C**

2 **A**, Centerpunch rod, nut and crosshead. When job is done, marks will show how far clearance had been out. (Cross-head mark used in Fig. 6 if scale isn't handy.) **B**, Loosen piston-rod locknut and run nut back to the end of threads

5 Wiggle crank on dead center and adjust piston rod so piston barely touches cylinder head. Then screw nut to cross-head, **A**. Replace cover. Turn engine to head end. Wiggle crank, turn piston rod, **B**, till piston just touches head end

3 Turn engine with bar or with flywheel so piston is on crank end. If engine is large and has jacking engine, use it. This is not accurate operation, but dead center can be guessed by eyeing position of connecting rod at crank-disk center

6 Caution: When adjusting piston rod at head end, nut must move with rod. Measure from crosshead to nut, **A**. Screw nut in half that distance, **B**. Without a scale, count nut turns. Turn rod so nut is against crosshead. Tighten nut

Dead Centering An Engine

1 Before setting steam valves, engine must be on dead center. Put crank 10 to 15 deg above estimated dead center. Scribe line on flywheel with hermaphrodite, **A**. Make trammel and mark flywheel as shown. Mark the crosshead and guide **B**

2 In placing some vertical engines on dead center, crank web may be better for markings. After placing crank past dead center, scribe line on web, **A**. Make trammel to reach from web to engine frame. Mark crosshead and guide

3 Turn engine over dead center so mark on crosshead again lines up with mark on guide. Now mark flywheel with trammel as shown. During all these operations flywheel must be turned in opposite direction engine runs, as explained in Fig. 5

4 Split distance between two trammel marks on flywheel with dividers, **A**. Turn crank until center mark on flywheel comes to trammel. When using trammel always place bottom end in same center-punch mark on engine housing or bolt

5 Most text books tell when finding dead center to turn the engine in direction it runs. That's wrong. Turn flywheel in the direction of rotation past crosshead guide mark, then back to it to take up bearing wear in the running position

$$X = a + b + c$$
$$Y = 4 \times X$$

6 Unless main-bearing, crankpin and crosshead-pin wear is taken up by turning wheel in opposite direction, wear of **a**, **b** and **c**, or **X**, add up to four times **X**, **(Y)** on flywheel. Then trammel marks false dead center; valves are set wrong

Today It Pays to Make

WEAR

ALL RODS WEAR from reciprocating in stuffing box. Shoulders wedge through the packing, making sealing a tough job

REMOVING PISTON

FIRST STEP in removing old rod is to clamp piston in vise. (Never clamp rod in vise.) Loosen nut one turn only. Then smack piston and rod assembly down hard against wooden plank. Nut protects threads and keeps rod from digging into wood

CUTTING TAPER

TURN DOWN shaft's end for small thread diameter, after both ends are centered and held between centers with lathe dog

MOVING TAILSTOCK half of total taper is one way to cut. Loosen setscrews, and nut, then jack over distance needed

COMPOUND REST is handy for cutting short tapers. Find right angle from the old taper, then feed the tool by hand

CUTTING THREADS

SET GEARS for right thread, place feed change lever in neutral, engage thread dial indicator against the lead screw

GRIND TOOL to 60 degrees to fit center gage. Place gage against straight shaft and set compound rest to 29 degrees. Cut thread by throwing in half-nut lever at any line on dial (for even threads). Draw the tool away at end with cross feed

New Piston Rods

CENTERING

Center drill

Recess

Dial indicator

TRUE new shaft with dial indicator by loosening and tightening the chuck jaws until dial needle shows zero all around

NEXT, run center drill against end of shaft slowly or drill tip will break. Be sure to oil the drill before using it

RECESS around center with lathe tool to protect center-hole edge from damage. Shaft will then be true between centers

FITTING

Remove cross feed bolt

Tighten these binding screws

A

Pencil

B

Turn

C

File

D

$\frac{1}{32}$ -in.

TAPER ATTACHMENT is the third method. Remove cross feed screw, tighten screws shown. Adjust to right inches-per-ft

NEVER FINISH taper with lathe tool only. *A:* First mark taper with pencil. *B:* Then insert taper into position and turn quarter turn. *C:* File end with fine mill file where the pencil marks are rubbed off the taper. *D:* Shoulder should not touch the piston

A

B

C

RUN CARRIAGE back by hand. *A:* Feed tool in desired amount by compound rest only Always use cross feed to remove tool and return to same point. *B:* Check threads with gage. *C:* Cut chamfer at end by placing the tool at an angle, as shown here

BUMP PISTON onto new rod against wood. Then tighten nut. Perfect taper will hold piston on without loading thread

Neck-Bushing And

1 Neck bushings are in many types of reciprocating machinery. Here they are shown in steam-engine stuffing boxes

2 Neck bushing is usually made of bronze. It helps seal stuffing box and provide rod bearing when **piston wears**

3 Worn cylinder and piston lowers rod. Neck bushing and gland wear at bottom. Packing squeezes in opening at top

7 Neck bushing can be made of babbitt or bronze. Cast plug by filling a can with metal

8 Chuck babbitt plug or bronze casting in lathe. Turn to snug stuffing-box-diameter fit

9 Then drill a hole slightly smaller than the piston rod diameter. Best way to center any drill in lathe is to run tool holder against drill. This steadies it for center

14 Now chuck the gland and rough-bore about ¼ in. out of hole. Rough tool marks help hold the babbitt metal best

15 Bolt gland to flange over gasket, to seal bottom. Then heat gland and pour it full of babbitt metal as shown

16 Chuck gland and drill babbitt. Bore for sliding fit over rod. Neck bushing and gland are ready to install now

Gland Tips

4 Piston and rod have worn cylinder and neck-bushing bottom, made packing **wear**

5 Neck bushing is held in place by making light press fit in cylinder opening through stuffing box. To remove bushing, **turn down round stock, drive out from cylinder end**

6 Cocked gland wears fast. Packing has hard time seal-**ing extra** clearance around rod

10 Bore smooth hole a few thousandths larger than rod, depending on engine service

11 Set the compound rest for 30 degrees and bevel casting's end as shown

12 Turn diameter for cylinder hole same size as hole. Only one chucking needed here

13 Cut with cutting-off tool and press into machine through stuffing box

17 With close fitting neck bushing and gland lined with babbitt metal, soft packing in stuffing box has easy job. Rod lasts much longer if packing is not tight. With neck bush-ing supporting piston, cylinder wears less. CAUTION! Be sure X equals X between crosshead guides. Take lower guide wear into account. Renew glands and neck bushings often

Aligning And Boring

1 First step is to stretch piano wire from aligning targets set on approximate center at both ends of the engine

2 Mechanic of Edw Purvis & Son, old time Brooklyn, N. Y. engine firm, mikes distance from 0.020-in. piano wire to machined boss inside crosshead-guide housing. He centers by adjusting the screws on alignment target at either end of long wire

5 Measuring cylinder all around **before** stretching wire tells mechanic cylinder's condition and how deep to set bore

6 After wire is shifted to exact center of engine, measurements are recorded from various points to wire. Then wire is removed and boring bar lined up with mike at same points. Allowance is made for wire size and boring-bar diameter in figuring

Engine Cylinder

3 Years of experience are needed for right "feel" in this delicate measuring step. Wire must be taut (it broke once on job). Measure from machined bosses on sides and from unworn crosshead-guide edges at top and bottom

4 Inside micrometer rests on edge of crosshead guide and is adjusted until it barely touches wire. Paper behind helps to see wire

7 Pipe supports boring-bar feedscrew handle from studs. String is weak and breaks if tool catches in cylinder

8 Blocks support small boring bar for corliss valve ports. Clamps hold bar in position at both ends of port like this

9 "T" slot in corliss valve is remachined for snug fit with valve stem to reduce lost motion in valve gear

Setting Piston Valves

1 Steam engine's piston valve admits steam at center, exhausts over ends. Inside rings (a) are guides. These valves can be set with sticks instead of using old marks, which are usually unreliable

2 One stick for valve and another for ports are needed. Soft wood of smooth grain is best. Use sticks about ½ in. by 1 in. and about 1½ times length of valve

3 Remove valve and place on bench. Label valve stick No. 1 at end. Lay alongside valve about two in. from numbered end. Use a sharp pencil or pen knife for marking the edges of valve

7 With dividers find center between B C on stick 1 and F G on stick 2. Use square to line up centers. With dividers on stick 1 find center of B C and mark across with square. Identify with X mark

8 Turn engine crank until top of valve L is flush with port edge M. Hold sticks in corresponding position, with A and F in line. Place the sticks on valve and mark stick 2 at cylinder head with pencil

9 Mark nail end of stick 2 with number 2. Saw off bottom end you marked at cylinder head. With stick 1 resting on valve, and stick 2 resting on cylinder block, A and F should now be in line

10 Hold sticks together and turn crank until valve reaches it's highest point of travel. Place mark J on stick 1 opposite center mark I of stick 2. Always turn the engine's crank in same direction

With Sticks

4 Mark port stick No. 2 at one end and drive tiny nail in the other end. Rest nail on bottom of the steam port in valve cylinder and make mark at cylinder head with a pencil or knife

5 Complete marking ports by moving stick to top of bottom port, **then** bottom of top port and finally the top of top part. Hold stick firmly against cylinder wall and make marks sharp

6 **(A)** Complete port and valve markings with square. **(B)** Turn stick 2 upside down. Shade sections AB and CD on stick 1 and outside sections EF and GH on stick 2, to show valves and ports

11 Turn crank until valve reaches its lowest position. Make mark K on stick 1 opposite center line I on stick 2. Valve is not set properly because center mark I is not in center of marks K and J

12 Distance K J is valve travel. Find half of this with dividers. Mark K' and J' on either side of center line. Hold sticks with center lines together and draw lines K' and J' across both, using the square

13 Distance from dotted line to K' in Fig. 12 is amount valve is to be lowered. Screw valve stem out of crosshead this amount. Place sticks as shown and turn crank. J' should line up with I

14 Turn crank until valve is at travels bottom. K' should be in line with I. Sticks show what's going on inside engine. Date sticks and keep them for your future reference for checking setting next time

The Alchemist

ENGINEERS VISITING the chemistry building of the University of Pennsylvania stop dead in their tracks on seeing this bas relief of the "Alchemist." Created by sculptor Donald De Lee, this work is highly symbolic of chemistry's dubious beginning.

The ancient alchemist tried without success to change base metals into gold, platinum, silver and other so-called "noble" metals. But where the alchemist failed, I figure that the modern metallurgist has succeeded.

Metallurgists have not only produced a metal alloy with noble properties, but at a cost far below those of the traditional noble metals. I'm talking about the familiar high-nickel-chromium steels, created by alloying iron with high percentages of nickel and chromium. Every power engineer knows how vital these metals are to our modern civilization, especially in the power plant.

My point is that many of our highly efficient tools and materials for turning out today's power at greatly lower rates had beginnings about as discouraging as the alchemists' experiments. But like the alchemist, there always were men who didn't know when they were licked.

All this proves that so long as you work toward improvement, you have a good chance of succeeding—if you stick to it.

6
Air conditioning, cooling

Cool Your Building Roof

TYPES OF SPRAYS IN ACTION _____

STATIONARY streams from jet *(below)* shower fine droplets. Roof thermostat controls water, keeps roof temp right

ROTARY cooler revolves slowly through gearing as wheel breaks up thin spray, uses no thermostat or electrical wiring

WHIRL JET throws fine sprays over roof, runs continually. Excess water, if any, runs into drain sump for recirculation

DOUBLE row spray head throws upper and lower streams. Upper sprays wide outer circle; lower inner area 10-ft diameter

Metal plate rests on roof

Slide

JET of fine spray is thrown over roof at 45 deg. Cooler runs continuously, revolves just enough to moisten roof

Spray angle

NOZZLE has no moving parts, but is designed so water whirls in jet, throwing hollow spray cone at very wide angle

CONTROLLING ROOF TEMPERATURE _____

160 F

Black asphalt

Insulation

INSULATED roof's temperature gets high as 160 F on hot day because heat can't get in. Roof deteriorates fast

150 F

UNINSULATED roof of same construction reached 150 F on same day. The spaces under this roof are hot, need cooling

103 F

Water 6 in. deep

FLOODED roof is cooler, but water stores heat, gets stagnant, mosquitoes breed— so must be treated. Roof may also leak

By Making It Sweat

THIS MUCH HEAT GOES IN

18 Btu per sq ft per hr

11.2 Btu per sq ft per hr

10.3 Btu per sq ft per hr

ASPHALT over concrete let in 18 Btu one hot day in August. Black roof absorbs most heat from hot rays of sun

DRY SLAG over asphalt of same concrete roof on same day absorbed 11.2 Btu. This is less heat, but too much for comfort

ALUMINUM paint over asphalt of same concrete roof on same day let in 10.3 Btu because light color reflects rays

ADVANTAGES OF SPRAYED ROOF

LESS air conditioning for top floor by 20%. Top floor cooler than floors below, smaller power bills, less upkeep

WALLS are cooler because heavy roof air spills down along walls, cooling them, so help make inside spaces comfortable

SAVES ROOF as volatile oils vaporize less. There's no thermal shock from sudden thunder storms because roof is cool

LESS WATER raises temperature slightly. But 108-F water blanket won't draw heat up through roof as will sweating roof

SPRAYED roof is cooler—spaces below are about 85 F—water evaporates fast as it hits roof, pulling heat from inside

SPRAYED roof cools in two ways. Water droplets intercept heat before hitting roof; roof film evaporates, pulling heat

Air Conditioning

1—NOT ENOUGH COOLING . . . INSULATE DUCTS AND CHECK AUTOMATIC DAMPERS

INSUFFICIENT COOLING in spaces when refrigerating system works right calls for a checkup outside the machinery room. Don't overlook anything

DUCTS WITH NO INSULATION often pass through hot areas, **A.** Quite often these areas are normally cool, but occasionally enough heat reaches ducts to transfer heat to conditioned air inside. Standard duct lagging is good investment. Also look for automatic dampers being out of order, **B.** Oil moving parts and work by hand to free

3—ODORS ARE BAD . . . CLEAN DIRT FROM INSIDE DUCTS AND CHECK DRAINS

ODORS are disagreeable when carried with conditioned air. They make the operating engineer unpopular and defeat purpose of supplying treated air

MOLD OR DIRT accumulations on evaporator fins, **A,** spoil air passing through them. If drain lines are poorly trapped or water in gooseneck dries up, **B,** sewer odors escape. Outside air improperly dampered, **C,** mixes with inside air, contaminates air supply. Look out for foreign matter such as bugs and dead mice in drain pan, **D.** Keep clean

5—LOW HUMIDITY . . . INSPECT SPRAY NOZZLES, STRAINER AND SOLENOID VALVE

LOW HUMIDITY is uncomfortable, it dries mucous membranes, causes headaches. Modern systems control humidity, keeping air's moisture right

SPRAY NOZZLES get plugged, **A,** or water pressure to nozzles is low or even shut off. Gage may not show right reading. Strainer, **B,** might be plugged. Check solenoid water control valve, **C.** It might be sticking or coils may be burned out. Inspect humidistat, **D.** Sensitive element is delicate, if insulated with dust it shows faulty reading

Trouble Shooting

2—MECHANICAL NOISES . . . TIGHTEN CABINETS AND FAN BELTS, SUPPORT PIPES

MECHANICAL NOISES in spaces are not only hard on nerves; but indicate that something is radically wrong with equipment. Get busy finding cause

FANS GET OUT OF BALANCE at times, **A.** Shut down and check. Rebalance in workshop. Fan belts slip, **B,** sending irritating noises through ducts. Belts are easy to tighten. Loose cabinet parts vibrate, **C.** It's good practice to tighten all screws at set periods. Pipes to refrigerating compressor vibrate, **D.** Bracket pipes right

4—NOT ENOUGH HEAT . . . VENT HEATING COILS, CHECK TRAP, BUY CIRCULATOR

LOW TEMPERATURES from insufficient heat call for action. Discomfort is hard to take and doesn't help production or operator's reputation

HEATER, A, may be too small for job or not spotted right. Drain lines from heater, **B,** are trapped improperly or trap doesn't work. Often fan, **C,** isn't running and no one checked it before complaining. Hot-water coils, **D,** get air bound. Open vent before looking further. Circulating pump is needed in line, **E,** or piping is poorly arranged

6—FLOODED FLOOR . . . BAFFLE AND INSULATE CHILL-WATER PIPING FROM AIR

WATER ON FLOOR is not only messy but creates expensive problems by soaking into floors, shorting electric equipment and even causing accidents

CHILL-WATER COILS leading into conditioning equipment, **A,** should be insulated to prevent sweating. Humidifier piping in cold fresh-air current, **B,** should be baffled or insulated from air. Water supply line, **C,** is in cold air stream or valves' packing leaks. Drip pans are usually a nuisance. Insulate lines and keep after all leaks

Temp, F	Vapor pressure in. Hg abs
32	.1803
40	.2478
50	.3626
60	.5218
70	.7392
80	1.032
90	1.422
100	1.933
120	3.446
140	5.881
212	29.92

Heat added

RELATIVE HUMIDITY is actual amount of moisture in air at any given temperature, divided by the greatest amount of moisture the same air could hold without condensation. A convenient way to look at this is in terms of pressures. That is, actual pressure exerted by moisture (*vapor pressure*), divided by highest vapor pressure possible under the same conditions, gives relative humidity. Let's look at a system that gives us the relation between this highest possible vapor pressure—the *saturation condition*—and temperature.

At left above, we have a sealed chamber with all air evac-

uated and water in the bottom. Now we get the system at exactly 32 F and let it settle. The pressure gage then shows 0.18 in. Hg abs. If we do the same for 40 F, the gage reads 0.25 in., and so on up to 212 F. Precise values of these gage readings for various temperatures are tabulated.

If we plot these pressures and temperatures, we get a *vapor-pressure curve, above*. For atmospheric air, this curve represents highest pressure that water vapor can have without condensation—it's 100% relative humidity. This pressure, plus that of dry air, makes up atmospheric pressure.

We often say, 'It isn't the heat, it's the

But what is humidity and how do we measure it? That's where the often misunderstood term "relative humidity" comes into the picture. To get it clear, start with explanations in the diagrams and captions, then read text to see how we find it

► IN THE DIAGRAMS and captions we've got a line on what relative humidity is, and on one of the items we need to measure it—saturation vapor pressure. We've also seen what wet-bulb temperature is. Now we need to know how to use wet-bulb temperature to get the other item—actual vapor pressure—involved in relative humidity.

If the air were saturated (100% relative humidity), we could measure vapor pressure directly below the saturation curve. But in our example, air

isn't saturated, so its actual vapor pressure is something less. As shown in the last diagram, actual vapor pressure is found under the curve at a point called *saturation temperature*. This is the temperature to which our air mixture would have to be reduced to bring it to 100% relative humidity.

To find the actual vapor pressure, we recall that at the wet-bulb temperature (*see captions*) the heat from the air to a water drop equals the heat taken from the drop to vaporize water. So we set

up heat-transfer relations for each condition and put them equal to each other.

For temperatures from 30 to 130 F, this gives a simple and fairly accurate rule: Actual vapor pressure equals saturation vapor pressure at wet-bulb temperature minus 0.11 times difference in dry- and wet-bulb temperatures. Then actual vapor pressure divided by saturation vapor pressure at dry-bulb temperature gives relative humidity. We usually multiply by 100 to get relative humidity as a percentage.

Let's see how this works for the diagram figures, taking wet-bulb temperature as 58.3 F. From the table, saturation vapor pressure at wet-bulb is 0.491 in. Hg abs. Then 70 minus 58.3 gives 11.7 as difference between dry and wet bulbs. Multiplying by 0.11 yields 0.129. Subtracting this from 0.491 we get 0.362 as actual vapor pressure.

Saturation vapor pressure at 70 F is 0.739 in. Now if we divide actual va-

70-F atmospheric air containing moisture approaching drop at 900 fpm

DRY-BULB TEMPERATURE. Curve shown here is a blownup reproduction of vapor-pressure curve on facing page. Water drop we've just been talking about is at 70 F, so it shows as a point on saturation curve at that temperature. Such a temperature, measured by ordinary thermometer, is called *dry-bulb temperature*

ACTUAL VAPOR PRESSURE. We now know highest vapor pressure possible for air at various temperatures *(table, curve, p 118)*. Let's call it *saturation vapor pressure*. For relative humidity, we also need actual vapor pressure.

We measure this by a *psychrometer*. To see how this gadget works, let's look at a water drop at 70 F, with moist air, also at 70 F, blowing on it at 900 fpm.

humidity'

por pressure by this, and multiply by 100, we get 49% relative humidity.

Wet-Bulb Thermometer. We've seen what relative humidity is and how we can figure it if we know dry- and wet-bulb temperatures. Dry-bulb comes from an ordinary thermometer, but how do we get wet-bulb? Even if you've never seen a wet-bulb thermometer, you've probably guessed by now that it's an ordinary one with its bulb surrounded by a wick wetted with water.

In what is known as a *sling psychrometer*, dry- and wet-bulb thermometers are mounted side by side. The instrument can be twirled around like a charm on the end of a watch chain to get air moving past the wet bulb at 900 fpm or so. This simulates the conditions of that drop of water we talked about in the diagrams and captions. Many psychrometers have a slide-rule gadget on which you set the two temperatures and read relative humidity directly.

WET-BULB TEMPERATURE. Now let's see what happens when air at 900 fpm starts blowing on the drop. Water at surface begins to evaporate. This vaporizing takes heat—latent heat of vaporization—so the drop is cooled. Temperature the drop finally reaches when cooling stops is what we call *wet-bulb temperature*

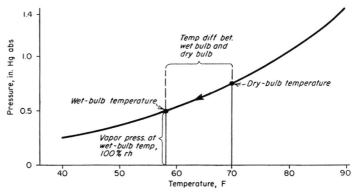

COOLING OF THE DROP stops at wet-bulb temperature because heat coming from the air to the drop eventually equals heat given up by the drop. As we'll see in more detail in the text, we can find actual vapor pressure of the air, and thus its relative humidity, once we've established the wet-bulb temperature

Here's how to figure humidity, wet-bulb temperature, air mixtures,

Humidity ...

Let's assume you know dry-bulb temperature is 80 F and wet-bulb temperature is 67 F. You want to find dew-point temperature (temperature at which moisture will condense out of the air for a given specific humidity and pressure, as air-moisture temperature is reduced) and % humidity.

Find 67 F on the 100% humidity curve and follow the line parallel to constant wet-bulb lines. Then trace the line from 80 F on the dry-bulb scale. Intersection shows dew-point temperature is 60.3 F, humidity is 50%. This point represents condition of the air.

Wet bulb ...

Now suppose you know dry-bulb temperature is 65 F and dew-point temperature is 55 F. You want to determine wet-bulb temperature and the % humidity. Approach is much the same as in our first example.

Follow the 65-F constant dry-bulb temperature line. Then, moving over to the vertical scale, locate 55 F. Intersection pinpoints air-moisture condition. Since chart doesn't have a fine enough grid, you'll have to estimate 58.9-F wet-bulb temp along the constant 100% relative humidity curve. The 69% humidity is also an estimate.

Air mixtures ...

Let's say 25% (point *A*) of a mixture of two air streams is at 96-F dry bulb and 75-F wet bulb. The balance (point *B*) is at 60-F dry bulb and 48-F wet bulb. Then, 0.25 × 96 = 24; 0.75 × 60 = 45. Resultant dry bulb of the mixture is 24 + 45 = 69 F.

Now draw a line between *A* and *B*. Intersection of 69-F dry bulb with the line between *A* and *B* establishes point *C* for the mixture. Wet-bulb and dew-point lines, running through point *C*, are conditions of the mixture. Its wet-bulb temperature is 56.2. Dew-point temperature is 46.6 F.

Get to know air-moisture data

It sets the groundwork for well-planned air conditioning

Psychrometric chart shown here in skeleton form is nothing more than a graphic portrayal of various air-moisture mixtures.

Lines of constant dry-bulb temperature are almost vertical. Constant dew-point temperature lines are horizontal. Constant wet-bulb temperatures slope downward to right. Constant % humidity lines are curved. Before using chart you must know any two values. You can then pinpoint the air-moisture condition. —*Courtesy The Trane Co* ➤

Heating, cooling...

The chart can also trace air-conditioning cycles. Remember, in tracing any cycle, dew-point temperature is constant as long as there's no change in moisture content of the air.

Thus heating air without changing its moisture content would take place along a horizontal line of constant dew point. The dew-point line would, naturally, be set by the initial condition of the air. Initial air at point *A* would be heated to point *B* along horizontal line *A-B*. If you cool air without condensing moisture, the condition of the air moves from point *B* to *A*.

1 This canvas water bag exposed to hot desert sun is cooler than the surrounding atmosphere because of evaporation

2 Electric fan evaporating perspiration carries away excess heat, cooling body to normal temperature despite the heat

Why Cooling Towers Cool

Here's the lowdown on why cooling towers cool below atmospheric temperatures.

▶ IN OPERATING refrigerating plants or condensing turbines, an important problem is removing and dissipating heat from the compressed refrigerant or the exhaust steam. This heat is usually removed from the "gas" by transferring it to water in a heat exchanger.

From there it may be dissipated in a number of ways. If the equipment is on the bank of a river or lake, the intake and discharge connections can be piped to prevent mixing the heated discharge water with the cooler inlet water. If the source of cooling water is a well or city supply the discharge may be piped back into the ground or into the sewer or open waterway. But such outright waste of this heated water is usually costly and in some localities illegal.

There are a number of reasons why cooling water should be reused. The most important is that very few plants are lucky enough to be near a limitless water supply. Another is that using city water for cooling is costly unless it's reused. The third is that most water contains dissolved salts, and using a continuous supply of such raw water quickly coats the heat exchanger with scale. In a cooling tower the water is cooled by exposure to air after each passage or cycle, allowing the same water to be reused a number of times.

Atmospheric Water-Cooling. Water, by having its waste-heat load transferred to the atmosphere, can be reused in a continuous cycle and thus conserved. The heat can be transferred by bringing the water and air together indirectly (non-contact) as in an automobile radiator. Another way is with evaporative-cooling equipment such as atmospheric spray ponds, cooling towers, etc.

Water, when cooled by the evaporative method loses about 1000 Btu (heat units) for each pound of water evaporated. This heat taken away in the water vapor produced is called the *latent heat of vaporization.* When air removes heat (in vapor) from water in this way, it can cool the water below atmospheric temperature.

This fact is important in transferring heat from water to air. It allows the water cooled by evaporation to serve plants having a great variety of temperature needs. It also allows a little water to cool a much greater heat load than if it were not below atmospheric temperature.

Cooling by Evaporation. Surface evaporation keeps water cool in the old-fashioned porous canvas water bags, Fig. 1. Likewise evaporating perspiration helps the human body, Fig. 2, to keep its normal 98½ F temperature in places where the surrounding atmosphere is much hotter.

This latent heat of evaporation is the primary cooling effect produced by blowing air over wet surfaces or through sheets of falling water in a cooling tower.

Cooling-tower engineers also refer to "sensible heat"—heat you can *feel*—as temperature. The higher the temperature of a substance, the greater its sensible heat. Where the air is cooler than the water there is some tendency (not counting evaporation) for the air to cool the water—that is for the air to get hotter (gain sensible heat) while the water gets cooler (loses sensible heat). On the average about 75% of the total heat removed is by evaporation (latent heat).

Wet- and Dry-Bulb. Cooling-tower engineers, like air-conditioning engineers, often speak of *dry-bulb* and *wet-bulb* temperatures. The temperature of air as read on the ordinary thermometer is called the dry-bulb temperature. Wet-bulb temperature is the final reading when the bulb of a thermometer is covered with wet cloth and the instrument is whirled around in a sling.

Wet-Bulb Is Lower. When the humidity is 100%, the wet-bulb temperature will be the same as the dry-bulb (ordinary thermometer). For any lower

3 Cooling tower lowers water's temperature because water drops spreading over laths are exposed to atmosphere. Slight evaporation carries away heat, cooling water

120 F — Hot water to tower

All temperatures used are illustrative only and subject to wide variation

Cooling range

90 F — Cold water from tower

Approach

75 F — Wet-bulb temperature

4 Schematic definition explains terms "cooling range" and "approach"

humidity some of the water will evaporate to cool the bulb, so the wet-bulb reading will fall below the dry-bulb. The drier the air the greater will be the difference.

Knowing the wet- and dry-bulb readings you can always find the relative humidity from a table or chart.

Other things being equal, cool air is better than warm air for a cooling tower. Most important of all is a low wet-bulb temperature, indicating either very cool air, very low humidity, or a combination of the two.

Here's a scientific fact worth remembering: No cooling tower, even in theory, can ever cool water below the wet-bulb temperature of the inlet air. In actual towers, of course, the final temperature will always be a few degrees above the wet-bulb temperature. Fig. 3 pictures water falling from one level to another in a cooling tower. The splashing droplets wet the wood laths giving a big exposure of water surface to the upmoving air.

Cooling effect can be speeded by (1) increasing air velocity over the wet surfaces, (2) increasing exposed wet surface, (3) lower barometric pressure, (4) increasing water temperature to the tower, (5) reducing air's humidity.

Cooling-tower designers must consider these properties of air: (1) wet-bulb temperature, (2) dry-bulb temperature, (3) humidity, (4) total heat in Btu, (5) pressure, (6) weight and (7) velocity.

The wet-bulb temperature assumed in designing evaporative water-cooling equipment is generally close to the average maximum wet-bulb for the summer months at the given location. The exact wet-bulb temperature selected will depend on operating conditions, temperature limitations and geographical location.

Operating Terms. Those who operate or manufacture specialized mechanical equipment gradually develop a somewhat specialized "glossary" of words and phrases that describe particular related functions, parts, or characteristics of their equipment. Here are some of the most common terms used with atmospheric water-cooling equipment.

Cooling Range is the deg F that water is cooled by the water-cooling equipment, Fig. 4. It is the difference in temperature between the hot water coming to the cooling tower and the temperature of the cold water leaving the tower.

Approach is the difference in deg F between the temperature of the cold water leaving the cooling tower and the surrounding air's wet-bulb temperature, Fig. 4.

Heat Load is the amount of heat thrown away by the cooling tower in Btu per hour (or per minute). It is equal to the pounds of water circulated times the cooling range.

Pumping Head is the pressure required to lift the water from the water's top in the basin to the top of the tower and force it through the water-distribution system, Fig. 5. Pumping head is equal to the static head, plus the friction loss in the distribution system and the velocity head (head required to maintain velocity of water).

Drift is the small amount of unevaporated water lost from atmospheric water-cooling equipment in the form of

Inlet to water distribution system

Side of tower

Static head

Pumping head is static head (as illustrated) plus velocity head and friction loss

Catch basin

5 The term "pumping head", shown here is used in cooling tower design

mist or fine droplets. It is water entrained by the circulating air. Drift is a water loss independent of the water lost by evaporation. The drift loss, unlike the evaporation loss, can be reduced by good design.

Blowdown is the continuous or intermittent wasting of a small fraction of circulating water to prevent a concentration of chemicals in the water. The purpose of blowdown in water-cooling equipment is to reduce the soluble solids, or hardness. This reduces the scale-forming tendencies of the water.

Makeup is the water required to replace the water that is lost by evaporation, drift, blowdown, and small leaks.

Delignification . . .

fungus attack and . . .

salt deposits on wood cooling-tower slats are three good reasons for topnotch water treatment. Fortunately, cooling towers don't usually end up in this condition. But it's good insurance to check for trouble symptoms. To aid you, here are . . .

18 Q & A's on treating cooling water

1 Is there an ideal treatment for all cooling waters?

A   Economics plays an important role in both selection of cooling-water systems and methods of water treatment. It's *possible* to prescribe an ideal water for each cooling job, but equipment and chemical cost is out of the question for most installations. So treatment must be tailored to fit the size and type of cooling-water systems currently in use.

COOLING SYSTEMS

2 What are the main types of cooling systems?

A   The two main classifications are "once through" and "circulating." In the simple once-through system water passes through the heat-exchange equipment only once. It is increased in temperature about 20 F before going to waste.

Circulating systems are more complex. Cooling water also passes through a heat exchanger or other apparatus where cooling occurs. Temperature increases from 6 to 20 F. Heated water is then cooled in another exchanger using an external cooling medium such as an air-cooled radiator.

3 Is there a trend to the use of circulating systems?

A   Since the once-through system calls for an almost unlimited supply of cooling water, today's trend is toward the use of circulating systems where the water is cooled for re-use. Cooling is brought about in the open type by actually evaporating a part of the cooling stream. In a closed system the cooling is usually done in an air-cooled radiator. A separate circulating or once-through setup can also be used.

4 Where are open and closed systems used?

A   Open type is most frequently used for cooling water in chemical plants, oil refineries and power-plant condensers. Closed system is usually limited to jobs where the quantity of cooling water isn't large. Typical applications include compressor cooling jackets and diesel engines.

EXCESSIVE SCALING

5 What causes scale formation?

A   Most common type scale found in heat-exchange equipment is calcium carbonate formed by the breakdown of calcium bicarbonate in the cooling water at high temperatures. Magnesium compounds and calcium sulfate are rarely the cause of scale in cooling systems. Scaling is affected by temperature, rate of heat transfer, calcium, magnesium, silica, sulfate and alkalinity concentrations and the pH.

6 How is the Langelier index used?

A   Langelier index measures the tendency of calcium carbonate to precipitate from water under given conditions of calcium hardness, alkalinity, pH, temperature and total dissolved solids. Charts, nomographs and special slide rules can be used to quickly calculate this index for any water supply when the analysis is known. Positive index means water has a tendency to deposit scale while a negative index shows a tendency to dissolve scale and corrode. Note that index does not measure amount or rapidity of scaling. This depends only on calcium carbonate or bicarbonate content.

Index is decreased by reducing calcium hardness, calcium bicarbonate, pH or a combination of all three. Total dissolved solids has a relatively minor effect on the index.

Blowdown can be used to reduce the calcium and alkalinity in the circulating water. But it increases makeup and is seldom used as the only means of control.

Other treatment methods used to adjust the analysis include: cold-lime softening, usually followed by an acid feed for the adjustment of pH and alkalinity; partial zeolite softening with hard water by-pass, also followed by an acid feed if needed. Some moderate-hardness high-alkalinity waters can be handled with an acid feed alone.

7   What is the "controlled scale" method?

A   It is a method of keeping index of the circulating water slightly positive, say in the range of 0.5 to 1.0 at the highest temperature in the system. This is usually enough to insure deposit of a thin impervious layer of calcium-carbonate scale on the surfaces of the cooling-water system. Careful adjustment of pH, alkalinity and calcium is needed to keep the desired index and protect the system against the ravages of excess scaling and corrosion.

8   What is the role of surface-active chemicals?

A   Surface-active materials prevent crystal growth and therefore scale formation. They increase the solubility of scale-forming salts so precipitation doesn't occur when solubility limits are exceeded. Chemicals used for this purpose include special types of phosphates, tannins, lignins, etc. They can handle the job of scale prevention alone and are also helpful in broadening the range of Langelier index control in the controlled-scale treatment.

CORROSION

9   How do you control cooling-water system corrosion?

A   Corrosion inhibitors such as chromates, phosphates, silicates and alkalies are used. Anodic inhibitors of the chromate and phosphate type decrease metal attack. But if the concentrations are too high, pitting and tuberculation result. Use of chromate in combination with phosphate has been successful in controlling pitting, tuberculation and other metal losses from corrosion, and dosages are at economical levels.

Since O_2 and CO_2 are primary cause of corrosion one of the more direct treatments is deaeration. With a vacuum deaerator, oxygen can be reduced to as low as 0.1 ppm depending on water temperature and equipment design.

SLIME AND ALGAE

10   What causes slime and algae to form?

A   Algae growths are composed of millions of tiny plant cells that multiply and produce large masses of plant material in a short time. Slime growths are a gelatinous mass of micro-organisms that cling tightly to secluded surfaces in the system, trapping organic and inorganic matter and debris along with scale-forming materials. Any appreciable buildup seriously interferes with heat-transfer efficiency.

11   What chemicals give effective treatment?

A   Chemicals that prevent slime and algae troubles are those that exert a toxic action on the micro-organisms.

Chlorine is fairly general in toxicity for most bacteria, algae and protozoa, but continuous application in a once-through system is too expensive. Intermittent treatment gives good results at much more reasonable cost.

Copper sulfate is toxic to simple algae, but is less toxic to slimes. Use is limited because copper sulfate precipitates at a pH of 8.5 or higher. Phenolic and other quarternary ammonium compounds are also used. They have the advantage of resisting precipitation at high pH and are not removed from the system by aeration in the cooling tower.

12   Why is an intermittent feed best?

A   Intermittent large doses of chemical rather than a steady feed of a much smaller quantity keep the micro-organisms from getting used to the treatment and building up an immunity to it. It also avoids the danger of a new strain of algicide- or slimicide-resistant bacteria developing.

13   How are accumulated slime and algae removed?

A   Best way to get rid of these growths is by mechanical cleaning. In the case of algae the algicide may only loosen the growth enough to set it free in the system to plug lines and damage pumps. After the system is completely cleaned, start out on the right foot with a good chemical-treatment program to prevent future buildup of these growths.

WOOD DESTRUCTION AND DECAY

14   What causes cooling-tower wood destruction?

A   This destruction is usually caused by one or a combination of: (1) chemical attack commonly known as delignification (2) mechanical rupture of wood cells from salt crystallization and (3) biological attack causing wood decay.

15   What is delignification and when does it take place?

A   It is the removal of tannins, lignins and cellulose products from the lumber, leaving long stringy fibers with greatly reduced strength. It apparently occurs in the presence of high concentrations of sodium carbonate in the circulating water. Oxidizing agents like chlorine speed up the process.

16   Why are salt deposits harmful?

A   Intermittent splashing on parts of the cooling tower produces concentrated salt deposits by evaporation. Salt-saturated water in these areas is absorbed by the wood. Then crystals form beneath the surface of the wood mechanically disintegrating it. Result is a mass of soft loose fibers susceptible to both chemical and biological attack.

17   What is fungus attack and how is it caused?

A   It is a form of biological attack that is caused by micro-organisms that use the wood for food. These wood-destroying fungi reproduce by means of spores that are normally airborne and alight on the surface of the wood. The attack is not confined to any particular section or make of cooling tower. It is most noticeable in the mist sections and takes place on both the surface and the interior of the wood.

18   How do you combat fungus attack?

A   Proper control of alkalinity, pH and solids content of the circulating water will go a long way toward preventing chemical attack on the wood. It will not *prevent* biological attack, but will decrease chances of damage from fungus.

One preventive today is use of properly impregnated cooling-tower lumber. Chromated copper-arsenate salts show much promise for this. A different approach is to try to chemically destroy the fungi in the circulating water by treating with sodium pentachlorophenate, zinc sulfate, and sodium chromate. Use intermittent dosing. It is often effective to spray a concentrated solution of toxicant directly on the cooling-tower wood in those sections most likely to be attacked by the fungus, including tower mist areas.

BRRR R R—avoid heating failures this winter by doing work now . . .   **Schedule maintenance at once . . .**

14 Q&A on preparing your heating system <u>now</u> for winter

1   What's best time to overhaul building heating system?
A   Most building operators do this job when equipment comes off the line in spring. It's best to clean boiler and sooty ducts as soon as system is shut down. Reason is that ashes and soot absorb moisture from humid air. Because of sulfur in these deposits, resulting sulfuric acid rapidly corrodes the metal.

2   Is galvanic action a problem in hot-water heating systems?
A   Galvanic action can be a very serious problem because of dissimilar metals in piping, fittings, heat exchangers, boilers, etc. Galvanic action precipitates iron into water, which lowers pH down toward acid condition.

Water going toward acid increases galvanic action, which in turn precipitates more iron, causes oxidation—corrosion. If water is not treated for five to ten years, look for serious trouble. If organic compounds are used, 1 part in 500 parts of water often prevents trouble.

3   What equipment in heating system should be checked?
A   Everything. Follow system through building and come back to boiler with return system. Take in all piping and check joints for leaks, corrosion and condition of insulation.

4   Should hot-water heating systems be drained in summer?
A   Many large buildings have hot water heated by steam in heat-exchangers. Hot water is then circulated through space

heaters. Many air-conditioning systems having chilled water are now being adapted for hot-water heating in winter. Chilled water is usually at 40 F, while hot water may be anything over 180 F. But slime and algae form in such systems when water is under 140 F. Because oxygen is gradually liberated from system while water is hot in winter, don't drain in summer, but treat water chemically.

5   Radiators: What maintenance is needed?
A   Check for physical defects—usually they need little maintenance. Some steam radiators have orifice plate on inlet side between valve and radiator. In that case they have no trap on outlet side. If it's a high-pressure system, you might take coupling apart once in five years or so and check hole size in orifice for wear. Radiators that don't heat properly usually have fouled valves. Make sure air bleeder valve on radiator works.

When trap is on outlet, renew valve disk if needed, or grind in. If the packing gland of valve is screwed down all the way, remove old packing and put in new.

6   Unit heaters of the ceiling type: What maintenance is needed?
A   Coils of small unit heaters often spring leaks. Check by placing pressure inside coil. Open steam trap and clean out, check valve seat and disk, also bellows for cracks. Check electric motor on unit. Clean out motor and grease bearings. Don't forget that more motor bearings are burned up from

Check returns and all the traps . . .

To keep your building warm on coldest days

overgreasing than from lack of lubrication. If in doubt, leave out plug of bearing and run motor so excess grease can run out. Then screw in plug again. If unit has fan, clean the blades. This allows them to run more freely and also keeps them from getting out of balance, which loads the bearings and shaft.

7   Valves: What attention is needed?
A   All valves in system, whether you buy or generate your own steam, should be checked for tightness, packing, good working condition. Best way is to lift valve bonnet. Then you can usually see if valve is tight by observing seat and disk condition. At times a valve needs a new stem, or packing may be hard, then leak steam during heating season.

8   Expansion joints: What attention is needed?
A   Expansion joints are often of slip type. Because there is constant movement in this joint when steam is turned off and on throughout zone, you should repack joints each summer. Before doing so, shine sliding part with fine grinding cloth to make sure no burrs or scale cut packing.

9   Reducing valves: What attention is needed?
A   Besides large reducing valve that steps down steam from utility firm as it enters building, there may be other reducing valves at various zones of either high or low pressure. One zone may operate on as low as 1 psi while another uses 50 psi. Open each reducing valve, check valve seat and disk, and other internal working parts.

Diaphragm may be critical part, although that depends on design and material. If there are any indications of failure from constant flexing, it's good insurance to replace diaphragm with new.

10   Condensate pumps: What attention is needed?
A   Clean electric motor and lubricate same as any motor in system. Because this unit can shut down entire system, have only responsible person do work. Check float for leaks and clean out condensate-return tank if needed. Float may leak, get gradually water logged.

Remove packing from pump shaft and check shaft for grooving. If heavy wear, check pump and motor alignment. Best way is to break coupling (remove bolts) and use dial indicator. Don't assume that equipment stays in line just be-

cause it was in line at one time. Misalignment is costly.

Most pumps of centrifugal type need little maintenance, especially on condensate-return system. But checking wearing rings and impeller often shows that parts need renewing.

11   Return piping: What attention is needed?
A   Leaks and corrosion are very common in return lines. Look for corrosion around piping connections, especially at threads of nipples. If portion of your return line is buried, check for leaks as that is where corrosion takes place. Put hydrostatic test on buried section of pipe. If possible, run piping in covered trench to avoid future headaches.

12   Relief valves: What attention is needed?
A   New type relief valves on hot-water boilers are tested and approved by National Board of Boiler and Pressure Vessel Inspectors. They are stamped showing they comply with ASME Heating Boiler Code. These valves, rated in Btu discharge capacity of either water or steam, are more reliable than older type valves. Insurance firms advise renewing old valves with this newer type.

13   Controls on boiler: What maintenance is needed?
A   Safety valve should lift easily when test lever is raised. Clean water column so water level is easily seen in glass. Experience shows that if automatic water column is attached to boiler and water column, draining weekly during heating season keeps it working right.

Summer is time to check low-water fuel cutout, emergency water feeder, pressure controls and firing controls. This includes any automatic firing device. Take low-water fuel cutouts apart and check carefully. If manufacturer has local service center, let him do this work. Or you might return equipment to factory.

14   What is common cause of radiators or other space heaters, which are supplied by one riser in system, not heating while other units do heat?
A   Building sections that support piping and hangers often settle, which may restrict circulation in riser. Heaters on other risers may have good circulation at expense of one that has settled. To avoid, check all possible places where settling might take place. Do this every summer when lines are cold and then elevate the risers to their original positions.

Vapor barriers combat leakage: Achilles' heel of low-temperature systems

Insulation for temperatures below 32 F is only as good as its vapor barrier. A vapor soaked material has very little insulating value. First let's see how water vapor gets into insulation and then examine various ways to keep it out.

Air has a measurable vapor pressure governed by its temperature and relative humidity. The lower the temperature and relative humidity the lower the vapor pressure. Water vapor is present in air as a gas and will penetrate any material that absorbs air. When an insulated body is at temperatures above ambient, vapor pressure of water is higher at the insulation's inner surface than in the surrounding air. This pressure differential drives vapor outward from the inner surface (drawing below), through the insulation into the outside air.

Under these conditions insulation need only be protected from water impingement on the cold outer surface with some form of weather barrier. Heat of the insulated body keeps insulation fairly free of water vapor and moisture.

When temperature of the insulated body is lower than ambient, vapor pressure at the inner surface is less than that of air at the outer edge. So direction of water vapor flow reverses bringing water vapor into the insulation. Amount of vapor flow varies with the difference in vapor pressure. This in turn is proportional to temperature difference between the insulated body and the outside air.

Sketch below shows what happens to water vapor inside insulation. At dew point temperature the water vapor will condense. Actual formation of water takes place at the first cooler vapor resistant surface within the insulation. In some cases this is at the cold surface itself. If the insulation absorbs and retains water, its effectiveness as a heat barrier is soon lost. If temperatures within the insulation are below 32 F, ice forms to buckle the material; allow still more water vapor to enter and cause additional destruction.

Icing and water damage don't show up immediately with some of the water impervious insulation used for cold lines. Water vapor in these cases may enter through cracks or unsealed joints. Any resulting localized freezing causes only minor damage. But during shutdown all accumulated ice in the system thaws and runs along the piping to collect at some low point in the system. When low-temperature

Moisture is the culprit that destroys low temperature insulation . . .

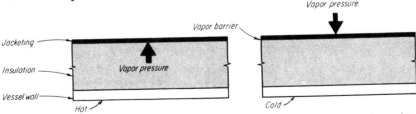

Higher vapor pressure at warm insulation surface drives out potential vapor leakage

Higher temperature at outer surface forces vapor laden air into the insulation

When the inside of a wall is insulated and interior space is held at a low temperature, moisture present will condense. This happens at the dew point — the critical temperature at which invisible vapor condenses to water. Dew point temperature varies with temperature differential and amount of humidity present. If insulation absorbs and retains moisture it soon loses insulating effectiveness. And if temperatures within the material are below freezing ice forms to destroy insulation. Solution: vapor barrier on warm side to stop moisture.

Trouble starts when vapor leaks in through damaged vapor barrier or through joints not properly sealed. Moisture travels along inner surface, freezes and destroys insulation

operation is resumed this large accumulation of water then freezes to severly damage insulation at this point.

What happens in those parts of the low-temperature system operating at temperatures above 32 F? Of course, the moisture won't freeze and visible harm may not occur. But water does collect in the insulation next to the inner surface with the attendant danger of severe metal corrosion.

Vapor barriers are the only practical answer to moisture problems. Aim is to stop movement of vapor at as high a temperature as possible so that dew point temperature will not occur where moisture is present or where condensation can take place within the insulation. Completely effective vapor barriers do not exist in practice either because of material's properties or defective application.

Since it is practically certain some moisture will pass the vapor barrier, what other provision can be made to keep the insulation dry and prevent its eventual destruction? In the case of refrigerated spaces this can be done by using insulation that readily permits water vapor flow plus a vapor-porous interior finish. Any small amount of moisture that gets by the vapor barrier then migrates through the insulation and deposits at the coldest spot inside, usually the refrigerant coils.

But when insulation is applied to cold tanks, pipes and similar equipment, there is no chance for moisture to escape on the cold side. So the vapor barrier must keep out moisture or the insulation will eventually fail.

Materials used include: (1) structural barriers (2) membrane barriers and (3) coating or surface barriers.

Structural barriers are the rigid sheets of reinforced plastics, aluminum and stainless steel jacketing. These come either as flat, corrugated or embossed sheets, often fabricated to exact dimensions and ready to install. They are usually fastened with bands, clasps or some other device. Joints may also be sealed with a plastic or tape.

Membrane barriers are the metal foils, laminated foil and treated papers, coated felts and papers and plastic films or sheets. One example is a new asbestos fiber coated with a polyvinyl choloride film. Today many of these barriers come attached to the insulation. Others are available in roll form. The joints of molded pipe insulation often come with flaps that can be sealed with an adhesive or tape.

Coating barriers come in fluid form as paint or in semi-fluid state. These are also known as surface coatings and often are the hot melt type used over insulation of cold storage spaces, piping systems and equipment. Material may be asphaltic, resinous or polymeric type. Pigments and solvents naturally depend on application and design. These are applied by brush, roller, trowel, spray or mop.

Rating of vapor barriers is according to vapor transmission rate expressed in perms. A perm of one, for example, means that only one grain of water will pass through one sq ft of material in one hour for every inch of mercury pressure differential.

In low-temperature insulation, a better vapor barrier performance than one perm is needed. For coolers operating between 32 and 50 F, 0.2 perm rating is sufficient. Most asphalt-saturated, shiny glazed papers qualify.

At still lower temperatures even greater efficiency is demanded. Practical rating in the freezing range below 32 F is about 0.01 perm. Laminated paper and aluminum foil products or aluminum foil alone, are used.

Great care must be taken in application to avoid damaging the vapor barrier material. Joints between sheets must be sealed and care taken to avoid rupture due to expansion and contraction of the structure itself. When the vapor barrier is punctured, the resulting hole must be sealed.

Adhesives used to laminate some vapor barriers of the membrane type; paper to paper, foil to foil, foil to paper, paper to plastic, etc, often have flame extinguishing properties. When material reaches combustion stage, gases or vapors released tend to smother the flame.

. . . but correctly applied vapor barriers effectively keep most of it out

Jacketing sealing these insulation joints is a 15-lb felt applied directly over the surface. In addition it serves as a protective cover

Plastic insulation finish being applied here is a combination vapor barrier and paint, used for cold and hot lines and equipment

Don't wilt when air-conditioning trouble hits you. Just use this handy chart to . . .

WELL-KEPT air-conditioning units give little trouble; keep 'em fit with right care

Track Down Air-Conditioning Ills Fast

SYMPTOM	POSSIBLE CAUSES OF TROUBLE; CURES
HIGH HEAD PRESSURE	1 Not enough condenser water, or water too warm; check water-regulating valve, look for plugged pipe 2 Noncondensable-gas or air in system; purge until head pressure becomes normal 3 Condenser fouled; clean tubes, tube sheets, heads 4 Too much refrigerant; bleed excess to a drum 5 Evaporative condenser faulty; check and correct air and water flow; clean surfaces 6 Evaporative condenser too small; install larger condenser or use small watercooled unit for peak loads
LOW HEAD PRESSURE	1 Too much or too cold condenser water; adjust water-regulating valve 2 Not enough refrigerant; look for leak in system; repair; charge 3 Leaky compressor valves; check to see if suction pressure rises more than 2 psi per min in Freon units. If it does, overhaul compressor 4 Oil return valve or trap leaks; spot by feeling for hot oil return; repair or replace valve or trap
LOW SUCTION PRESSURE	1 Evaporator load light, causes compressor to short cycle; run at lower speed or fit capacity control 2 Clogged strainer; clean or replace strainer 3 Low-pressure switch setting too low; adjust 4 Compressor oversized; fit with a capacity control or run at lower speed 5 Hiss in expansion valve. This is caused by flash gas in liquid line. Cure by subcooling liquid 6 Evaporator pressure drop too large; fit an external equalizer to stop high superheat 7 Not enough refrigerant; look for leak in system; repair; charge
HIGH SUCTION PRESSURE	1 Too much load on evaporator; check conditioned-space insulation, fresh-air leakage, infiltration 2 Compressor suction valves faulty; overhaul 3 Expansion valve stuck open. With this trouble suction line will be too cold; repair valve 4 Evaporator oversized; run compressor faster or buy a large compressor 5 Low-pressure switch set too high; adjust 6 Expansion valve oversized; use smaller valve

SYMPTOM	POSSIBLE CAUSES OF TROUBLE; CURES
SYSTEM CAPACITY LOW	1 Evaporator pressure drop too large; use an external equalizer on expansion valve 2 Superheat adjustment wrong; fit external equalizer or adjust expansion valve 3 Clogged strainer, valve or pipe; clean or replace 4 Dirty cooling coil; clean dirt off coil; defrost to remove excess ice; check air flow to see it is normal after cleaning and defrosting
COMPRESSOR WON'T STOP	1 Temperature high in conditioned rooms; look for oversized fan, wrong damper setting or too much Infiltration. See that rooms are insulated right 2 Discharge pressure high; check cooling water flow and temperature, clean condenser tubes or evaporative-condenser coil, purge excess Freon, air and noncondensable gas 3 Too low a temperature in conditioned rooms; check thermostat settings in the rooms, adjust to needed temperature; look for cold drafts that may be striking thermostat 4 Not enough refrigerant; look for leak in system; repair; charge 5 New heat loads in conditioned spaces; these may come from new or different production methods; increase compressor capacity or reduce heat load
ODORS IN ROOMS	1 Refrigerant leak; find, repair, charge, test 2 Brine leak; find, repair, adjust concentration and test piping 3 No trap in air washer drain to sewer; fit with suitable trap 4 Cooling-coil surface dirty; shut down air washer and clean coil; check intake-air filters for dirt 5 Odor-producing units near air intake; relocate odorous units or air intake
COMPRESSOR OIL LEVEL TOO LOW	1 Dirty strainer, tubing or valves; clean 2 Liquid slugs in suction; reset superheat, check to see thermal bulb is securely fastened 3 Not enough charge; look for leak, repair, charge system, and test with a halide torch 4 Suction line traps oil; repitch and fit loops 5 Oil separator faulty; repair, add oil, test

7
Refrigeration

► UNLESS A NEW air-conditioning refrigeration system is charged right, it will work poorly. When there isn't enough charge in system, high- and low-side pressures drop below normal. Then system capacity drops too. If too much Freon is added, high head pressures result. This also cuts the systems capacity for cooling.

Charging the new system for the first time is important. Here's where the operating engineer first gets acquainted with his new unit—while it's warming up.

As Freon first enters system, check for leaks. Read "How To Test New Freon System" (following page) for complete details on finding leaks. If you find major leaks, pump out system right away and repair before going on.

When charging, go slow so you can check for the minor leaks. Feed Freon gradually. Periodically, stop compressor and weigh Freon drum to see how much charge you're feeding in. Mark minor leaks as you find them. No new system can be expected to be leakproof. It's best to find all leaks now.

After charging to full pressure, check again for leaks. Mark them. Then pump out system, fix all leaks and recharge.

Check liquid strainer. Freon-12 is a good cleaner and loosens scale and dirt in system during first charging.

1 First step in charging new refrigeration system is attaching an accurate compound gage to the suction-gage connection. Gage should show pressure above atmosphere as well as inches vacuum. Attach pressure gage to discharge connection. Install these gages permanently

2 Place Freon-12 service drum on scale and connect to charging valve on liquid line. Use 3/8-in. flexible copper tubing or a flexible metallic hose. This is important, because solid piping strains as scale lowers and causes leaks or a break. Open the drum valve slightly to test charging connections for leaks. Leaks bubble when connections are coated with soap-and-water solution. Open compressor discharge and suction valves, condenser hot-gas and liquid valves and liquid solenoid valve. Then open charging valve, and crack drum valve enough to raise 40 psi in entire system

3 Test every joint in system with exploring tube of a halide torch. Pay particular attention to all connections at compressor. Slightest trace of Freon-12 turns torch flame from blue to a green color. Repair leaks and recheck with torch to make sure entire system is tight

4 Turn on air-conditioning fan to put load on evaporator. Close condenser liquid valve. Note weight of service drum. Lay drum on scale and block so valve end is lower. Scale won't handle drum, lay it on floor and weigh from time to time. Open charging valve and drum valve. Circulate water through condenser and run compressor intermittently until system is charged. Freon should show in bottom of condenser gage glass. Close charging valve and open condenser liquid valve. For air conditioning, keep suction pressures 30 to 45 psi and head pressures 115 to 140 psi

TEST NEW FREON SYSTEM

Combination pressure and vacuum gage

Purge valve

Condenser hot-side valve

Discharge valve

Suction valve

Liquid solenoid valve

Condenser

Charging valve

Thermostatic expansion valve

Low-side gage

Refrigerator compressor

Strainer

External equalizer connection

Evaporator

▶ BEFORE PLACING a new air-conditioning plant in service, the refrigerating-system piping must be thoroughly checked for leaks. It makes no difference how well the piping joints have been made up or how reliable the workmen are, the only sure way to know if the system is tight is to test for leaks. What do leaks mean to the operating engineer? Just this; he will lose his Freon charge and the system won't work right. Even the tiniest leaks will cause him headaches day after day. Every joint in the system is a possible source of leaks. The contractor should make a thorough search for leaks before turning the plant over to you. Since it's your baby after it's put in, you'd better know how to make this test right so you can check the work while it's being done. The following steps show the right way to test for Freon-12 leaks in any refrigeration system.

1 Close the discharge and suction valves to the compressor. Open the condenser hot-side valve **C**, charging valve **D**, liquid solenoid valve **E** and thermostatic-expansion valve **F**. Disconnect the external equalizer line to the thermostatic-expansion valve to keep it from closing. Then attach pressure gages to the high and low side of the system. The high-side gage can be attached to the purge valve. The low-side gage can be attached to the external equalizer connection after the piping is removed

Charging connection

CO_2 drum

2 Connect a drum of CO_2 or dry nitrogen to the charging valve. Raise pressure in system to about 15 psi. Check suspected joints by coating with solution of soap and water with a few drops of glycerine. If no bubbles show, raise pressure to 50 psi and soap again. Resweat defective joints and test again until no bubbles appear and pressure in system is maintained

Flexible exploring tube

Flame

Sweated joint

Halide torch

3 Replace the CO_2 drum with a Freon-12-service drum. Raise the pressure in system to about 10 psi. Test all joints with the exploring tube of a halide torch. The slightest trace of Freon-12 will turn the torch flame from its normal blue color to a characteristic green color. A gasoline torch can be used for this—also denatured alcohol or acetylene gas

Charging connection

CO_2 drum

4 Disconnect the Freon-12-service drum and reconnect the CO_2 cylinder. Remove any relief valves that might raise and cap openings. About 10 psi Freon is still in system. Boost this to 175 psi with CO_2. Check all joints with the halide torch. Repair any leaks and repeat procedure until leaks are stopped. Keep 175 psi for 24 hours. If pressure holds, system is tight

Charging valve

Vacuum pump

5 Release pressure from system. Connect the suction side of a vacuum pump to charging valve. Discharge pump to atmosphere, running it for 24 hours. If the system does not lose more than about 1-in. of vacuum, it can be considered dehydrated and free from leaks. That's because this is a triple check for leaks and the only method to use before accepting the job

1 A simple refrigerating system consists of a cooling coil where the liquid boils; a compressor that compresses the vapor to a higher pressure, a condenser that cools the vapor and changes it back to a liquid and a liquid-storage vessel

How Machines Make Cold

Knowing how liquids act at different temperatures and pressures helps you spot refrigerating troubles in a jiffy.

▶ REFRIGERATION is nothing more than removing heat from a substance we wish to cool. All we have to do to coax heat out of a warm object is to place it near or in a cold substance. But we can't get very far with natural cooling in the summer because we don't have enough nature-made cold substances. So the only answer is—make your own cold with a mechanical system, Fig. 1. Knowing *what* has to be done, let's see how we go about building such a system.

Energy Needed. To make artificial cold, we'll have to get heat to flow "uphill" from a cold body or area to a hot one (relatively speaking). This means working against nature and it's going to take energy, somewhere along the line, to haul the heat away. This costs money.

For the time being let's forget the energy part of our problem and start by letting the heat move downhill, naturally. We can do this by taking advantage of (1) what happens when liquids change to vapors and back and (2) the way these changes are tied to the pressure acting on the liquids.

How Liquids Act. To get our thinking straight about the characteristics of liquids, let's start with a few we know because we can see and feel them. Water is certainly a familiar one and we know it is still liquid right up to 212 F as long as we keep it in a vessel open to atmospheric pressure. We also know that if we keep on adding heat to the water it will boil and turn to vapor at this temperature and pressure, Fig. 2A.

Other substances that we can see as liquids open to the atmosphere are chloroform, which boils at 140 F, and ether, which boils at 95 F.

Now that we have a few boiling points off our chest, let's see what they have to do with moving heat. Take water for example; if we want it to boil, we add heat. For every pound we have to add one Btu to raise the temperature one degree. Starting with water at 32 F we need only 180 Btu to get one pound to 212 F. But, to turn a pound into steam we must add 970 Btu (latent heat of vaporization).

Effects of Pressure. So far we have been talking about how a few liquids act at atmospheric pressure and we can see that the temperatures involved will not do much in the way of keeping things cold. Fortunately we can take advantage of a liquid's high latent heat of vaporization and get it to boil at the temperature we need by varying the pressure acting on it. Increasing the pressure raises the boiling temperature and decreasing it lowers the boiling point.

But to do this we must use a closed container which puts the liquid out of sight. A good example is a boiler feeding a condensing turbine. Here the water does not boil until it is heated considerably above 212 F, Fig. 2B, because the boiler confines the vapor given off as the water gets hot so pressure is much higher than atmospheric.

Cool Steam. Steam from the boiler drops in pressure as it passes through the turbine to enter the condenser, which may be under a 29-in. vacuum (a pressure much lower than atmospheric). You can prove that steam leaving the turbine is considerably below 212 F by laying your hand on the exhaust hood. And it can't be a liquid

2 Pans above show how pressure acting on water surface affects boiling

Atmospheric pressure

Ⓐ 212 F

100 psi (abs)

Ⓑ 328 F
Higher pressure means higher boiling point

29 in. of vacuum

Ⓒ 79 F
Lower pressure means lower boiling point

3 Cylinder of liquid ammonia vented to atmosphere holds —27 F in box

Vent

Atmospheric pressure

Insulation

Ammonia

—27 F

at this point because the turbine blades wouldn't last long if water passed through them.

Last, but not least, you can refer to the steam tables and find that water can exist as a vapor at the low-temperature and low-pressure conditions of the turbine exhaust.

Now we're getting somewhere. If water can exist as a vapor in a vacuum of 29 in., and we know that it is cool enough for us to lay our hand on the turbine exhaust, we have proof that water will boil into steam at a temperature below 212 F if we keep the pressure below atmospheric and add heat, Fig. 2C. Another thing in our favor, from a refrigerating standpoint, is that it still takes about the same number of Btu to vaporize the water at this lower temperature and pressure.

If we have to talk about liquids penned up out of sight, let's see how a common refrigerant like ammonia acts. Checking the tables of ammonia characteristics we find that it is a gas at atmospheric pressure and temperature; so we have to confine it in a closed system to keep it liquid permanently if the temperature is higher than its boiling point of —27 F. By the way, we have to expend energy to get it penned up.

Looking at it another way, suppose we have a cylinder of ammonia whose temperature is 80 F. A pressure gage connected to it will read 138 psi. The more we cool the cylinder (at 70 F pressure is 113 psi) the less the gage will read until we get it down to —27 F. Then the gage reads zero.

How Ammonia Cools. Suppose we put the cylinder inside a heavily insulated box and open the valve so ammonia can escape, Fig. 3. As the pressure falls, the ammonia starts to boil furiously because it is at 80 F. The only source of heat for boiling is from the liquid and its metal container. So what happens. The large body of liquid cools rapidly, and as it cools its vapor pressure keeps in step with the temperature until the tank and liquid reach —27 F. Pressure gage then reads zero and only a small whiff of ammonia gas comes out occasionally as heat leaks in from outside the insulation. We now have a tank of liquid ammonia open to the atmosphere, but we lost about 19% as vapor or gas in the boiling process to cool the cylinder down to —27 F. As soon as we opened the ammonia valve and released the pent-up pressure, the liquid started to boil because it contained too much heat to remain a liquid at the resulting pressure.

Heat Haulers. Boiling absorbs heat, in this case from the liquid itself, so the liquid ammonia soon gets so cold it no longer contains an excess of heat units (vapor pressure and temperature are in balance) and boiling stops. The same thing holds true with other refrigerants except they have different boiling temperatures at atmospheric pressure. For example sulphur dioxide boils at 14 F, methyl chloride at —10 F, Freon 12 at —20 F, Freon 22 at —44 F and carbon dioxide at —109 F.

Real Refrigeration. Suppose we need a lot of room. We insulate the space, cover one or more walls with pipe coils,

to get lots of cooling area, and connect them to a large cylinder of ammonia. Open the valve, feed liquid ammonia into the pipe coils where heat picked up from the cooling area causes it to boil rapidly, and what happens. The resulting gas escapes to atmosphere and we have a refrigerated space at —27 F. If this is too cold, put a valve on the vent to atmosphere. Then throttle the discharge and build up pressure in the coils until you get the temperature you want.

Fine, you say, but that ammonia blowing to waste costs money. We can't afford to get our refrigeration by wasting ammonia but saving and using it again requires investing a lot more money than our simple gadget cost.

Starting Uphill. So far we have been letting heat flow downhill in its natural way. But reclaiming the ammonia is another matter. Here is where it takes energy, because we have to buck nature to get the heat out of that —27-F ammonia vapor and turn it into a liquid. And we have to remove the heat with some substance available to us at normal temperature—say water at 70 F. Let's follow through without worrying about pressure drop caused by pipe friction.

From a practical standpoint the best we can hope to do with 70-F water is to cool the gas to 80 F. Now, how are we going to cool gas that has a temperature of about —27 F, with 70-F water? Maybe the ammonia tables will give us the answer; so let's see what they say about 80-F ammonia. Well liquid ammonia at 80 F has a vapor pressure of about 138 psi gage. But our gas is at atmospheric pressure. Sure, but you said you wanted to reclaim it, so dig down and buy yourself a compressor. And while you're at it, get one rated for at least 150 psi to cover yourself if the cooling water should push well above 70 F on some hot summer day.

While in the buying mood get a good condenser and some pipe and you're ready to build a closed refrigerating system. So you want to hook it up like Fig. 1. OK, now vapor from the cooled space (evaporator) enters the compressor, which has to do work on the gas to get it up to 138 psi. The gas gets hot—in fact it jumps to about 280 F in the compressor cylinder. We get the pressure we need all right but end up with a lot of hot gas that has to be cooled down to 80 F before it will start condensing back to a liquid. With the correct size condenser our 70-F water should do the job nicely and prepare the ammonia for another journey around the circuit.

Pressure-Temperature Chart for Refrigerants

1 Use this chart to check your evaporator and condenser pressures. If they don't match the respective temperatures, check your instruments before looking further

2 Compressor takes vapor away and raises it to the condensing pressure

How Machines Make Cold

*Here we learn how many Btu a pound of refrigerant absorbs
when changing to vapor, Part 2. A further step compares heat
content per cu ft of vapor, which determines compressor capacity*

► ON PAGES 134 and 135, we learned how the boiling temperature of a liquid depends on the pressure acting upon it; so let's see how this affects the various common refrigerants. But first let's learn how refrigeration is measured and machines rated.

Standard Ton. Large pieces of refrigerating equipment are rated in tons. Now just what is a ton of refrigeration? *It is the cooling effect obtained by melting a ton of ice in 24 hr.* Notice that the capacity is based both on the amount of heat absorbed and time—a ton of refrigeration is not just a quantity, it is a rate of cooling.

Btu Per Ton. Let's see what one ton of refrigeration means in Btu. The latent heat of fusion (freezing) of ice is 144 Btu per pound; the amount of heat that must be removed from 32-F water to turn it into ice at 32 F. In other words the process takes place without a change in temperature. So if we have to take 144 Btu out of every pound of water to freeze it into ice, the ice is

going to absorb 144 Btu when melting to water.

A ton of ice, therefore, must absorb $2000 \times 144 = 288{,}000$ Btu in melting. And if the process *must* take 24 hr to complete, then it occurs at $288{,}000 \div 24 = 12{,}000$ Btu per hour and $12{,}000 \div 60 = 200$ Btu per minute.

Now we can define a ton of refrigeration as the removal of 200 Btu each and every minute in a 24-hr period. But don't jump to conclusions and say one ton of refrigeration will produce a ton of ice in 24 hr.

This may sound confusing after using a ton of ice to define a ton of refrigeration, but remember that a ton of ice absorbs a constant and never varying number of Btu as it melts. But if we attempt to freeze this ton of ice in 24 hr, we have mechanical losses and heat flow into the system from the surrounding atmosphere. Both of these use up refrigerating capacity. It takes about 1½ tons of refrigerating capacity to freeze a ton of ice in 24 hr.

Equally important when figuring capacity is the rate at which the refrigeration must be done. A one-ton compressor (rated on 24-hr basis, remember) is not large enough to handle a 288,000-Btu load if the cooling job has to be done in 5 or 6 hours.

Machine Rating. Just stating that a one-ton compressor will produce one ton of refrigeration in 24 hr is not the whole story. We have to be more specific about the operating conditions to be able to compare machine performance. These conditions are defined as (1) an evaporator pressure corresponding to a saturated-vapor temperature of 5 F with a 9-F allowable superheat at the compressor intake and (2) a condensing pressure corresponding to a saturated-vapor temperature of 86 F with an allowable subcooling of 9 F at the evaporator expansion valve.

Now that we have the standard ton streamlined down to 200 Btu per min at 5- and 86-F pressure conditions, we can compare the refrigerants.

TEMPERATURE RELATION AND VAPOR VOLUME

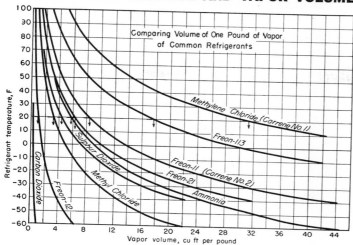

Comparing Volume of One Pound of Vapor of Common Refrigerants

3 Here is how the vapor volumes of the common refrigerants compare. Before using this information to select compressor capacity, figure Btu content per cu ft

BTU NEEDED TO BOIL ONE POUND OF LIQUID REFRIGERANT

(Latent heat of vaporization)

Temp. deg F	Ammonia	Methyl Chloride	Sulphur Dioxide	Carbon Dioxide	Freon– 2I	Freon– II	Freon– I2	Freon– II3
−40	598	190	178	137	114	87	73	..
−30	591	189	177	134	113	87	72	72
−20	585	186	175	129	112	86	71	72
−10	576	184	173	125	111	85	71	71
0	569	182	171	120	110	85	70	71
10	561	179	168	114	109	84	69	70
20	553	177	165	109	107	83	68	69
30	545	175	163	102	106	82	67	69
40	536	172	159	95	105	82	65	68
50	523	169	156	87	104	81	64	68
60	518	167	152	77	103	80	63	67
70	509	164	149	64	101	78	62	67
80	498	161	145	45	99	77	60	66
90	489	158	141	...	99	75	59	66
100	478	154	137	...	97	76	57	65

Boiling Temperatures. Take a look at the simple curves in Fig. I. They represent vapor pressures of the common refrigerants at different temperatures. If you prefer gage pressures, use the right-hand scale at left, otherwise use the absolute scale at left. Refrigerants all act pretty much alike although each one has its own pressure-temperature range. To explain the curves, suppose we use the small compressor setup in Fig. 2 with its suction line connected to an insulated drum of ammonia. To change drum pressure we can run the compressor at any speed we like.

Suppose the ammonia has a temperature of 75 F. What should the pressure gage read? Tracing up the 75-F line to the ammonia curve, follow the arrow to the right and read 135 psi. Let's run the compressor at full speed for several minutes. A quick look at the gage shows the pressure to be 65 psi, and we can hear the ammonia boiling furiously. What happened? Well we took away part of the ammonia's vapor pressure, so the liquid gets busy trying to restore conditions to a balance.

Temperature Lag. At this stage of the game the thermometer may not have changed, but if we watch it closely it will show a drop in temperature by the time the liquid stops boiling. The boiling uses up some heat in the liquid and it has to be a little cooler as explained last month. If pressure rises, say to 85 psi, and remains steady, the curve shows that thermometer should read about 58 F.

To get the temperature you need, run the compressor and pump the ammonia pressure down to the value corresponding to the temperature shown on chart. Then run it just fast enough to carry the vapor away as it forms and still hold the pressure steady. This is the principle of any refrigeration system regardless of what refrigerant is used. Now let's see how refrigerants differ when they boil.

Vapor Volume. The curves in Fig. 3 show the volume of one pound of the different refrigerants after they have been changed to vapor at the temperatures given on left-hand scale. The vapor in each case has a saturation pressure corresponding to the temperature so we can compare the refrigerants in terms of volume per pound.

For example, suppose we need a room or space held at 20 F. We find (arrows) that carbon dioxide has the smallest volume. Freon-12 is 1 cu ft, methyl chloride is almost 3½, sulphur dioxide 4½, ammonia almost 6, Freon-21 over 6, Freon-II is 8½, Freon-113 almost 18 and methylene chloride 31½. On this basis alone it looks like compressor size, needed to do a given refrigeration job, varies all over the lot.

But just a minute. Suppose we examine things a little closer. Remember we learned that cooling and freezing equipment is rated on a ton-per-24-hr basis, equal to 200 Btu per min. Fig. 3 shows how much vapor a pound of each liquid forms. But take a look at the table. At 20 F ammonia latent heat of vaporization per pound is 553 Btu. Even though its vapor volume is almost 6 cu ft per lb, the ammonia vapor does contain 553 ÷ 6 = 92 Btu per cu ft.

Compare this with Freon-12 having a volume of 1 cu ft and heat content of 68, which means that it contains 68 Btu per cu ft. So you see compressor capacities for the two refrigerants do not vary much for the same cooling job at 20 F. On the other hand there is quite a difference between ammonia and Freon-113 capacity requirements.

11 steps to foolproof refrigerant-leak testing

▶ PLANNING TO INSTALL more refrigeration capacity to handle summer air-conditioning loads? Having trouble playing detective with small leaks in your existing system? Then the latest guide used by The Trane Company is for you. It's your road map to a refrigerant system that's free of even the smallest leaks. Here, in 11 easy-to-follow steps, is the gist of it.

1 If test pressures exceed relief-valve setting or safe-pressure limits on bellows used in pressure controls, remove these items before building up pressure. (*Test before covering pipes or turning on condenser water.*)

2 Plug the thermostatic-expansion-valve inlet to make sure both high- and low-pressure sides of the system are fully tested. Common way to do this: Remove cap screws and insert a plug in valve's flange body. Hold plug in place with a solid diaphragm over the inlet. And don't forget to disconnect the external equalizer line from the expansion valve.

Remove the thermostatic expansion valve's power assembly during leak testing to avoid damaging it.

3 For high-side test connection: Close compressor suction, discharge valves to make sure you won't be testing the compressor itself. Connect a drum of dry CO_2 to the liquid charging valve. A word of caution: Don't use oxygen or acetylene instead of CO_2. You'll be inviting a possible explosion. And be sure to have a gage and regulator on the CO_2 drum, to control pressure.

CO_2 pressure can build up to 1000 psi or more if it isn't watched. Second gage, between regulator and liquid charging valve, reads test pressure. Install this gage with a flare tee.

Admit CO_2 to system's high side until gage reads 225 psig or pressure specified by local codes. Close regulator and charging valves. Then disconnect drum just ahead of the tee. To read high-side test pressure, cap tee and open the charging valve.

RECOMMENDED METHOD of connecting your vacuum pump to the refrigeration system before charging it with refrigerant

4 For low-side test connection: Admit CO_2 to the suction line through the external equalizing pipe that was disconnected from the expansion valve (see step 2). Drum should be upright and fitted with standard gage and regulator. Gage, installed with flare tee, between regulator and equalizer line reads test pressure. To isolate drum from system, install a packless diaphragm valve between regulator and test gage.

After admitting CO_2 slowly until test gage reads 125 psig (for Freon-12 systems), close regulator and packless diaphragm valve. Then disconnect drum at the packless valve and cap the valve's open side.

Don't forget to check with local codes for test pressures used with different refrigerants.

5 With system under test pressures, tap the sweat or solder connections with a rubber or rawhide mallet. Hit hard enough to start faulty connections leaking, but not hard enough to break a sound joint.

6 A drop in gage pressure shows up a major leak. You can spot such leaks by tracking down escaping gas noise.

To find smaller leaks brush all possible leakage points with soap solution and watch for bubbles. A few drops of glycerin in the soap solution insures good bubbling. Using distilled or rain water for the solution yields best results.

7 After bubble-testing of high- and low-side systems, break charging-valve and external equalizer-line connections; allow CO_2 to escape. Repair all leaks spotted by gage pressure loss or by bubble testing. A word to the wise: Don't try to repair leaks while the system is under pressure. Also, take bad joints apart, clean and remake them as a new joint. Don't add more brazing material over a poor connection.

8 After repairing leaks, charge high- and low-side of the system with a small quantity of Freon-12. Use five pounds of F-12 for each ten tons of refrigeration, adding larger part of the F-12 to the low side. (For charging, follow the same steps used for CO_2, above). Next, disconnect refrigerant tank, hook up CO_2 drum and build up to original test pressures.

9 Check the entire system with a halide leak detector, photo on facing page. Make sure you run the exploring tube over all parts of the system under pressure. Small amount of escaping refrigerant colors the flame green. Big leak may put out the flame or turn it dense blue with a reddish tip. If F-12 in the air colors the flame regardless of the searching tube's position, ventilate the room thoroughly.

If an open flame can't be used, re-run the bubble test where leaks have been repaired. If additional leaks show up, you'll have to run through repairing and testing steps again.

10 After you've tracked down all system leaks, let the equipment stand 24 hours under test pressures. If gages show pressure changes—taking into account room-temperature changes—you're up against more leak testing. But remember: Pressure will rise or drop about three psig for each 10-F rise or fall in temperature.

11 After final leak testing and repairing, release pressure, remove all expansion-valve plugs and reassemble the equipment for normal operating conditions. You're now set to pull vacuum on the system, see drawing at top, to prepare it for charging.

139

INDICATING REFRIGERATOR compressor shows what's going on inside machine

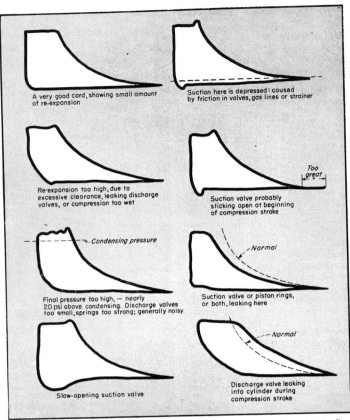

A very good card, showing small amount of re-expansion

Suction here is depressed: caused by friction in valves, gas lines or strainer

Re-expansion too high, due to excessive clearance, leaking discharge valves, or compression too wet

Suction valve probably sticking open at beginning of compression stroke

Too great

Condensing pressure

Final pressure too high, — nearly 20 psi above condensing. Discharge valves too small, springs too strong; generally noisy

Normal

Suction valve or piston rings, or both, leaking here

Slow-opening suction valve

Normal

Discharge valve leaking into cylinder during compression stroke

THESE CARDS show the experienced engineer that his compressor is either in first-class shape (top left) or that a number of things are wrong, and exactly what they are

The Indicator-Card Story

▶ You may not know it, but James Watt invented the steam-engine indicator, as well as the double-acting steam engine. As a young man, Watt was an instrument maker. One day he repaired a small model of the Newcomen-type atmospheric engine for the University of Glasgow. The model fascinated him so much that it awakened his first interest in steam power.

After building the first steam-powered double-acting engine, Watt developed the indicator. He needed it to find out what went on inside the cylinder at all times.

High-Speed Engine. Modern high-speed steam engines don't need to be indicated as often as old slow-speed corliss machines, with their many adjustable parts. The more valves and valve linkage, the more chances for engine to get out of adjustment.

In the same way the enclosed-type refrigerating machine of today generally operates satisfactorily for years, without needing to be indicated. For this reason most compressors today are no longer fitted with indicator connections, unless so ordered.

But the indicator does provide a check on what is occurring within the machine. Taking indicator readings may prevent serious waste in your plant. This is especially true if the

compressors are opened for inspection only after long intervals.

Large Compressors. Large ammonia and Freon compressors seldom run at speeds above 360 rpm. Taking an indicator card under such conditions is not difficult. Easiest way is to fit a home-made eccentric to the shaft, between stuffing box and belt wheel or motor, to reproduce the motion of the pistons.

Caution. *On ammonia machines, use an indicator with steel cylinder and piston. Ammonia corrodes brass.*

Set the eccentric so it is at top of its stroke when piston of cylinder being indicated is at top of its stroke.

KEEP THESE SAMPLE CARDS FOR REFERENCE
TO COMPARE CARDS FROM YOUR COMPRESSORS

Normal

Compression too wet:

Poor card caused by restricted passage between indicator and cylinder

Discharge valve lifts too high, or has poor action

Discharge valve sticks open due to weak spring, causing leak. Or piston of indicator sticks.

Card from a horizontal machine at 206 rpm. Valves have begun to "dance" at this speed.

VALVE AND OTHER mechanical faults in refrigerating compressors are often not known without opening machine, so cards are taken. Here's what they tell you

You can use a steam-engine indicator to check refrigerating-compressor performance if it can be connected to the cylinders. Here is how to interpret the more common cards taken

Best way to do this accurately is to set crank on dead center with trammel

Then set eccentric with trammel. You don't need to cut a keyway on shaft or keyseat in eccentric. Just hold in right position with setscrew. Then place permanent guide mark on shaft and eccentric for future settings.

What Cards Tell. The refrigerating work being done by each cylinder and the indicated horsepower required to drive the machine, can be found from the cards. Any other method of measuring the cooling load is likely to be much harder than with cards.

The cards shown here carry notations

that tell their own story. They will help any engineer find out how to fix what's wrong. It pays refrigerating plants with large slow-moving compressors to take these cards at set periods. It's the only way to know what's going on inside the machine.

Then operating engineer can place his finger on the fault the first time without having to guess at trouble. I've known of compressors that ran with major efficiency losses for years before trouble was discovered accidentally.

Cards not only help to make adjustments but keep engineers on their toes.

The attention you give your new refrigeration system just before putting it into operation may make all the difference between a lemon and a peach.

These easy-to-follow steps tell how to make . . .

START AND STOP the compressor several times at 10-second intervals before running in. Read compressor pressure gages. Correct operating difficulties before proceeding

Final checks and adjustments on

▶ THE BEST WAY to avoid costly and annoying shutdowns after a refrigeration system is operating is to go over it with a fine-tooth comb before starting. Then check it again after running 72 hours and once more after a week.

Before operation:

Before starting your new refrigeration compressor for the first time, check the entire system. Put all controls through their complete cycle of operation.

Adjust the low-pressure switch so it starts the compressor whenever suction pressure rises above the desired setting. Adjust the differential or cutout points as low as the job will stand to prevent short-cycling.

Check cutout point for the high-pressure switch, too. This is generally 175 psi.

See that all interlocks work right. For example, the compressor should be "locked out" if the evaporative condenser fails to run properly.

Check the thermostatic expansion valve for proper superheat adjustment. Put a thermometer on the suction line near the remote bulb of the thermostatic valve (sketches, facing page). Read refrigerant temperature at a point as near the remote bulb as possible, but on the compressor side of both the remote bulb and the external-equalizer-line connection.

You get the most accurate temperature reading by using a thermometer well. If no well is provided, attach thermometer with duct seal or use a clip-on-type thermometer. In these last two methods, reading accuracy depends on the bond between thermometer and pipe.

Read the suction pressure at the gage installed in the back-seat port of the compressor suction valve. Difference between thermometer reading on the suction line and the temperature calculated from the suction pressure is the superheat.

This method does not consider any

pressure drop in the low side of the system. If you suspect abnormal pressure drop, take steps to measure it and determine its effect.

One method for measuring pressure drop is to install a tee at the connection between the thermostatic expansion-valve external-equalizer line and the suction line proper. Connect a refrigeration compound gage to this tee. This gives the true pressure reading at the evaporator outlet. Gage readings can then be used for obtaining superheat measurements.

Most air-conditioning installations have superheat settings of about 10 degrees.

Also, check head pressure. If it is too high, adjust the water-regulating valve for increased flow. If head pressure is too low, adjust valve so that the water flow will be reduced.

Set thermostat to maintain desired temperature in the area to be cooled.

Just before starting the system, check for the following conditions:

1. Check temperature at evaporator outlet

2. Thermometer well construction

3. Attaching thermometer without a well

USE THERMOMETER WELL to get most accurate reading of refrigerant temperature. Otherwise, tie thermometer on line, use putty to insulate bulb from the atmosphere

refrigeration systems

1. Shutoff valve on water line to shell-and-tube condenser open.

2. Oil in compressor at or above center of sight glass.

3. Compressor suction and discharge valves open.

4. All moving parts free to move. (Check by turning compressor over by hand several times from the drive coupling.)

5. All shutoff valves open except by-pass valves installed for other purposes.

6. Solenoid stop valve on magnetic coil control.

7. All compressor suction, discharge and oil-pressure gages connected and open to read operating pressures.

8. Readings on compressor pressure gages OK when compressor is started and stopped several times at 10-second intervals. (Correct any operating difficulties before proceeding.)

After making these checks, let system run normally for at least 72 hours before making a final check.

During 72-hour run-in:

Mechanic should stay on the job continuously for at least 48 hours of the 72-hour run-in period. Here are the things he should do:

1. Keep a close watch on oil level in the compressor, never permitting unit to run short of oil.

2. Immediately after starting, take amperage and voltage readings on all motors to determine that they are running at nameplate conditions.

3. Check entire system for leaks several times during the 72-hour run-in.

4. Check all controls for proper operation; reset if necessary; see that they are calibrated.

5. Observe and record all operating pressures and temperatures.

6. Check superheat setting of the thermostatic expansion valve; adjust if necessary.

7. Adjust and set compressor capacity control system. Even though this may be preset at the factory it should be reset if necessary.

After 72 hours:

Here are the points to check after system has run 72 hours:

1. Compressor oil level. If low, do not add oil immediately, but operate the system three or four hours and check frequently to see that returning oil does not restore the proper level. If oil has not returned at the end of four hours, add some. If oil is at the proper level on starting, check frequently during the 3- or 4-hour run to see that level stays constant.

2. Refrigerant flow at the sight glass. Flow should be solid with no bubbles. These indicate a refrigerant shortage. If bubbles are present now, there is probably a leak, and system should be checked completely with a halide torch.

3. Liquid line from receiver to expansion valve. See that there is no appreciable temperature change over the length of the line. If a strainer or stop valve has a warm inlet and cold outlet, there is restricted flow or inadequate subcooling. If liquid is properly subcooled, there is an obstruction. Remove and clean the obstructed part.

4. Superheat setting of the expansion valve.

5. Adjustment of the evaporator pressure regulator valve.

6. Head and suction pressure. If correct, the suction and discharge shut-off valves should be back-seated. Remove test gages (unless they are permanent) and plug gage tappings.

7. Compressor seal. Stop system and test seal with a halide torch to be sure it is holding tight.

8. Motor alignment. Check compressor motor for alignment and adequate lubrication. Avoid over-lubrication.

9. Clean the water strainer ahead of the water-regulating valve.

10. Air-handling units. Adjust belts for proper tension and check alignment. Check for adequate lubrication. Avoid over-lubrication. Check air filters to be sure they are clean.

After one week:

1. Drain compressor crankcase. Remove inspection plate and thoroughly clean crankcase, swab out with carbon tetrachloride or similar cleaner. Then recharge with clean, new compressor oil, as specified by the manufacturer.

2. Tighten bolts in the coupling between compressor and motor.

3. Clean all filters in the entire system both in the compressor and liquid line.

Do not take these steps at the time of the 72-hour check. More regular running time is needed to clear the system of foreign material.

—*Courtesy, The Trane Company*

Purging your
refrigeration system

The knowledge is common among refrigeration engineers and system operators that air in the condenser causes a rise in head pressure. But what they're apt to overlook is *why* air's presence increases head pressure and *how*, in many ways, this increase can add significantly to overall costs.

First of all, let's briefly review air's effect on condenser performance, then see how this adverse effect makes itself felt throughout the system. When air molecules spread themselves across the condenser's heat-transfer surfaces it becomes extremely difficult for the refrigerant-gas molecules to reach these surfaces to give up their heat energy. Thus air's presence has the effect of reducing amount of heat-transfer surface available to the refrigerant gas. The only way to offset this "smaller-condenser" effect is by increasing temperature—and pressure —of the refrigerant gas so that total amount discharged by the compressor will be condensed.

As discharge pressure rises, a number of effects are apt to follow. One, certainly, is an increase in power input. Accepted rule: for each 4-lb increase in head pressure—due to air or other noncondensable gases in a refrigeration system — compressor power costs go up 2% and compressor capacity drops 1%. This 4-2-1 formula for increased power cost and decreased capacity applies to all common refrigerants, under wide differences in suction and head pressure. It's obvious that such power-cost increase means a profit decrease some-

where along the line. Too, decrease in compressor capacity means operating the unit more hours to get the same refrigeration effect. In many cases, other losses caused by lowered compressor capacity may add up to far more than the extra cost involved in operating the compressor against the higher head pressure.

Added cost factors: Cooling-water use increases and heat-transfer efficiency drops off. Using more cooling water may mean much or little to the individual plant, depending upon its location, pumping system, water source and cost, recovery factor and so on. However, it's a safe bet that unnecessary water used in any plant ups operating costs to some degree.

Excess head pressure takes its toll from the compressor itself. Bearings (this applies to the driving motor also) are subjected to greater loading. Higher temperatures accompanying increased discharge pressures tend to shorten valve life, promote lubricating-oil breakdown, And gasket failure often goes hand-in-hand with excessive head pressure.

From a practical standpoint, it's impossible to keep air and other noncondensable gases from sneaking into the majority of refrigeration systems. If system operates with suction pressure below atmospheric, air leakage is most likely through valve packing and compressor seals. But, whether suction pressure is above or below atmospheric, air will find additional ways of getting into the system: (1) when system is opened for cleaning

or repair (2) when you're making or breaking hose connections to discharge refrigerated cars or trucks (3) when you're adding compressor lube oil (4) by refrigerant or lube-oil breakdown (5) from impurities in the refrigerant itself.

A sure sign that air or other noncondensables have crept into your system is a discharge-pressure-gage reading higher than normal for any given refrigerant temperature leaving the compressor. Table 1 lists temperature-pressure relationships for three widely-used refrigerants: saturated ammonia, Freon 12 and Freon 22.

Example: If ammonia temperature leaving the compressor is 84 F, theoretical condenser pressure is 149 psig. If the discharge-pressure gage reads 169 psi, 20-psi excess pressure is increasing power costs by 10%.

Liquid temperatures are easy to read if the liquid line from the condenser is equipped with a mercury well. Without the well, reasonably accurate readings can be obtained by taping the bulb of an immersion thermometer to the liquid line. Cover the bulb with about 1-in. of asphalt sealing compound or similar protective substance, then insulate with suitably heavy cloth. After a lapse of 3 or 4 minutes it should be possible to take a reading accurate to within $\frac{1}{2}$ to $\frac{3}{4}$ of a degree.

Remember, however, that air and other noncondensables are not the only causes of excessive condenser pressure. If condenser simply isn't big enough for the job, or if it's permitted to operate with fouled or scaled tubes, head pressure will go up. But in most cases air is likely to be the culprit and must be purged before head pressure can be reduced to the proper level.

Purging, like everything else, can be done in more than one way. To be effective it must be done slowly, giving air molecules time to migrate to purge connections.

It's possible, of course, to purge manually, though it isn't practical from an expense standpoint except on very small units. Table 2, for instance, shows that manually purging 1 cu ft of air from an ammonia system at 90 F and only 4-psi excess pressure due to air wastes 51 cu ft of ammonia. Or, from Table 3, manually purging 1 cu ft of air from an F-12 system at similar temperature and excess-pressure conditions allows 32.3 cu ft of Freon to escape.

At best, then, manual purging either takes a lot of time or not all the air is removed. In the latter case, adverse effects of air in the condenser continue, though at reduced rate. Summing up the case against manual purging: (1) It wastes refrigerant. (2) It takes a man's time which might be invested to better advantage elsewhere. (3) Air is almost never completely eliminated. (4) Released refrigerant gas may be disagreeable,

could be dangerous. (5) Purging may be postponed or ignored because of time and trouble involved in doing it.

Automatic purging, then, appears to be the logical way of handling the problem. Let's refer again to Table 2 and contrast the 4216 cu ft of ammonia lost by manual purging at 80 F and 4-psi excess head pressure due to air, with the 0.25 cu ft of refrigerant lost at a temperature of 0 F in an automatic purger with head pressure of 157 psia. Under these conditions automatic purging is 42.6/0.25, or 170 times, as efficient from a refrigerant-loss standpoint. Checking Table 3, we note that automatic purging of a Freon-12 system with same temperature and excess head pressure is 26.3/0.31, or 84.8 times, as efficient as manual purging.

Automatic system purging, commonly referred to as "refrigerated" purging, is applicable to practically any system at relatively low cost. It involves nothing more than installing commercially available, simple, dependable refrigerated devices. These automatically and completely separate air by condensing the refrigerant gas. Then they release the air to the atmosphere. Advantages derived from these devices, which require almost zero maintenance, are obvious in view of today's constantly rising operating costs and dependability requirements.

1—Too-high gage reading means air in the system

If tempera-ture, F, is:	Gage pressure should be: Ammonia	Freon 12	Freon 22
44	64.7	40.70	75.04
46	67.9	42.65	78.18
48	71.1	44.65	81.40
50	74.5	46.69	84.70
52	78.0	48.79	88.10
54	81.5	50.93	91.5
56	85.2	53.14	95.1
58	89.0	55.40	98.8
60	92.9	57.71	102.5
62	96.9	60.07	106.3
64	101.0	62.50	110.2
66	105.3	64.97	114.2
68	109.6	67.54	118.3
70	114.1	70.12	122.5
72	118.7	72.80	126.8
74	123.4	75.50	131.2
76	128.3	78.30	135.7
78	133.2	81.15	140.3
80	138.3	84.06	145.0
82	143.6	87.00	149.8
84	149.0	90.1	154.7
86	154.5	93.2	159.8
88	160.1	96.4	164.9
90	165.9	99.6	170.1
92	171.9	103.0	175.4
94	178.0	106.3	180.9
96	184.2	109.8	186.5
98	190.6	113.3	192.1
100	197.2	116.9	197.9
102	203.9	120.6	203.8
104	210.7	124.3	209.9
106	217.8	128.1	216.0
108	225.0	132.1	222.3
110	232.3	136.0	228.7

2—Ammonia lost through manual purging

Temp, F	NH₃, psia	Air, psia	Total psia	Cu ft lost per cu ft air purged
90	180.6	4	184.6	51.0
		8	188.6	25.5
		14.7	195.3	13.9
80	153.0	4	157	42.6
		8	161	21.3
		14.7	167.7	11.6
70	128.8	4	132.8	35.3
		8	136.8	17.7
		14.7	143.5	9.6
30	59.74	97.26	157	0.64
20	48.21	108.79	157	0.45
10	38.51	118.49	157	0.33
0	30.42	126.58	157	0.25
—10	23.74	133.26	157	0.18
—20	18.30	138.70	157	0.13

3—Freon 12 lost through manual purging

Freon, psia	Air, psia	Total psia	Cu ft lost per cu ft air purged
114.3	4	118.3	32.3
	8	122.3	16.1
	14.7	129	8.8
98.76	4	102.76	26.3
	8	106.76	13.1
	14.7	113.46	7.15
84.82	4	88.82	23.2
	8	92.82	11.6
	14.7	99.52	6.3
43.16	59.60	102.76	0.77
35.75	67.01	102.76	0.55
29.35	73.41	102.76	0.41
23.87	78.89	102.76	0.31
19.20	83.56	102.76	0.23
15.28	87.48	102.76	0.18

Refrigeration Oil Level

1 Oil level shows incorrectly in refrigeration compressor after system's been shut down for a long time, A. That's because there's no way to know how much liquid refrigerant is mixed with the oil. After a long shutdown, oil contains a lot of liquid refrigerant, so the level naturally is higher.

The best time to check oil level is after the machine has been running for some time, B, then there is only a little refrigerant mixed with the oil. While compressor is running, refrigerant is evaporated out until only normal quantity remains in solution. Then oil level shown in the sight glass is correct.

2 If oil level is lower than normal after long shutdown, A, it's a sure thing that actual running oil level is much too low. If compressor is started without adding oil, it won't be long before refrigerant is pumped out of oil and machine may be ruined, or at best the bearings may be harmed.

In this case, be sure to add oil until level reaches ¾ full in sight glass, B, before starting compressor. Then after machine runs a while, check level again. It should be normal after refrigerant is pumped out of oil. Never let oil get out of sight in glass, or you won't know how high it is. Keep crankcase oil at right level.

3 To add oil to crankcase, close suction valve, A, and open discharge valve. Then run machine a few revolutions. This will reduce crankcase pressure to a few pounds above atmosphere. *Don't pump down to vacuum.* Idea is to have slight pressure so moist air does not enter the refrigerating system.

Then remove oil filling plug slowly to bleed off the slight pressure in crankcase. Add oil through a clean, dry funnel, B. Pour oil steadily so no air enters through funnel. Screw in plug as soon as funnel is empty. Caution: Never leave a Freon system open for long, close it as soon as possible. This goes for other refrigerants also.

4 Most Freon-12 systems use copper tubing, A. Usually great care is taken when installing tubing to keep dirt out of system. If system is originally clean, chances are crankcase oil will not be contaminated enough to need removal for filtering or replacing with new after the breaking-in run.

When steel or iron piping and fittings are used with Freon-12, be sure to draw a sample of oil from crankcase every six months, B. Check oil in glass jar. If tiny particles of foreign material show up after foam has settled, replace all oil. It is a quick way to test oil. Be sure to check and renew the oil every six months.

8
Pumps

FOUR REASONS AGAINST PERIODIC OVERHAULS

1 High cost

2 Pump unavailable for emergencies

3 There's chance of faulty reassembly

4 It is uneconomical with this type pump

1 **CALIBRATED ORIFICE** on this barrel-type centrifugal boiler-feed pump is used to measure leakage past balancing device, indicates degree of internal leakage

How often should you overhaul your centrifugal boiler-feed pumps?

► WHILE PUMP OPERATORS have long argued over the advisability of opening feed pumps at regular intervals for inspection and overhaul, pump designers are almost unanimously united in believing there is no advantage gained until performance shows a maintenance need.

Modern designs. Table, above left, shows four good reasons against periodic overhauls—cost, availability, damage and poor economy. Modern designs and materials-selection practices are primarily directed at lengthening the time between shutdowns for overhaul and maintenance.

Higher investment for equipment, keynoted by greater use of stainless-steel alloys, would make it extremely uneconomical to subject these units to unnecessary overhauls, if they were chosen to run for extended periods.

Schedules. The cardinal principle in scheduling complete overhauls is that a centrifugal boiler-feed pump need not be opened for inspection unless (1) factual or (2) circumstantial evidence indicates it is necessary. There is but one exception to this rule.

Factual evidence constitutes (1) a

fall-off in pump performance justifying the expense of an overhaul (2) pump noise (3) driver overload indicating trouble (4) excessive pump vibration. In extreme cases, the pump may have seized at the internal clearances as a result of foreign matter, overheating or a mechanical defect.

Circumstantial evidence is accumulated from past experience with a particular pump or from similar equipment on the same service. For example, if a group of chromium stainless-steel alloy pumps run 60,000 hr continuously without needing a complete overhaul, an additional duplicate unit can be run for the same period.

Exception. This rule does not hold for corrosion-erosion troubles. These are rare, however, because much attention has been devoted in recent years to proper materials for high-pressure high-temperature boiler feed pumps. Today, 5% chrome stainless casings and 13%, or higher, chrome stainless fittings are established practice for all feed pumps serving 850-psi and higher-pressure plants. The practice is beginning to spread to plants operating at lower pressures.

But since ordinary materials (cast iron for casings, bronze for fittings) are still used in many low- and medium-pressure plants, corrosion-erosion troubles are still a potential danger. So when pumps of these less-resistant materials are used with feedwater that may cause severe and rapid deterioration, it is advisable to open the unit for inspection soon after initial operation. It should also be opened at intervals thereafter to establish an experience pattern that proves the equipment is satisfactory for its job.

Causes of wear. Erosive action of the water flowing past wearing rings and interstage bushings causes part of the wear at all internal clearance joints. Another portion is due to infrequent momentary contacts, which occur on relatively rare occasions during pump operation.

No matter how rigidly a pump is designed, certain conditions arise during transient loads (partial cavitation or similar sudden hydraulic impacts) that impose a temporary vibration on the rotating element sufficient to cause a slight momentary contact at the running joints.

This contact may not be enough to cause permanent damage or seizing, but it does affect the running clearances. A similar condition may exist if the shaft deflection under stationary condi-

2 EFFECT OF WEAR on head-capacity curve of a centrifugal boiler-feed pump. Leakage is nearly constant at all heads

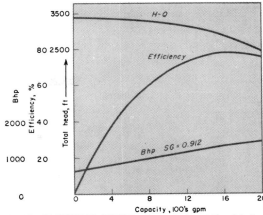

3 CHARACTERISTIC CURVES of a typical centrifugal boiler-feed pump, which is rated at 1645 gpm with a 2840-ft head

tions exceeds the existing clearances. This can occur even though the deflection is reduced to less than the running clearance by the supportive effect of internal clearance joints acting as additional steadying bearings. In these cases a slight amount of wear occurs every time the pump starts.

It is impossible to state the exact amount of wear attributable to erosion or to momentary contact. But the more rigid the pump construction, the less the effect of transient conditions on wear of the running joints.

Varying wear in exactly similar pumps operating under apparently similar conditions indicates existence of a varying frequency of transient conditions and vibrations.

Effect of wear. As running clearances increase with wear, a greater portion of the gross capacity of the pump is short-circuited through the clearances to a lower-pressure stage. The effective or net capacity delivered by the pump against a given head is reduced by an amount equal to the leakage increase.

While in theory the leakage varies with the square root of the total head, it is sufficiently accurate to assume the leakage increase remains the same at all heads. Fig. 2 shows the effect of increased leakage on the shape of the head-capacity curve. Subtracting the additional internal leakage d from the initial capacity at each head in the pump range gives a new head-capacity curve, as shown in Fig. 2.

Performance. If the pump whose characteristics are shown in Fig. 3 delivers, when new, 1645 gpm against a 2840-ft head, while absorbing 1370 hp when handling 314-F water at an efficiency of 78%, let's see what an in-

crease in internal leakage of 2% of the initial capacity will do to performance.

With increased leakage the pump delivers 1610 gpm against the same head and requires the same power input. Efficiency is reduced by the ratio of the initial capacity to the present capacity, or to 76.5%. At the original capacity, 1645 gpm, the pump now develops a 2825-ft head at 77% efficiency.

Since a capacity margin is always provided in feed-pump design, the unit in Fig. 3 may normally be required to discharge only 1480 gpm at 2840-ft head, with the excess pressure throttled by the feedwater regulator if the pump operates at constant speed. So the additional leakage does not impair normal performance. The decision to overhaul the pump at this point depends solely upon evaluating the reduced economy against the cost of the overhaul.

In high-pressure feed-pump operation it is good practice to renew internal clearances when the leakage doubles. At this point the added power to meet station loads becomes appreciable and usually justifies the overhaul cost.

Performance tests. An accurate test of the pump head-capacity curve gives a reasonable indication of the internal condition. Proper instrumentation for determining pump capacity and pressure is a must, both for routine operation and for tests.

A rigid schedule should be set up for pump head-capacity tests so the performance can be compared with the original values when the unit was new. This comparison, not the mere passage of a certain time interval, should establish the degree of internal wear and the need for an overhaul. Tests, which are cheaper than overhauls, can be made

without taking the unit out of service.

Power-consumption increase for a given flow rate, change in pressure drop across the feed regulator and speed measurements on variable-speed units are approximate means of determining performance when full head-capacity tests cannot be made or are not accurate enough.

Measuring leakage. A calibrated orifice, Fig. 1, fitted with a differential pressure gage or manometer can be installed in the balancing-device leakoff line if the leakage is returned through external piping. By inference, the orifice readings can give an accurate indication of the condition of the internal running joints. They give a direct indication of the condition of the balancing device.

If the pump design is such that the pressure drop per inch of wearing-ring and balancing-device length is about the same, the wear rate of the balancing device can be taken as indicative of that of other parts.

For example, if after 50,000 hr operation, balancing-device leakoff doubles, it is likely that wearing-ring clearance has about doubled.

Spare parts. To insure restoring a pump to service promptly after an unexpected overhaul, the following spare parts should be kept on hand: one set of casing wearing rings, one set of impeller rings (if used), one set of interstage or diffuser bushings, one spare balancing device made up of rotating and stationary parts (if used), two sets of shaft sleeves, several sets of packing, one set of bearing bushings, and parts for the thrust bearing. A completely assembled spare inner assembly should also be kept on hand.

Suction

Water-jacketed
stuffing box

Bearing bracket

Forged-steel
barrel

Sleeve
shaft-bearing

Stuffing-box
bushing

Stainless-steel
impellers

Two-piece stainless-
steel twin volutes
and stage pieces

Now let's open, inspect and overhaul

▶ IT WILL HELP YOU understand all pumps so much better if you know construction details of a high-pressure boiler-feed pump, and how to inspect and assemble it, even though you don't have this type of unit yourself. Modern double-case barrel-type centrifugal pump is today's answer for pressures up to 6000 psi, to 600 F. In the common 2600-psi range, this type runs at about 3600 rpm, has up to 12 stages.

Long service without major overhaul is common for these pumps. But when you do open one, do your job thoroughly because this unit is vital to plant operation. Photos show every important step. They should guide you to do the job right, even if you have never opened your pump before.

Pump overhauled here is a Worthington, 9-stage unit, with discharge diameter 8 in., design capacity 657,000 lb per hr, suction pressure 102 psi, discharge pressure 2582 psi, suction temperature 320 F, speed 3500 rpm, driven by electric motor through hydraulic coupling.

We start right out by overhauling pump parts shown in cross section at top of page. Of course, oil lines and other piping connections are disconnected. Here we show only major pump parts that you will have to give special attention.

1 Here's how to open

1 Remove coupling, gland, packing, then bearing cover on inboard end. Next, take weight off lower bearing bushing as shown, roll out the bushing by sliding it up over the shaft. With load off bearing housing, you can unbolt and remove it

Forged-steel
discharge head

Bearing
bracket

Kingsbury thrust bearing

Wearing ring

Combination balancing
drum and disk

Thrust shoes

Direct-driven
oil pump

this barrel-type feed pump in 3 steps

2 Now let's switch to the outboard end. Remove direct-driven oil pump and bearing cover. Remove thrust shoes, roll out leveling plate and the base ring

3 Screw off the thrust collar-nut by first loosening with special wrench. Now slide thrust collar off shaft. Next, remove the bearing so shaft is clear

4 Remove nuts from outboard bearing bracket. Fasten wire sling to eyebolt and swing bearing bracket clear of shaft. Guide so it does not touch shaft

5 Special tools come with all barrel-type feed pumps. Set tool up on shaft to remove the shaft sleeve as here. Tighten each nut a little until the sleeve starts to move on shaft

6 Remove shaft-sleeve packing with packing hook. Aim hook at stuffing-box bore—not at shaft as hook will scratch it. Always remove all the old packing from any stuffing box

10 Move rotor carriage, raise or lower with hand wheel to align with casing studs. Lock it in place with two nuts

11 Place the carriage crosshead over the shaft at discharge end of pump. Screw crosshead onto the locking nut

12 Next, slide the special inboard shaft support-head and shaft protecting sleeves over the shaft, bolt in place

2 And here's how to inspect_____

15 Heat each impeller to 300 F to expand over shaft. Keep moving torch. Use special clamp to lift the impeller

YOU MUST WATCH THESE DETAILS WELL
A: wearing rings for wear, clearance
B: metal-to-metal joints for mating
C: impeller for a good shrink fit
D: 2-piece twin volutes and stage pieces' condition
Compare clearances with those of manufacturer

7 Remove balancing disk from supported shaft with special wrench. *Never turn rotor if bearings are out*

8 Next, remove nuts, swing discharge head clear. Remove stay-bolt nuts (small nuts in cylinder), washers and keys

9 Remove nuts from suction head at other end, take weight with chain fall. Guide carefully, swing clear of shaft

13 With carriage bolted to two cylinder studs, slide out pump's inner assembly. Two mechanics can easily do this job by turning pinions over the pinion rack on each side of carriage. Heavy rotor is easily moved thus to any part of your plant

14 Lift rotor assembly off the carriage with crane and place vertically on wooden blocks. Remove the stay-bolt nuts

1 Examine thrust shoes carefully. Wear should be very little as relocating balancing device takes the thrust load

2 Mike thrust collar for wear. There is usually little or no wear here as softer babbitt on shoes takes wear

3 With rotor assembly removed, inspect the inside of the barrel thoroughly. Look for signs of erosion, dirt, scale

4 Measure ID of interstage bushing, balancing-disk, wearing rings and OD of mating parts. Renew parts if clearance is about double that in the manufacturer's instruction book

5 Place impeller in stage piece, move laterally against a dial indicator to see diametral clearance. Compare this reading with data given in your manufacturer's instructions

6 Now install twin-volute diffuser and repeat the process to determine diametral clearance in interstage bushing. Compare this with the data given by the maker of your pump

7 Inspect impeller bore carefully. Since the impeller is shrunk on the shaft and keyways do not cut through, there should be no sign of leakage here. But check carefully anyway

8 Check the tongue of the volute carefully for any signs of erosion. Volutes are of stainless and usually this is not a problem. But don't overlook anything while the pump is open

9 Check metal-to-metal joints between stage pieces for signs of erosion. Since stage pieces are stainless, this is usually not a problem either. But, again, don't miss anything

HERE'S DISCHARGE END

A: 9th-stage impeller   **B:** sleeve seal
C: water-jacketed stuffing box bushing
D: packing set   **E:** last-stage diffuser
F: balancing-disk head
G: shaft sleeve   **H:** bearing bracket

Please note: This is the high-pressure end
of the pump. Make sure you know these
details. Give each part your careful attention for long troublefree service

3 Now let's close it up right

1 To expand over shaft, heat impeller
with flame if lacking an oil bath.
Here hot plate distributes heat evenly

2 Slide heated first-stage impeller in
place. Make sure that bearing surface
is coated with molybdenum disulfide

3 Blow compressed air through hot impeller to cool. Parts are numbered.
Fit into the right position on the shaft

4 With first-stage impeller shrunk on shaft, mount assembly vertically in first-stage stage-piece, as shown here

5 Next, install first-stage twin-volute diffuser in stage piece. Shaft is machined slightly smaller for each successive impeller. Joints are metal to metal, so be sure to get them perfectly clean before mating or you might run into some trouble

9 Thoroughly clean metal-to-metal joint, between discharge head and casing barrel, with crocus cloth. Tight joint here is important, depends on perfect contact of mating surfaces

10 As Frank Anderson, *right*, installs balancing disk with special wrench, Editor Steve Elonka and Ted Edwards of Worthington Corp, *left*, look on. Shaft is on hydraulic jack

BEARINGS ARE IMPORTANT

A: bearing bushing   B: thrust-bearing shoe
C: thrust collar   D: oil pump on shaft
E: jacking screw for centering the shaft
F: thrust-collar nut   G: thrust sealing ring

6 When replacing wearing ring in stage piece, always lock Allen setscrews by upsetting with center punch as here

7 Put molybdenum disulfide on bearing surface before shrinking impellers in place to prevent galling next overhaul

8 Install subsequent stage pieces, impellers and the twin-volute diffusers until the complete rotor is assembled

11 With rotor assembly in place and pump casing ready to close up, install shaft sleeve and tighten shaft sleeve-nut with spanner wrench. Sleeve takes all the packing wear

12 Check the vertical movement of shaft in wearing rings and diffuser bushings by jacking up on the shaft against dial indicator. Refer to recommendations for the maximum clearance

13 Bolt thrust-bearing housing in place, install lower journal bearing-bushing. Adjust jackscrews on housing sides for total vertical clearance, split in half to center shaft

14 Pack pump, assemble oil and cooling-water lines, drains. *Caution:* Before starting overhauled pump, always roll pump by hand while cold, also afterward warm with suction open

For Long Life, Check

1 Centrifugal pumps take a beating from continuous operation. Open boiler feed—and other vital pumps—once a year for a good look-see inside. Check for wear, corrosion, fractures and cavitation. Poor circulation during the off-pumping periods in high-pressure pumps causes plenty wear. Boiler feed pump's arch enemy is low head of water on suction side from sudden head drop, causing water to flash. Then the pump overheats, wearing sealing and channel rings, or bending shaft

4 Check shaft sleeve through stuffing box for wear from packing. Turn down worn sleeve or replace with new. Look for shaft corrosion at ends of the sleeve

5 Renew shaft packing, align lantern ring with sealing hole. When running, lubricate shaft with slight gland leakage. This keeps air out if under vacuum

6 Water-flinger ring protects thrust bearing from water leakage. See that ring is snug fit on shaft. If not, seal between shaft and ring with sealing paste

9 For shell-type bearing, check clearance by placing lead wires on shaft and screwing down bearing cap. Mike wires. Remove shims for right clearance

10 On shell-type bearings, take bridgegage readings to find wear on lower half of bearing. Insert feelers between gage and shaft. Rebabbitt if low bearing

11 Examine impellers carefully for wear at wearing rings. Check vanes for erosion, cavitation and corrosion. Vanes at eye and discharge are trouble spots

Centrifugal Pumps Annually

In these times of peak loads, high prices and slow deliveries,
it's cheaper to look inside the pumps now than be sorry later

2 Break coupling and align pump and motor shaft while pump and lines are hot, as hot piping tends to throw pump out of line. Align pump again after it's cold. Mark halves, place rubber, rotate halves together and make rim and angular checks

3 If coupling is flexible pin-type, remove all pins and bushings. Clean wearing surfaces of bushings and pins, check for roundness. Lubricate pin and replace elastic bushings that are shot. Check wearing surfaces of all type couplings

L-shaped wearing ring

Labyrinth wearing ring

7 Keep clearance between impeller and casing wearing rings close to design recommendations. This opening affects pump's efficiency. Usual clearance should be no more than 0.003 in. for each inch wearing-ring diameter. Never let this opening double the original clearance. Labyrinth type leaks less than L-shaped shown here

8 Clean oil or grease out of bearings. Check for water leakage that knocks bearings out. Clean out all the old lubricant and inspect bearing for wear

12 Tap key in impeller for being tight. Shaft twisting while under load, expansion, corrosion, or original poor fit makes impeller progressively loose

13 Check thrust-bearing clearance by breaking shaft coupling and barring shaft both ways against thrust bearing. Measure shaft movement from each side of impeller, near tips, housing. Consider expansion of shaft from cold to hot running condition. Before closing pump, renew plastic sealing rings as old ones harden from heat

Do You Have Inside Dope

ROTARY feed pump in this tannery plant pinch-hits for old reciprocating duplex

LOBE PUMP is form of gear type, must have gears to drive each shaft, or lobes would slip. Liquid travels around lobes and casing. No valves needed in pump

INTERNAL GEAR has stationary crescent to prevent losing liquid back into low side. Liquid from suction fills spaces between the gear teeth as they unmesh

SCREW PUMP has one power rotor and two idlers. Liquid from suction enters at rotor's ends, is trapped by meshing screws and carried along between threads to discharge. Pump works against pressure of 1000 psi while handling oil at 7000-rpm

SINGLE-SCREW pump has one screw turning in double-threaded helix. Rotor is metal while helix is hard or soft rubber, depending on service. Softer helix bears against rotor, doing away with clearance. Use either end of the pump for suction

SEPARATE-BEARINGS pump is for abrasive liquids. Helical gears don't build up pressure at teeth ends as spur gear, so holes needn't be drilled. Outside bearings prevent abrasive wear by liquid. Watch shaft deflection by not overloading pump

UNIVERSAL-JOINT pump has two shafts connected in casing at angle through universal joint. Joint is spherical and, while revolving, pumps because of closing and opening action of joint. Spherical parts fit snugly, preventing liquid slipping back

on Your Rotary Pumps?

GEAR TYPE shown is simplest rotary pump. Liquid is moved between gear teeth and casing. Meshing teeth act as valve, preventing liquid sliding back to low side. Trouble is high pressure builds up from liquid trapped between ends of the meshing teeth. That forces gears apart against casing, second sketch. Solution is drilling holes in idler gear, third sketch. Then machine slot in the pin so pressure can escape to high side. Helical gears do not need this change as oil is not trapped

SLIDING-VANE—throws vanes against casing bore. Vanes move out from off-center rotor, sealing casing bore, and prevent slippage while pumping liquid

SWINGING VANE is much like sliding vane but vanes are hinged, and offer less friction. Vane is sturdier; so chance of damage is less if the discharge is closed

SHUTTLE-BLOCK pump has sliding block in rotor slot. Hole in block fits over off-center stationary pin fastened to pump cover, which slides block in the rotor

Rotary and Reciprocating Pumps Are Positive Displacement

► MOST PLANTS have rotary pumps of one kind or another. These sturdy work horses move oil, water, chemicals and many other liquids. Because they are compact and do their work so well, tendency is to forget all about them until they start giving real trouble. A quick look at the insides of the more popular rotary-pump types will help you understand them.

Principle. Because outside of these pumps resembles the centrifugal type, that's what many people take them for. But they are next in line to reciprocating pumps because both are positive-displacement pumps. The water is not "thrown," as by a centrifugal pump, but actually trapped and pushed around in a closed space as in a plunger-type pump. Rotary type combines constant-discharge characteristic of centrifugal with positive-discharge feature of reciprocating. Only difference from reciprocating is that discharge pulsates

with piston, while rotary's discharge is constant. And, of course, one rotates while the other reciprocates.

Construction. Casing of all rotary pumps is fixed, or stationary. But insides have tightly fitted gears, vanes, pistons, cams, segments or screws, depending on design. Working parts of most common types shown have two external spur gears; two spiral gears; two herringbone gears; external-internal gears; screws, swinging-vane, 2-lobe, 3-lobe or cam types; special lobe type; guided-vane types, cam and piston; rotary piston; and universal joint.

Maintenance. Packing trouble is cut down to minimum where only low pressure or suction end of pump has packing gland.

For best results, don't wait until these pumps lose much of their capacity from slippage. Inspect insides at set intervals and check up on clearances that allow liquid to slip.

Piston Centering and Valve

FIRST CENTER PISTONS

1 Open pump steam drains. Relieve liquid pressure. Bar crosshead knuckle until piston bumps cylinder head

2 Make mark A on piston rod against the gland. Use scriber, pencil or soapstone as convenient

3 Bar in opposite direction until piston bumps cylinder head. Make mark B against steam gland as before

4 Find center mark C between marks A and B with dividers or scale. Avoid burring at B or C rod

5 Final operation is barring piston rod until center mark C is brought back to the steam-end packing gland

6 Steam piston is now in exact center of cylinder regardless of rocker arm angle. Center the other piston

Setting for Duplex Pump

THEN SET VALVES

7 With piston in center of cylinder, valve ends should just cover steam ports at A and B as shown

8 A, Move valve against nut and measure steam port opening. B, Move valve in opposite direction and measure

9 If port openings at both ends are equal, valve is set. If unequal, remove pin from valve link for adjustment

10 Hold nut and turn valve stem to correct the inequality, screwing stem into or out of nut as shown

11 Measure port openings at A and B at ends of valves as before. If they are equal, valve is properly set

12 Before closing valve chest, move valves in opposite directions to uncover ports, or pump won't start

Overhauling Duplex

(To set valves, see "Operators' Notebook I" Jan, 1948)

1 This steam duplex pump is so reliable it's often given no attention. Then it stops, refusing to budge. Overhauling steam chest and valves is answer. Any OE can do it himself

2 Shoulders wear at ends of valve travel on valve chest, **A.** Slide-valve face wears thin and ends round. Steam enters cylinder under rounded ends, stopping piston on midstroke, **B**

5 Scrape away shoulders if they are light. If heavy, remove chest studs and use 12-in. mill bastard file. First chalk file. Never rub fingers on file's cutting edge or it slides

6 Filing is an art. A good mechanic can file a valve-chest face level by checking with straightedge like this. Keep checking and file only the high spots until no light is seen

9 After valve chest is level, grooves cut with chisel give raised-valve-surface effect. This avoids shoulders, saves valves from hitting shoulders, prevents short pump strokes

10 When old pumps are faced often, false face may be needed to make up for lowered face. Cast iron, ½ in. thick, fastened with flathead screws on gasket like this is answer

Pump Valve Chest

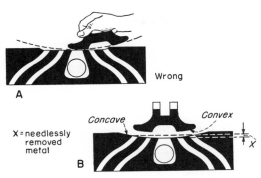

Wrong

A

X = needlessly
removed
metal

Concave   *Convex*

B

3 Never grind slide valves until valve-chest shoulders are removed. Grinding makes chest face concave and valve convex, **A. B** shows the metal needlessly removed when grinding

Rounded corners

Face to dotted line

4 Right way to overhaul valve is to place in lathe chuck and face it down until rounded corners are machined away. Use pointed tool, then blunt tool for smooth finish

Medium grinding compound

A

Normal valve travel

---Oil

B

7 After face is finished level with single-cut smooth file and valves are faced in lathe, grind with medium grinding compound, **A.** Remove compound, give "oil rub" for finish

Raised wearing surface

8 Some steam pumps with slide valves have a raised valve face like this. Since valves travel to ends of raised surface, no shoulders form. Light grinding is all that's needed

Cast-iron piece

Exhaust cavity

11 D-valves must be long enough to cover outside steam port edges. If right-length valves aren't in stock, attach cast-iron pieces. Make valve cavity cover ports as shown

Valve stem

A   *Nut*   *Amount worn*

Valve lowered

B   C

Screws

12 Worn valve lowers and valve moves away from the valve-stem nut, **A. B** shows how lowered valve allows nut to turn. To hold nuts, drill and tap for roundhead screws, **C**

Don't Forget to Renew

TYPES

PRESSED liner is thin sleeve that is a pressed fit into cylinder bore. Material is bronze, monel, cast iron or others

REMOVABLE liner is held by four bolts and is centered by four lugs. Easy to renew because it is not a pressed fit

REMOVABLE cylinder with bronze liner is held by binder bolts through cylinder head against gasket to prevent leakage

REMOVING PRESSED LINER

WEAR is on bottom of horizontal liner. Unless renewed, the piston rides on casting

SPLIT liner with chisel bar along bottom, where the metal is already thin from wear

CAUTION! It's easy to damage cylinder casting under the liner. Split only the liner

COLLAPSE the split liner with chisel bar. This isn't easy, also depends on the material

PRESSING IN NEW LINER

PRESS new liner into cylinder gradually like this. First clean cylinder and oil liners outside. If liner is tight after first few inches, remove it before it gets stuck tightly, for good

POLISH liner outside with coarse grinding cloth by holding it in lathe like this. If inside and outside micrometers are handy, polish liner so it is only about 0.0005 in. larger than cylinder

Your Reciprocating Pump Liners

REMOVABLE liner with clamp (ported distance piece) between cylinder head and liner. Cylinder bolts support the liner

REMOVABLE liner with double shoulders, held by distance piece against cylinder head. Small bearing surface on cylinder

NO LINER. Whether outside-packed (like this opposed plunger pump) or inside-packed type, plunger takes all the wear

PULL worn liner in one piece. It's much easier but you must have strongback shown to hook in back of liner. Turn groove in lathe to fit liner; then grind back edges round as shown

STRONGBACK against open cylinder keeps steady strain on liner while small sledge pounding against bar bumps it out. This method is much quicker and easier than splitting liner

SHRINK liner with Dry Ice. Place in wooden box and pack with sawdust. Break up Dry Ice and fill inside of the liner

WEAR GLOVES, slide liner in fast. You may have to tap liner with mallet. Bag of Dry Ice inside liner keeps it shrunk

WHAT IS DRY ICE?

This is solid carbon dioxide, (CO_2), and is 109 F below zero. It's solid in 10-in. cubes, weighing 50 lb. Count on about 0.001-in. shrinkage per inch of liner diameter. To keep liner expanding while fitting into cylinder, place bag of broken-up Dry Ice inside liner. *Caution!* CO_2 displaces air, so ventilate working spaces well. Also, avoid frostbite by covering bare skin.

INTEGRAL MOTOR PUMP has small portion of pumped fluid recirculated through motor section (can) and oil-filled stator cavity

ANOTHER ZERO LEAKAGE DESIGN has conical journal-actuated bearings to maintain minimum bearing clearance at all times

Integral motor pumps or separate motor pumps, how do they compare?

One hears more and more about the canned pump because it has made some impressive inroads into several areas long dominated by the separate motor centrifugal pump. One firm alone claims to have sold over 100,000 canned pumps since this design first made its bow a few years ago. Thus, a number of manufacturers are jumping on the bandwagon and producing the unique design.

Because of the confusing and often conflicting statements one hears when canned pumps are mentioned, we asked pump manufacturers to give us their views on eight areas of controversy that keep coming up. Most of these manufacturers make both integral and separate motor-type pumps. They are primarily interested in seeing that you apply the right pump. We hope that you put the following summary of their comments to constructive use.

1—The motors of canned pumps are low in efficency because of the can material on the rotor and stator. Conversely, separate motor pumps are "as efficient as modern technology can make them." Also, no material is required for covering the rotor and stator.

It is true that electromagnetic efficiency of a canned motor pump is partially reduced because of increased air gaps and the presence of nonmagnetic liners. But these losses

are somewhat compensated for by the elimination of stuffing boxes and mechanical seals. Don't forget that seals also contribute to the inefficiency of conventional pumps through friction losses.

The efficiency of a separate motor pump is normally based on the brake horsepower input to the pump versus the work done by the pump. No consideration of electric motor efficiency is made. Since it is impossible to separate pump efficiency and motor efficiency in a canned pump, the normal method of evaluation is to look at *power input* versus *work done* by the pump, or *wire-to-water* efficiency.

The same figure is obtained for the separate motor pump by multiplying motor efficiency by pump efficiency. When all of these factors are considered, the difference in ratings is frequently negligible.

Also, because a zero leakage pump is liquid-cooled, heat is removed more efficiently from the motor area. This cooling obviously permits the motor to work at higher amperage levels than would be possible in a standard air-cooled application. In other words, more work can be actually obtained from an electric motor of a given size when applied in a canned pump.

Amperage draw in canned pumps is usually insignificant in the smaller sizes up to about 7 hp. But, for the larger rating units, amperage draw can be important. While most zero leakage pump motors may be less ef-

Canned pump has zero leakage, does a specific job

Originally, the integral motor pump (also known as *sealless, zero leakage* and *canned*) pump, was applied only on highly specialized applications, where absolutely NO leakage could be tolerated. An example is the pumping of liquids subjected to radioactive contamination in nuclear energy electric generating stations. This unit has now invaded the general process field, and some other industrial areas as well.

Basically, the canned pump differs from other centrifugal pumps in that the pump and motor are combined in a single unit, and completely enclosed within a corrosion-resistant, nonmagnetic alloy cylinder (can). Part of the pumped liquid circulates through the motor section to cool the motor and also lubricate the bearings. One design has been simplified so that now it has only 14 basic parts. This compares to 50 basic parts required for a standard centrifugal pump of separate motor design.

However, all canned pump motors suffer losses because of the larger air gap required to accommodate the stator and rotor cans. But most manufacturers measure these losses and take them into account when writing amperage draw ratings for their canned pumps.

Bearings are the single, most critical parts of a canned pump, primarily because they are in the pumped fluid. The choice of bearing material often presents a complex problem. Why? Because the choice must be made from hundreds of available materials.

And when it comes to bearing life, it may depend on the clearance between shaft and bearing. This was pointed out in a study of submerged bearings conducted by Battelle Memorial Institute some   years ago.

An automatic thrust balance feature by one manufacturer has completely eliminated the axial bearing wear that was a serious source of bearing failure in earlier designs. Another improvement is the use of an oil-filled stator which extends motor life by keeping the windings cooler while protecting them from condensation damage problems.

Since the pump and motor are built as a single unit, without an external rotating shaft, no coupling, seals, or stuffing boxes are needed. Elimination of mechanical seal or packing maintenance and lubrication, and housekeeping problems from fluid leakage, is dramatically extending the range of potential applications for this pump.

For most applications, canned pumps are lighter and smaller than separate motor pumps. This makes their foundation cost considerably less. Smaller sizes can even be mounted directly in the piping without additional support.

---

ficient, there are many applications, such as those for hot oil, when the *can* barrier is not necessary. The motor efficiency of the zero leakage pump then approaches that of the conventional motor.

Pump manufacturers will agree that all industrial-type pumps can be built with efficiency as high as modern technology can make them. The only question is: Will the customer pay the price?

2—Standard canned pumps are designed for 3500 rpm only. This is critical because 3500 rpm requres more NPSH (net positive suction head) than 1750 rpm, for example. Also, 1750-rpm canned motors are special, which means more money and longer delivery. In contrast, separate motor pumps have motor and pump combinations available with a variety of speeds to suit pumping requirements.

Nothing could be further from the truth. Canned pumps are available in both 1750- and 1150-rpm speeds (from one firm), as a completely standard pump with price and delivery on the same basis as a standard line of 3450-rpm zero leakage pumps (from the same firm). And there are many lower-speed pumps furnished as a complete line.

While the 1750-rpm canned pump (for example) is not special, it is only more expensive if a larger pump is required to do the job at 1750 rpm than at 3500 rpm. But deliveries in either case are the same. Also, there is no good reason why these pumps cannot be made for various other speeds. It depends on the definition of *standard*.

3—The direction of shaft rotation is difficult to establish in the integral motor pump because no moving parts are visible. On the other hand, the rotation is easily determined in the separate motor pump by observing the shaft or coupling.

While it is true that the direction of shaft rotation cannot be observed in canned pumps, the pressure gage on the discharge side of any centrifugal pump will indicate correct rotation. The reason is that a centrifugal pump develops only about 60% of its head when running in reverse. Thus, it is a simple matter to run the pump a few moments against a closed discharge valve as soon as it is wired into the system. Then reverse the electrical connections and again run the pump for a few moments. The connections which result in the higher discharge pressure are obviously the correct ones.

While observing pressure is easy enough to do, it may not always be practical, especially during the installation period of new, larger plants. The wiring is usually completed long before the process equipment is ready for initial startup. By the time the piping is flushed and the plant is ready for its trial run, the operators may naturally assume that all equipment is properly hooked up.

To make sure pump rotation is correct, a pressure check would obviously have to be scheduled for all pumps as soon as flushing or process fluid is available in the plant. Conversely, the wiring in an existing plant is usually done by the plant electrician, who can check rotation as soon as the pump is hooked into the system.

Another method is to use a relatively inexpensive phase-sequence meter that checks actual connections against the phase sequence indicated on the pump nameplate.

Also, a shaft rotation indicator is now available which quickly gives the answer. It can be used on any type of centrifugal pump or other rotating machinery to insure initial operation in the proper direction, instead of the usual "hook up and try" approach.

Don't forget that while there are no external rotating parts in an integral motor pump, this feature is basic to its main advantages. Under some conditions, pumps can be checked by looking into the suction opening, but no

Separate motor pump is the workhorse of industry

In contrast to the integral motor pump, the separate motor type comes in numerous designs. Regardless of the design, they all have a separate motor and a separate housing for the pump (see illustrations on the opposite page). All separate motor pumps require a seal around the shaft to prevent the pumped fluid from leaking out of the casing. It may be a mechanical seal (face type), or jam-type packing in a stuffing box around the pump shaft. The stuffing box may have a connection for pumping a clean protective fluid into the lantern ring to keep abrasive or corrosive liquids away from the packing.

Although leakage from a mechanical seal is usually so slight that it may not be apparent, it still cannot be depended upon as zero (see *Mechanical seals*, 24 pages, *Power* magazine). But where zero leakage is not the main consideration, the separate motor pump with seals has been widely used for years.

For one thing "standard" centrifugal (off the shelf) pumps cost less initially, and there is usually a standard size, design and speed for almost every application. As one pump manufacturer told us, "The only reason some commercial centrifugal pumps may not be as efficient as modern technology can make them is because they have to be sold for prices which are not very much higher than the cost of the raw materials they are made from." Thus, they are popular.

Just as the canned pump (because of its more compact design and lighter weight) may be installed in the line in smaller sizes without a foundation, so may a new design of separate motor pump. This is known as the in-line pump, which is mounted in the line like a valve. Thus, the ability to locate an in-line pump where it is needed, instead of in a fixed bay, avoids long piping runs and cuts installation expense.

For example, one recently introduced in-line design lets you take out the entire pump cover and rotating assembly for service without disturbing the pump casing, piping connections, or driver. The mechanical seal in this design is also serviced, or replaced, by removing the rigid coupling and gland, again without disturbing the rotating assembly or driver. Features of this type keep the separate motor pump in competition with most special designs introduced from time to time.

In a way, it is unfair to compare the integral motor pump with the separate motor design, since the canned feature was originally developed only to prevent leakage. But it is a fact that the canned pump is invading the general pump market in some areas because it has proven to be highly reliable in respects other than zero leakage. Our aim in this report is to help bring the outstanding features of both types of pumps into proper focus.

---

manufacturer would like to see this information in his installation instructions for obvious reasons.

4—Clearances in the plain bearings of the canned pump must be maintained to assure close running clearance between the rotor and stator. But wear, destroying the bearing clearance, will result in short pump life due to contact of the rotor and stator. On the other hand, the motor bearings of the separate motor pump are isolated from the pumpage, or any liquid carrying abrasive particles. Thus there is no problem from abrasive wear.

It is true that bearings and journals are two of the most critical components in a canned motor pump. In addition, proper materials must be selected for each application to insure long service life. But today, manufacturers of canned pumps select the proper bearings and journal combinations for a broad spectrum of applications.

Although bearing tolerances must be maintained, there is ample allowance for reasonable wear. Therefore, there is no reason to state that "wear will destroy the bearing clearance, resulting in short pump life." As a single, isolated example, one paper plant has had over 35,000 hours of trouble-free operation from a canned pump. In fact, this pump has required much less attention and maintenance than could normally be expected from their older-design pumps with stuffing boxes or mechanical seals. That is why the canned pump was initially installed since, in this case, zero leakage was not the main problem.

As for bearing life, it can be lengthened in abrasive service by inserting a filter in the external piping for circulation from pump cavity to motor cavity. And the filter can be cleaned periodically without opening the pump.

One new type of rotating bearing, for instance, claims to reduce bearing wear over stationary bearings and increase its life by a theoretical factor of four. In actual practice, bearing life has been increased as much as 10-to-1. Wear resulting from thrust loads is also claimed to be minimized by the use of a thrust control design.

Another new design of bearing for canned pumps is a conical bearing (photo, right side of p 171) spring loaded against a tapered journal. This feature is claimed to assure bearing journal contact at all times and compensates for wear on the bearing faces.

Again, it is generally true that bearing wear is critical on canned pumps. But remember that close-coupled, separate motor pumps may develop leaks around the shaft which can damage motors. And, in some cases, special enclosures may be required to prevent this damage. So, if bearings are considered the Achilles' heel of canned pumps, you might say that seals are just as critical in separate motor pumps.

5—Repair of the zero leakage pump is expensive because it requires replacement of the rotor and stator *can*, the bearings and the shaft. By comparison seal component replacement for the separate motor pump, when necessary, is inexpensive.

There is no good reason why a proper inspection and preventive maintenance program will not eliminate expensive canned pump repairs, just as with other types of pumps. Bearings are usually the only items required when the pump is serviced intelligently. Canned pumps used by the AEC, for example, have run for years without repairs in critical nuclear energy electric generating applications. These large canned pumps have chalked up an enviable record. Not only that, but replacement of bearings in canned pumps usually takes less time than the replacement

SEPARATE MOTOR PUMP is sealed with jam-type packing and is connected to driver through flexible coupling on end of shaft

ANOTHER SEPARATE MOTOR PUMP DESIGN is sealed with packing backflushed with a clean fluid to prevent excessive wear

of a mechanical seal. And if a mechanical seal fails, pumped liquid may be sprayed over the surrounding area. This is a potential hazard which usually cannot be tolerated when zero leakage pumps are installed.

Replacement frequency of bearings and gaskets in canned pumps depends on the application. The best way to determine frequency is to monitor bearing wear over the first few months of operation. Wear can then be extrapolated in terms of time based on bearing tolerances, and tabulated in instruction books.

Of course, there have been failures. But the misuse or misapplication of any complex equipment can cause catastrophic failure. One big saving with a zero leakage pump is in installation cost. By comparison, separate motor pumps require special technical care in leveling, alignment, packing break-in, and mechanical seal adjustment.

Complete failure of the canned unit requires expensive replacement of bearings, journals and cans. However, there are records of maintenance-free design life-ratings in excess of 75,000 hours.

6—If the *can* material thickness on the rotor and stator must be increased to combat the effects of erosion, corrosion, or pressure, motor efficiency will drop as the gap between the windings increases. In contrast, the separate motor pump has no problems with the effects of erosion, corrosion, or pressure on the motor since the motor is not in contact with the pumped fluid.

Basically this is correct. But can thickness and air gap are standard and fixed throughout in some manufacturers' lines of canned pumps. To combat corrosion, a variety of materials is available. Also, special backup plates around the can are used to increase the pressure rating of the

pumps as required. Thus, the academic truth of this statement has no bearing on pump's practical applications.

7—Compared to canned pumps, the separate motor pump costs much less.

There is a differential in original cost, although one must look at the complete picture. When compared on a quoted price basis, canned pumps are almost invariably more expensive. However, one must not overlook their compact size, lighter weight, and elimination of the need for heavy bases, complicated alignment, and leveling procedures. Thus the overall cost of an installed canned pump may be less than that for other types.

Even when the initial cost figure is higher, don't forget that the zero leakage feature commands a premium price. Then again, as the canned pump gains widespread acceptance and production runs increase in quantity, this price differential will probably be reduced.

8—The auxiliary equipment that must be included to protect the motor from abrasives, temperature or corrosion, must be added to the cost.

Once again, the integral motor pump has never been offered as an overall solution to all centrifugal pumping problems. As a result, it is rarely offered for abrasive applications. The only exceptions are when a toxic or explosive fluid cannot be trusted to mechanical seals—no matter how elaborate they might be.

Heating or cooling jackets and similar accessories are available for special applications, but such applications normally require similar auxiliary equipment for the separate motor-type pump. An example is the use of steam or cooling water for mechanical seals. ∎

Casualty experience gives centrifugal-pump users a good guide to preventing failures—shows areas of design, installation and operation that are most apt to cause trouble

EMERGENCY SERVICE-WATER PUMPS should be protected by suction strainers. Strainers keep foreign material from getting into the impeller and damaging it

To guard against centrifugal-pump failures, know their primary causes

Probable failure, the sword of Damocles in any centrifugal-pump installation, depends on factors ranging from application and pump type to degree of preventive maintenance. But a statistical review of past failures can guide users by showing them the sensitive areas. Forewarned is forearmed; emphasis on the weak links will go a long way toward catching potentially serious troubles before they get a chance to become actual.

A recent statistical survey of centrifugal-pump failures, made by The Hartford Steam Boiler Inspection and Insurance Company, provides a good basis for users to predict trouble areas and guard against them.

Table, right, breaks down initial failures of 132 centrifugal pumps by general service area and part that initially failed. Further analysis shows *why* the majority of these failures happened.

Pump-shaft failures predominate—they represent

Based on information supplied by The Hartford Steam Boiler Inspection and Insurance Company, Hartford, Conn.

50%—so let's look at them first. Many feed-system centrifugal pumps (including boiler-feed) operate at relatively high speeds. So the wear and tear of misalignment is magnified. Failure percentages reflect this.

On the other hand, the possibility of foreign objects entering pump suction is far greater with other energy-system pumps (which include service-water pumps). But since a service-water pump usually runs at lower speed for higher capacities, it has generally larger internal passages. Many solids will pass through without causing enough internal stress to actually break the shaft, so here, the percent of failures tracing to foreign objects is relatively small.

Progressive cracking because of corrosion or vibration is a disease common to all pumps; this area needs maintenance emphasis regardless of pump type. At the design stage, proper material selection can minimize corrosion failures. First step here is to know all properties of the liquid pumped; then you're on reasonably safe ground as you go ahead with material selection.

Vibration can come from hydraulic cavitation or mechanical sources. Former should be a function of correct original application—but if you have a pump in operation which is cavitating badly, take immediate steps to relieve it. The process may not be cheap, but in the long run it will be far less expensive than letting the pump run to destruction. There are two ways to attack cavitation: First decide whether your system can be altered to increase suction head, by reducing pipe friction, increasing static head or perhaps cutting pumping temperature. Second, if these steps are not practical, perhaps the pump manufacturer can give you a different impeller design with lower npsh requirements.

Progressive shaft cracking due to mechanical vibration is inexcusable. All rotating equipment should be inspected frequently to be sure that vibration is well within allowable maximum set by the manufacturer. If it exceeds these limits, the pump should be taken off the line immediately and the cause of vibration determined and corrected.

Bearing failures are next on the list. The vast majority stem from lubricant difficulties, regardless of pump service. Loss of lube occurs when oil level falls too low or, in pressure systems, when the oil pump fails. Former is an inspection fault; latter can be overcome by providing a backup pump to cut in on low oil pressure.

Contamination of lubricant takes many forms. It may just be dirt. Even if an oil system is not exposed to the air, vents are provided to equalize air pressure with changing lubricant temperatures—letting the system breathe. These vents alternately draw in and expel air which may be loaded with dust or other plant impurities. Then, over a period of time, the impurities collect and form sludge. Answer here: inspect the lubricant as frequently as necessary to assure yourself that it remains satisfactorily clean. In critical equipment where bearings are pressure lubricated, a filter with bypass arrangement may be a good investment to continuously clean system oil.

Water is a contaminant too. It can enter a lubrication system through cracks in cooling jackets and oil coolers. But another water source is less obvious than these—condensation. As we have said, any lubrication system breathes—and to some degree it inhales water vapor. Temperature within the bearing housing dictates how much of the intake water vapor condenses out. When bearings are overcooled—to a temperature below the ambient wet-bulb—water vapor will condense out within the bearing housing. Since this water vapor has no other place to go, it will collect in the lubrication system and build up. Water contamination then promotes corrosion and ultimate bearing failure.

Loss of main-pump suction is another source of bearing troubles, especially in relatively high-speed feed-system pumps. Vibration, caused by loss of suction, overloads the bearings, and they fail prematurely. Defective bearings show up more in feed-system pumps, also because of higher operating speeds which accelerate failure and will not long tolerate a marginal bearing.

Internal pump damage again breaks sharply between feed-system pumps and others. Condensate in feed systems is normally clean from a physical viewpoint—at least there are no sticks and stones in it. But these pumps

Pump-failure study shows you what to watch out for

	Feed-system pumps	Other energy-system pumps	Total
Pump-shaft failures	14	52	66
Reasons for shaft failures			
Misalignment or poor shaft installation	50%	26%	
Progressive cracking (vibration, corrosion)	42%	54%	
Foreign objects	—	12%	
Other causes	8%	8%	
Bearing failures	23	32	55
Reasons for bearing failures			
Loss or contamination of lubricant	45%	71%	
Loss of main-pump suction	14%	3%	
Misalignment	9%	10%	
Defective bearings	14%	6%	
Other causes	18%	10%	
Damaged pump internals	4	7	11
Causes of internal pump damage			
Loss of pump suction	72%	8%	
Foreign objects	14%	80%	
Other causes	14%	12%	
Total failures	41	91	132

are more prone to suction-side disturbances. Their high speeds accelerate the effects of suction troubles so that internal damage to impellers, wearing rings and interstage bushings can be extensive.

On the other hand, other energy-system pumps are frequently not blessed with such physically pure water. They're the ones that do catch the sticks and stones, so their greatest internal damage stems from foreign objects. The cure for feed-system pumps lies in proper operation and design—plenty of suction pressure margin. In other energy-systems pumps, suction screens will keep out trash that might cause internal damage. Be careful of suction strainers, though. Unattended, they can be worse than none at all. If suction strainers are permitted to load up, pressure drop across them increases with attendant loss of suction pressure. And so, out of the frying pan into the fire—you have kept out the trash only to ruin your pump because of loss of suction pressure.

Of course, the list of things that can go wrong with a centrifugal pump is endless, but these failures top the list. Odds are that special attention to these potential trouble areas will pay off in reduced maintenance and pump-replacement costs. And downtime will be cut too—a real money saver.

Pump Suction Do's & Don't's

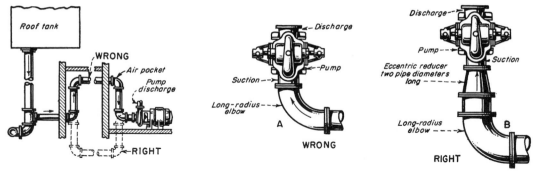

UPWARD LOOP. Never put an upward loop in the suction line of a centrifugal pump. It may form an air pocket and reduce capacity. Full lines show how a pump was installed. Even with pressure tank, pump would not deliver expected flow. Swinging the loop downward (shown dotted) corrected the trouble

HORIZONTAL ELBOW. Never place elbow in horizontal plane at suction nozzle, **A.** Elbow throws flow too much to one side. Connect straight pipe, 2 to 7 pipe diameters long, between elbow and pump suction to straighten out flow. You can use reducer, 2 pipe diameters long, between elbow and intake, **B**

PIPE SIZE CHANGES. Avoid sudden changes in suction-pipe sizes. They produce air pockets, **A** and **B,** that reduce pump's capacity or cause it to lose prime. To change the pipe size, use tapered increasers or reducers instead of flanges shown

PIPE REDUCERS. Don't use a concentric reducer between suction pipe and pump's inlet as it forms an air pocket

PIPE SIZES. Where suction pipe is larger than pump's intake, use eccentric reducers as in **A, B** and **C.** Where suction pipe branches from a header make connection same size as the header and use an eccentric reducer as at **B**

LONG SUCTION LINE. When suction line is long and water level in suction well is below centerline of pump, have line slope up to the pump as shown in full lines. If the pipe slopes downwards, as indicated by the dotted lines, an air pocket forms at the high spot and reduces pump capacity

VERTICAL SUCTION LINE. When suction line rises up to pump, a long-radius elbow connected to suction is permissible, but it's better to insert a piece of straight pipe between the elbow and pump intake. On discharge side, a concentric reducer or increaser is satisfactory for most installations

9
Air compressors, systems

Piping Tips for Air

INTAKE AIR LOCATION

INTAKE at compressor in machinery room is common practice in many small units. This is not best hookup because air is usually warm close to the compressors

BASEMENT intake doesn't help much because air is usually as warm there as in machine room, especially if taken from basement ceiling. Basement air is damp

OUTSIDE intake is usually best because cooler air takes up less volume and machine compresses greater weight of air on each stroke. Machine stays cooler

INSTALLING INTAKE LINES

DISCHARGE and intake lines should never be placed in same duct. Intake-air temperature is raised. Then there's loss of about 1 percent for each 5-F intake rise

INTAKE AIR lines on large compressors need supports. Otherwise weight is on cylinder, and compressor is forced out of alignment. Expensive repairs follow

LONG INTAKE lines must be increased 2 in. in diameter for every 15 ft of piping. Always make pipe at least same size or larger than cylinder intake opening

OUTSIDE INTAKE POINTERS

INTAKE FILTER on outside is protected from rain by hood and from solids by screen cage. This prevents paper and other large particles plugging the filter

OUTSIDE INTAKE close to drain or exhaust lines is bad. Vapors are naturally drawn into intake line. It's best to locate intake where there's only clean, dry air

WHILE PLACING air intake on outside, give some thought to selecting shady side of building, and overhanging roof. This is usually the north side of the building

Compressor Systems

DISCHARGE LINE HOOKUPS

DISCHARGE line to receiver must be same size or larger than opening on machine. Make long radius bends to reduce friction or compressor will be working too hard

WITH SHUTOFF valve in discharge line, install safety valve between it and compressor. Set safety 5 to 10 psi above discharge pressure. This is important

DISCHARGE LINES shown in this notebook enter receivers near top or bottom, as in most plants. There's only one right way. You write us—we'll print answers

RECEIVER TANK LOCATION

RECEIVER or tank stores air and acts as an aftercooler. Be sure to place it as close to compressor as possible. Also use large radius bends and large piping

OUTSIDE receiver on building's shady side radiates more heat. Then cooler outside air helps knock out moisture, furnishing drier air at same cost

TANK outside must have safety valve and gage INSIDE to keep from freezing. Lift safety valve every week by hand ,or by raising pressure to check if it works

AFTERCOOLER PIPING

FACE THIS fan-cooled aftercooler to blow away from receiver. Cooled compressed air and moisture go to tank. Aftercoolers assure drier air needed for any service

WATER-COOLED aftercooler condenses moisture in hot, moist air from compressor. It prevents water hammer, freezing, leaky joints. Hookup to discharge

PIPE-TYPE aftercooler is usually used with separator on large jobs. Aftercooler condenses moisture and separator removes it from air. These two make good team

Checking & Starting

CHECK BEFORE STARTING

1 First hours of new compressor are important. Go over all bolts and screws that might have loosened during shipping and check for any loose wiring

2 Belt tension is ok if 2 lb force pushes belt down distance equal to the thickness of a belt

3 Fan belt tension is best with about one-quarter-in. slack from a straight-line path shown

START AND OBSERVE

7 Unload the compressor by hand before starting. On some machines this is done by small hand unloader. On other types you unload machine by turning a switch to **off** position

8 Good practice calls for turning the compressor over by hand a few times before starting. This tells you if the machine is free to turn, also lubricates the internal parts well

LOW OIL-PRESSURE CAUSES

Keep level up here

Oil pump

11 Low oil level in crankcase causes low oil pressure. This is serious. Oil drops below pump; air is pumped

12 Lube-oil-strainer plugging prevents oil from entering pump. Then oil pressure drops gradually in machine. Remove strainer and clean it at set periods. Never fill crankcase above normal or rings carbonize, stick, and the machine loses **power**

New Air Compressor

4 Air filters protect compressor's internal parts from being worn much by dirt. Fill with right oil and tighten the wing nut before starting the machine

5 Never start compressor without filling crankcase with lube oil of recommended type first off

6 Oil cups on compressor need oil before machine is started. Be sure to find all of the cups

9 Arrow cast on most air compressors shows right direction of rotation. This is important. If machine rotates in reverse direction, oil pump fails and machine burns up very fast

10 Oil-pressure gage is most important instrument on any compressor. Watch it closely. Pressure above 35 psi at first means oil is cold. If it falls under 15 psi, stop compressor

13 Low oil pressure is also from loose bearings and loose wrist-pin bushings. Only cure is to take up this wear

14 Check bypass for being cracked open if oil pressure doesn't build up in a hurry. Valve regulates pressure

15 If pressure still doesn't build up, remove bypass valve bonnet and check for foreign matter stuck to seat

Money burns when you let pressure fall below that for which air tools were designed. This article, based on new "Compressed Air Handbook", of Compressed Air & Gas Institute, shows how much you're losing and what to do about it

Stop Air-Dollar Waste in . . .

▶ COMPRESSED-AIR POWER is a wonderful thing, rightly used. It's too bad many users miss out by letting air pressure get low. By low air pressure, we mean low at the point of use when the tool is running.

Most air tools are designed to run at 90 psig (pounds per square inch, gage). When pressure falls below this, the tool can't do the work for which it was designed.

Causes of Low Pressure—Main causes of low air pressure are: (1) Compressor is too small (2) Piping is inadequate (3) Too much air leaks out of system.

Don't jump to the conclusion that low pressure shows the need for more compressor capacity. Maybe so, maybe not. To find out, first see whether the compressors are running full blast when the pressure is low at tools. Even if they are, don't overlook two other possible causes of low air pressure—leakage and inadequate piping.

If pressure drop from air receiver to point of use is more than 10% average, or 15% at the very worst point in system, correct it before you buy more compressors. Chart on next page will show how much air your present or planned piping can handle properly.

Air Leakage. It is just common sense that every cubic foot of leakage eliminated is another cubic foot available for useful work. Eliminating leakage cuts the power bill and restores the system capacity.

Low Pressure. With portable pneumatic tools you can boost production 37% for about 30% increase in air consumption if you run up the air pressure (at tool) from 70 psi to 90 psi.

What to Do. The cures for low air pressure are just as definite as the causes and effects. Faulty air-power conditions usually result from poor planning or increased air uses without corresponding expansion of the system. Make a survey to find out the extent of low-pressure conditions and the present air-flow requirements. Then try to estimate your future demands.

Pressure drop varies roughly as the square of the velocity of the air going through the pipes. So if the flow is doubled, the pressure drop is $2 \times 2 = 4$ times as great. This works the other way also. Enlarging the line reduces the velocity and very much reduces the pressure drop. For example, a 3-in. line, 1,000 ft long, will handle 500 cfm with a 2.5 psi pressure loss, while a 4-in line of the same length will pass about 1,000 cfm with the same drop. Remember, too, that the installed cost of a 4-in. line won't be much more than that of the 3-in. line because most of the cost is labor.

But what do you do if your line is already in and the pressure drop is excessive? One remedy is to run a new line parallel to the original with frequent interconnections. Or you may install a loop system with some outlets taken off the new line to relieve the old. Sometimes it will pay to put in a com-

plete new air-distribution system.

Since everything in this country has a tendency to grow bigger, it's always smart to install oversize lines for branches and manifolds for tool attachments.

And don't forget the hose. Losses of 15%—even 25%—in the air pressure are frequent in the hose alone. What a pity to waste all this in the last few feet after you've spent good money to produce enough air and deliver it to the hose at the proper pressure. For one thing, don't use a hose longer than you need to make the tool reach the work and to give proper freedom of movement to the operator. Above all, make hose big enough (see chart).

Every cubic foot of leakage salvaged is pure gain. Some systems leak as much as 10 to 20% of the total air compressed. Individual air leaks may be small, but a lot of them add up to real money. Run an air-leak test on the whole plant and inspect every section of hose. Remember that a single $\frac{1}{8}$-in. hole wastes 182,000 cu ft of air per month, at a cost of $9.

The most likely locations of small leaks is around valve stems, in hose connections, unions, drains, home-made guns and lines leading to tools that are non-operating. When you have eliminated leakage, inspect system regularly to keep it tight.

Only after you've done all these things, do you have time to consider new compressor capacity if you can't get along with what you have.

Cubic-feet free air per minute

Nominal pipe sizes-std weight pipe

▶ THIS CHART HELPS you figure pressure loss in any length pipe. In example (dotted line) initial pressure is 100 psi gage, quantity is 1,000 cu ft of free air per min and piping is 4-in. standard.

From 100 on left scale run *horizontally* until *under* 1000 cfm on top scale. Then move parallel to *slanting lines* until *over* pipe size on bottom scale. Now move *horizontally* to right scale and read pressure drop per 100 ft of pipe. Chart redrawn by permission, from original by Walworth Co.

.. 1-PIPE FRICTION

EQUIVALENT FRICTION OF FITTINGS

With factor below, to figure out how much pipe would have the same friction as the given fitting, multiply factor by nominal pipe size in inches and get equivalent pipe length in feet. For example, a 3-in. open angle valve would be equivalent to 3 × 14 = 42 ft of straight pipe:

Fitting	Factor	Fitting	Factor
Globe valve (open)....	26	Close return band......	6
Angle Valve (open)....	14	Standard elbow........	2.5
Swing check (open)....	7	Medium-sweep ell.....	2.3
Gate valve (open)......	0.5	Long-sweep ell........	1.8
Standard tee, through side		45-degree ell..........	1.2
outlet.............	6	square ell............	5.5

▶ THE FOLLOWING example shows convenient way to work out and total the pipe equivalents of all fittings. Add their total to actual piping length to get total equivalent piping. Then, in this case, multiply 3.18 by pressure drop per 100 pipe ft found from above chart.

Fittings (for 3-in. pipe)	Factor	Equivalent feet	No. of fittings	Total Equiv Feet
Standard Ell......	2.5	7.5	4	30
Tee, Side Outlet..	6	18	1	18
Angle Valve.....	14	42	1	42
Globe Valve.....	26	78	1	78
		All Fittings		**168 ft**
		Actual Piping		**150 ft**
		Total Equivalent Piping		**318 ft**

.. 2-FITTING FRICTION

PRESSURE LOSS IN 50 FT OF HOSE

Hose Size Inches	Line Pressure psi gage	Air flow in cfm (cubic feet free air per min)										
		50	60	70	80	90	100	110	120	130	140	150
½	50	18										
½	100	8	13	19	27	37						
¾	50	2	4	4	6	9	11	14				
¾	100	1	1	2	3	4	4	5	7	8	9	11
1	50	x	x	x	2	2	3	4	5	7		
1	100	x	x	x	x	x	1	1	2	2	2	2

▶ TABLE AT LEFT lists pressure drop (in pounds per square inch) to nearest pound. All drops of less than 1 psi are indicated by X. Empty spaces indicate hose size is too small for given air flow. Air flow (cfm) is in cubic feet of free air per min. Note how much the pressure drop is lessened in each case by jumping to the next larger size hose.

.. 3-HOSE FRICTION

Let's take a look at

TWO-STAGE COMPRESSOR runs continuously, supplies 125 psi air to automatic extrusion machine in sewer pipe plant. Intercooler is in the base of this machine

SLIDING VANES move in slot; are held against the cylinder by centrifugal force

How compressor works

▶ POPULAR 2-STAGE AIR COMPRESSOR is the rotary, sliding-vane type. Air is trapped between the vanes as the eccentrically mounted rotor passes the inlet opening, making it a positive displacement machine. As rotor turns toward discharge port, volume of cell between any two vanes decreases and air pressure rises to rated discharge value. Centrifugal force holds vanes against bore, preventing trapped gases from slipping by. Thus vanes act as pistons and valves.

Machine is designed for common electric motor and internal-combustion engine speeds—needs no speed reducer. Since it's a rotary unit, air flow is free from pulsations.

Each unit or stage has a cylinder with heads and is water jacketed. Steel rotor, with integral shaft, is arranged eccentrically within the cylinder on same vertical center line. Shaft is supported on radial roller bearings in each head.

Blades, made of a thermosetting plastic composition or metal, slide freely in slots. Eccentric mounting of rotor inside cylinder forms a free space, crescent-shaped in cross section, divided into compartments by the blades. Volumes of compartments vary from minimum to maximum on intake or suction half of revolution; then from maximum to minimum on compression half.

Lube oil is force-fed to cylinder and to bearings, forming a thin film of oil on the running surface of cylinder wall. This lubricates blades and cylinder.

Single-stage machines are OK up to 40 to 50 psi. Use these units in zones requiring pressures lower than carried in central system. Zoning saves network of distribution sys-

tem and need for reducing pressure from a central plant.

Pressure regulator. Pressure in system moves spool valve, Fig. 4, from bottom to top seat, admits air behind small piston. Piston acts against main valve, moves it to its seat. At same time, relief valve opens, venting air trapped in cylinder to atmosphere. Thus machine runs continuously, as regulator is controlled by pressure rise and fall in air receiver or discharge lines. Operation is automatic, needs no attention once it's set. Two simple adjustments allow variations in pressure or range setting if needed.

Multiple unit hookup. Have one unit on start-and-stop, while additional units run continuously. Two-stage machines are for pressures over 50 psi, as they are more economical at higher pressures than single-stage unit. Final air temperatures are also much lower. For capacities through 229 cfm (actual free-air delivery) at 100 psi, machines have intercooler tube nest in base. Larger units have separate intercooler. Two-stage machines are normally direct-line coupled to prime mover; no speed reducers are needed.

Maintenance. Open machine about once a year. Examine rubbing surfaces for wear. New vanes are same width and length as slots. Keep suction filter clean to keep contamination out of cylinder. Adjust mechanical lubricator feeds properly and see that reservoir always has enough oil so it is visible in the lubricator gage glass. Keep enough cooling water flowing through the cylinder jackets and intercooler to hold temperature where needed (usually about body temperature at final water outlet funnel).

Courtesy, Fuller Company, Catasauqua, Pennsylvania

the sliding-vane compressor

1 Rotor has fixed roller bearing at one end, floating bearing at other to take care of the longitudinal expansion and contraction differences

2 Blades automatically compensate for wear; check degree of wear when opening unit

A   Compressor packing set

4 Regulator is controlled by pressure rise and fall in air receiver or discharge line as compressor runs continuously

B   Vacuum packing set

3 Typical packing arrangements for air service only. As you can see, V rings are not installed in the same direction in the vacuum and compressor machines. Experience shows that these packing sets run for long periods, with little or no gland adjustment required

Used as a vacuum pump

Vacuum pumps (sliding-vane type) are very similar to the compressor just described. Single-stage machines create vacuums up to 29.5 in.; compound machine up to 29.95 in. (referred to 30-in. barometer). Above 28 in. vacuum, 2-stage vacuum pumps are recommended because of their high volumetric efficiencies. Curve at left shows typical characteristics of these compound vacuum pumps between 28 in. and 29.95 in. mercury. Absence of vibration and air pulsations are often an important factor for application of these pumps.

Photo by Ewing Galloway

What Do You Take Me For, a Fathead?

THIS HIPPOPOTAMUS IS A DEAD RINGER for a chief engineer I worked for many years ago. It was an industrial plant and I had the day shift.

Shortly after starting, I got interested in the old dc switchboard. The fuses and knife switches were pretty dirty, so I started checking all of them.

Some fuse clips were badly burned. And I found double links in many cartridges. It was the old "penny in the fuse" practice, but I was surprised to find that in a plant.

"If this board's so sloppy," I thought to myself, "there must be plenty wrong with it." First off, I checked the generator's field rheostat settings. They were pretty high. Then I noticed the main-circuit-breaker trip arm was tied down. I untied it in a hurry. Then I set to work polishing the fuse clips, cartridge ends and knife switches with fine sand-paper. Next I got rid of all the double fuse links, and there were plenty.

Two days later, on my morning watch, the breaker kicked out. "What's going on here?" thundered the chief, tearing into the engine room.

I explained about untieing the breaker because I didn't want the generator to burn up.

"I tied that down because I don't want it kicking out," he yelled. Then he warned, "When I tie down a breaker, I want it left that way. WHAT DO YOU TAKE ME FOR, A FATHEAD?"

I've learned there are plenty of his kind around even today. They're heavyweights, all right. About as heavy as a hippo, and about as fast . . . mentally. They don't know why protective devices are built, and they don't take the trouble to find out—until it's too late.

10
Diesel engines

Let's time your dual-fuel engine and tune it up in 5 careful operations

1 Find top dead center ...

2 Mark flywheel ...

3 Time fuel pump ...

1

Establish your engine timing marks

► Suppose you have to time a dual-fuel diesel that has no marks on the flywheel, then tune up the engine from indicator cards and make all adjustments so the engine is ready to roll. Can you do it? If not, read on.

To get this story right, I trailed my old friend, the capable erection superintendent for Worthington Corp. Engine shown is a 16x20 dual-fuel job, 8-cylinder, 4-cycle, 360-rpm and 130-bmep.

1. First, place No. 1 piston top dead-center (TDC). For stationary engines,

place protractor (water level) against crank web. With web vertical, piston is on TDC. For marine engine, use a trammel (as shown in Chap. 5, on steam engines) as these engines aren't level.

If flywheel is already marked with TDC, see that pointer over flywheel aligns with this mark on it when crank web is in true vertical position. *Note:* This check must be right because all initial fuel-injection settings are made from these TDC marks, then used as a base for fuel-injection measurement.

OIL AND GAS—they're both the same to this engine. But for top performance, know how to keep your engine in tune

4 Check port closing marks . . .

5 Find fuel-port closing spot . . .

6 Set valve clearances

2. With crank on TDC, place square against flywheel rim to check wheel marking with pointer on engine frame. If there's no pointer or mark on engine frame (new engine), scratch mark on frame where scale in photo touches the frame opposite flywheel. Check remaining TDC marks for other cylinders on flywheel in same way for accuracy.

3. With TDC marks on flywheel correct for all cylinders, check timing marks on fuel-pump window frame, aligning with pump-cup guide. When marks line up as in photo, fuel port is at closing position.

4. Check distance of port closing marks before TDC on flywheel rim. Make this check for each cylinder.

5. Next scribe fuel-port closing position on flywheel rim opposite centerline mark on engine housing.

6. Now set inlet and exhaust valve clearances at their initial "cold setting." You can do this at same time you make TDC checks. Be sure to set valves on combustion stroke while they are in closed position. This is very important.

To help this work and other checks, firing order of engine is stamped on plate attached to engine. After setting clearances on one cylinder, follow firing order for remaining settings by jacking flywheel to cylinder TDC marks and repeat for each cylinder.

Make the final "hot settings" of inlet and exhaust valves later while the engine is running at full load and gets to normal operating temperatures.

(Continued on next page)

1 Test the spray valve . . .

2 Take the spray valve apart to make adjustments . . .

5 Reassemble spray valve with a torque wrench . . .

6 Test spray again . . .

2

Check fuel injectors and fuel pumps

1. Remove spray valve from cylinder head and test for opening pressure. Gage shows pressure of 3600 psi, which is 100 psi too high for this engine.

2. Dismantle spray valve so you can adjust the opening pressure setting. Torque wrench shown is not needed to take injector apart; use any wrench.

3. Remove shims from inside spray valve to lower opening pressure to 3500 psi. Shims vary from 0.002 to 0.0365 in. for pressure settings. If opening pressure is low, add shims. There is about 100-psi change per 0.010-in. shim when setting is close to specified 3500-psi opening for this engine.

4. Clean spray holes in nozzle to assure an even spray pattern. Holes are 0.018-in. diam—clean with 0.016-in. wire supplied by maker. Do this job at regular intervals if possible.

5. Reassemble spray valve with torque wrench, using 400-lb torque pressure. This wrench is a good investment. It

helps place right tension on valve casing and working parts. Your engine maker will give you right torque pressures for various engine parts. Hold spray valve firmly in vise.

6. Test spray-valve opening again now that you have removed a shim. Gage shows 3500 psi needed, which is pressure when valve "breaks" to atomize fuel oil.

7. Dismantle dual-plunger pump and check seats of main delivery and pilot-plunger delivery valves. Seat must be especially right for pilot-plunger valve as it meters tiny amount of fuel oil.

8. Check metering edge of plunger for erosion and scratches or wear grooves that cut down pump efficiency. This is a ticklish job and parts must be absolutely clean. When reinstalling barrel and plunger in pump body, make sure both are free of tiny dirt particles. Don't use rags in handling parts. Threads often get in pump, causing trouble.

3 Remove shims . . .

4 Clean spray holes . . .

7 Check dual-plunger pump . . .

8 Look for erosion . . .

9 Align fuel control-rod . . .

9. Now check the pump control-sleeve and the metering fuel control-rod teeth for fitting. Tiny white timing marks on both control sleeve and the fuel control-rod must align with one another as shown. Fuel pump is timed from these reference points.

10. Next check fuel pump's timing-mark alignment with control sleeve and control rod after reinstalling in pump body. Dots must line up as in step 9.

11. Measure width of pilot-oil timing washer and record. This washer times pilot oil when engine goes into gas operation. Thinner washer advances pump timing, thicker retards it. Every 0.002 in. makes about 4-psi difference in pressure.

Raise or lower final gas-operating combustion pressures needed with these washers. Usually you won't have to change washer as pump maker installs the right dimension for your engine.

(Continued on next page)

10 Check inside of pump . . .

11 Measure the timing washer

1 Take exhaust temperatures...   **2 Adjust pump rod's length . . .**   **3 Check cylinder temp again . . .**

6 Study cards for unbalance as Fred Carter is doing . . .   **7 Check lines with rule . . .**

3

Take indicator cards and balance engine

1. Now that you have made oil-operation checks, start engine, apply the load. Next use selector switch, record exhaust temperature of each cylinder.

2. Adjust fuel pump metering-control-rod so temperature is same from all cylinders. First loosen lock nut and adjust setscrew as shown. To raise temperature, screw in—to lower, screw out. When temperature for a cylinder is right, tighten lock nut as shown.

Each pump's control rod has reading in mm. If amount is more than 2 mm on any rod, check that pump's timing and corresponding cylinder pressures. Reading on all control rods should be as nearly the same as possible.

3. Check cylinder temperature again. Here it is exactly right, 820 F. All cylinders must be same.

4. With exhaust temperatures all the same, take indicator cards. Always blow out indicator cock before attaching indicator. Unless you do this, tiny carbon particles may plug pressure passage in line, giving faulty readings. Take cards of cylinder-combustion pressure at full load only.

5. Combustion-pressure balance card shows if each cylinder is doing its share of work. Remember that cylinder temperatures may balance perfectly, yet combustion pressures be out of balance. This is the most important of all balancing operations.

6. Study pressure card for unbalance by checking against engine maker's instruction book. This book also tells you right indicator spring to use. Ruler for measuring the pressure lines must have same pressure rating as on indicator spring to tell the right story.

7. Here No. 7 cylinder's firing pressure is low, while ruler shows other pressures fairly well balanced around 800 psi.

8. Check to see that aligning mark does not go out of window at top or bottom of fuel cam stroke. Photo shows that alignment mark is low but not out of sight for No. 7 cylinder.

If the required shimming won't cause this line to go out of the window at top or bottom, you can add shims. It is important that shims you add do not raise pump too high because that will jam the pump and cause damage.

What do you do if shimming does raise the marks high enough to leave window? Answer is that you must make needed adjustment by camming individual cylinders of the fuel pumps.

9. Insert shims under pump plunger to advance timing and to raise combustion pressure. Here pump has been removed from its base and fuel pump's push rod is protruding from pump opening (not shown in closeup). This push rod makes up contact between pump and camshaft's fuel cam. Cap for fuel pump's push rod that houses shims is shown at right of shim being miked. Mike shims, then add 0.040-in. shim

4 Blow indicator cock . . .

5 Take combustion cards . . .

8 Check pump-plunger mark . . .

9 Insert shims under plunger . . .

10 Adjust fuel cam . . .

11 Check temp again . . .

12 Compare indicator cards . . .

13 Take "pull" cards at full load

to raise pressure and also to bring up pump plunger's aligning mark. Do this by taking difference between desired combustion pressure and actual recorded pressure in psi. Then add one half that figure in thousandths of shim thickness to raise pressure.

Follow same method of removing shims to lower the pressures. This adjustment gives approximate correct pressure. Make final setting by caming.

Check pump aligning mark in pump window at stroke's bottom again after shimming. You will see that mark is now higher in window. So any danger of damaging pump from too low an aligning mark (out of sight) has been removed because of shimming.

Now check pump plunger's aligning mark in pump window at top of stroke. Aligning mark at top of pump stroke is about same distance from top of window frame as mark is from frame bottom with pump stroke at bottom. Check each cylinder's fuel pump like this.

10. Make final adjustment by fuel cam if combustion is still slightly low with pump timing marks centered in pump's window frames.

Use indicator card for this operation. To raise pressures, advance adjustable cam in photo—to lower, retard. Raising pressures gives you lower temperatures while retarding pressures gives higher temperatures.

Look at pump window, while moving cam to see that line on plunger and window are in line. For changing engine
(Continued on next page)

timing as a unit, advance or retard the camshaft gear. Original camshaft setting is checked by placing camshaft key on No. 1 cylinder in level position, then having crank throw of that cylinder read 48 degrees BTDC (before top dead center).

When over-all adjustment change is made to raise or lower pressures, the inlet and exhaust valve—as well as gas cams on some camshafts—have enough overlap so operational settings are not affected.

11. Check cylinder temperatures after caming adjustments. Note that temperature indicated is somewhat lower then reading before caming change. That's because adjustments raised pressures.

12. Take combustion-pressure indicator cards again and compare them with ones taken before making adjustments. Note that pressure balance is much better on AFTER card.

13. Take one set of "pull" cards at full load to further check individual cylinder performance. Cards show timing and firing point. Pencil points to firing point here. This operation is also an important check for fuel-oil performance of any type of diesel engine.

1 Examine relief doors . . .

2 Check linkage lengths . . .

3 Examine for tiny gas leaks

4

At this time make these checks too

1. With engine checked through for oil operation, check for gas operation. Then inspect the relief doors on alternate rear engine-covers for working freely (once a month). Also check diaphragm for possible rupture.

2. Refer to instruction bulletin and check linkage length measurement. This measurement is very critical as governor opening, gas shutoff position and amount of gas-rack lead are all affected by accuracy of these settings. You also determine transition to gas from oil and oil to gas—resistance on gas-rack movement.

3. Keep gas-piping connections tight. Check for leaks with a soap-and-water solution. Bubbling effect of soap quickly helps you find the smallest leaks.

1 Align top of gas plug . . .

2 Open all indicator cocks . . .

3 Roll engine on air . . .

4 Place on gas operation . . .

5 Adjust gas metering-valve . . .

6 Adjust gas-valve clearance

5

Now adjust engine for gas operation

1. With engine tuned up for oil operation, check top of gas-metering plug valve for alignment with metering edge of inlet sector of metering valve's casing. Do this by resting end of scale on both edges to find if flush. Have gas-fuel rack and governor in their "closed" positions during this check—and follow through on all metering valves on all cylinders. As soon as settings are made, make sure gas rack, metering valves and governor open and close as a unit. It's especially important that they close *all the way*.

2. Next open all indicator cocks before rolling engine over on air to purge system of gas. This is a precautionary step. With fuel pumps' racks in their closed position, pilot oil is being metered into the combustion chambers of

the cylinders and any gas leakage or build-up will ignite and cause possible detonations.

3. Now roll engine over on air to purge system of any accumulated gas. Check the gas pressure to engine's metering gas-valves. It should be 12 psi on gage, which is average setting for natural-gas operation.

4. Engine is now ready to run on gas —so place on "no-load" operation. First place the manual control-lever on its gas position. Only other adjustment needed to change to gas operation is to open the master gas valve. Do this before placing the lever in its gas-operation position while engine is running on oil.

It's best always to start on oil and then change to gas operation at no load. When shutting down, it's also good practice to change to oil from gas operation. That allows fuel pumps to operate on oil, which they would not do otherwise.

With selector switch, check full-load cylinder's temperature balance.

5. Next, adjust gas-metering valve's reach-rod length to balance the cylinder temperatures at full-load operation. Loosen lock nut, then turn reach rod *into* clevis to lower temperature—screw *out* to raise temperature. Make needed adjustments in all cylinder metering-valve reach rods to balance the temperatures.

To raise or lower all cylinder temperatures as a unit, lengthen or shorten the adjustable-rod connection between gas rack and governor. Caution! When shortening this connection too much to raise temperatures, the *close* position of gas rack might affect shutting off the gas because metering valves are partly open.

You are now ready for final cylinder temperature-balance at full-load gas

operation. With temperatures balanced, take indicator cards at full load to balance-out combustion pressures. Use same procedure as shown for oil operation.

6. If final combustion-pressure balance is slightly off, you can adjust gas-injection valve clearance. Tappet clearance has adjustment range of 0.012 to 0.020 in., *less* clearance to raise pressures and *more* to lower them.

Caming adjustments can also be made for pressure settings and balancing. If you must make over-all pressure adjustments, use pilot-oil fuel pump's timing washer shown for setting fuel pump.

When temperatures and combustion pressures have been balanced out according to instruction book, engine is ready for service on either gas or oil operation.

Be sure to test all alarm and shutdown devices on engine. You must also check and actuate the auxiliaries at set periods to make sure you have maximum protection at all times.

This engine has many exclusive features. They are: throttling air at light loads to improve operation and fuel consumption on gas operation; changeover when on gas or oil operation that's affected only by gas-pressure failure or manual control. These devices must also be checked from time to time if you want dependable operation from this or any engine.

Now that engine is timed for both oil and gas, work out a preventive-maintenance program as recommended by the maker. Keep a careful log and study it for possible warnings of trouble. The most successful diesel operators know their engines thoroughly, thereby keeping them at peak performance, with minimum shutdown. It's the only way.

Let's Take Over Watch

1 — Read logbook carefully to see what happened on the previous watch . . .

YOU ARE RESPONSIBLE for the plant the minute you take over a watch—your reputation as an operator is at stake. Study the logbook, and make sure you know of any unusual conditions that arose on the previous watch. A smart operator can often prevent breakdowns from telltale entries.

Above is a photo of George Green, first assistant engineer, studying the log in a smoothly running plant.

Use selector switch to check each cylinder's exhaust temperature. If temperature to any one cylinder is high, check log sheets to see if it has been high before. If not, adjust fuel oil to that cylinder until temperature is the same as in others.

If that cylinder knocks, it might be overloaded. If all its adjustments are right and it still knocks, the cause might be that the cylinder firing before the knocking cylinder is dead. Then this one knocks because it does the work of both. If dead, check fuel pump, also valve mechanism for that cylinder, starting with camshaft. Check lube-oil temperature going into engine. Adjust cooling water to lube-oil cooler if oil temperature isn't right.

Next, check fuel, lube and cooling-water pressures. Try to keep them as nearly ideal as possible. Lube-discharge temperature from engine often warns of overheated bearing or badly overheated parts inside. Try to locate cause by feel, or by listening carefully with ear to flashlight placed against various parts of engine housing.

Cooling water from engine tells the same story as lube oil from engine. If temperature has been climbing gradually over several weeks, the cause is usually scale building up inside engine jackets. Or it could be a fouled heat exchanger.

Make sure you know where lube oil is in governor by looking at oil-level sight. Keep level right. If it is a self-cleaning lube-oil strainer, give handle one turn as in photo. That scrapes strainer clean. Then check oil pressures to see if they have changed because of clean strainer.

Crankcase oil level should be up to mark. Check it with bayonet, and make sure you read it correctly. Lubrication is the engine's lifeblood. You can't always depend on the engine's tachometer readings. Good investment for your plant is a hand-type revolution indicator. Place it on end of shaft at pedestal bearing. If engine's critical speed is known to be within operating range, make sure engine runs slightly over or below this speed. While at the pedestal bearing, feel the bearing and, if it is oil-ring lubricated, look inside for good measure to see that rings are turning and level is right.

Trouble with modern engines is they run well with little attention for long periods. Then operators get lax, and they forget to make their rounds as they should.

While on watch, keep on your toes during peak-load periods. Get the habit of listening to each engine when taking over watch, and while making each round. Listen at each cylinder until you get so expert that you can detect any unusual noise. After a while you will know what every little sound means. Before turning watch over to your relief, be sure to tell him about anything that might give him trouble.

In a Diesel Plant

2 — Check: Cylinder exhausts . . .

3 — Lube-oil temperature . . .

4 — Fuel, lube, cooling pressure

5 — Check: Lube discharge . . .

6 — Cooling discharge temp . . .

7 — Governor's oil level

8 — Check: Crankcase-oil level

9 — Shaft revolutions

10 — Clear: Lube-oil strainer

Let's Start Your Diesel

Getting Ready

1 Open the inlet and outlet of engine's jacket and the cylinder-head cooling water, make sure water is flowing through

2 Pump up the lube-oil pressure by hand to about 5-10 psi. This prevents scored bearings caused by oil draining

3 Open fuel valve and pump up fuel by hand to make sure lines are filled, and also to help clear the lines of air

Starting Up

7 Open starting air-valve to crank the engine, close valve as soon as engine is firing and rolling under own power

8 Test cylinders for water by holding a dry rag under open-cylinder blow-off valves. If moist, check for cracks

9 Bring speed up slowly. If cold, wait till lube is 100 F. Observe oil, fuel, cooling pressure, and then throw on load

Keeping on Your Toes

13 Check bearing and other temperatures every hour. Do this by feel as shown, and by reading the various gages

14 Every hour, oil starting air-distributor (this engine). There are various oil cups that need hourly hand oiling

15 Check voltage and load every hour, jot down on log sheets. This practice is often skipped in small plants

Engine and Keep It Rolling

4 Open blowoff valves of each cylinder to avoid any damage if there is water in the cylinder from a cracked liner

5 Open starting-air valve on storage tank if the pressure is right for starting. If not, start the air compressor

6 Set the governor to starting speed. This engine is now ready to start as it has been properly prepared to roll

10 Check exhaust temperatures of all cylinders with selector switch as shown. Do this before leaving engine

11 Exhaust of this cylinder is very low because of air in the fuel-oil line. Crack connections and let air bleed out

12 If venting the fuel-line won't help, shut it down and replace the fuel injector because the unit is probably faulty

16 Every hour, log exhaust temp, cooling-water temp in and out, lube-oil temp in and out, lube-oil pressure in, out

17 Once a day, fill wick oilers for rocker-arm rollers, exhaust and also intake valves, etc, depending on engine

18 Enter time engine was started, or shut down and other important info in log book. Sign when going off watch

BEARING FAILURES TELL OPERATOR ABOUT ENGINE'S CONDITION

1 Tiny particles of cast iron are embedded in the bearing surface here

2 Foreign matter caked to bearing-shell back caused high pressure and heat

3 High pressure from foreign matter back of thin bearing caused failure

4 Fatigue and corrosion teamed up in high-speed high-pressure bearing

5 Bond failure is from faulty tinning or no tinning. Lining breaks away

6 Ridge around crankpin or journal may spoil bearing if not stoned down

Your Bearings Talk . . .

▶ YOUR DIESEL BEARINGS can tell you a lot about the engine's condition and operation, if you understand their 'language'.

To interpret the bearings' story, you should know lining materials, their characteristics, main causes of bearing failure and how to identify them, bearing-design features and lubrication.

Various lining materials are used in heavily loaded bearings for resistance to fatigue corrosion and for greater life.

Babbitt Linings. Babbitt makes fine bearing linings within its load limitations. Its frictional coefficient is low; melting point is low and it's soft and easily worked; it bonds well; and conforms to shaft. Also, foreign particles embed into it. Big advantage is that it's less abrasive and works well with unhardened crankshafts.

WHY BEARINGS FAIL

Dirt and other abrasive materials cause most trouble. They scratch and score surfaces.

Particles are often embedded in the babbitt. With a harder material, these particles might remain on the surface to score the crankshaft. If particles removed by light scraping are metallic and ferrous (shown by a magnet), look for excessive engine-parts wear, Fig. 1, such as liners, pistons and rings. If dirt, check air filters.

Mechanical conditions cause failures also. Bearing misalignment is common. Then crankshaft is distorted. Such distortion deforms the bearing surfaces and causes localized high pressure areas so bearings get hot. Bearing-shell distortion is from foreign matter on shell's back, Fig 2. This causes high pressures and temperatures, but only in bearing effected, Fig. 3.

Modern engines with higher cylinder pressures and higher speeds have two common bearing failures. These are fatigue and corrosion of lining material.

Cracks become deeper, until they alter direction and break parallel to the bearing backs just above the bond layer, Fig. 4. Then as load increases on surrounding areas and oil fails, bearing burns up completely.

OTHER FAILURES

Corrosion. Corrosion is from acid components of combustion, together with condensed water vapor, collecting in lube oil. Build-up of harmful acids is speeded by excessive oil temperatures. Acids etch bearing surfaces. Then area is reduced, until in extreme

* Photos, Federal-Mogul Corp.

cases results are similar to fatigue. At times it's hard to tell if failure is from fatigue or corrosion, and not only because appearance is often similar. But both causes may have caused final failure.

Most babbitts resist corrosion well, but resistance varies with different types. With tin-based babbitts, for example, the high oil temperatures promote corrosion, cause fatigue failure before corrosion is serious. Lead-based babbitts with over five percent tin or one-half percent silver aren't effected much by corrosion. The SAE formulas and a number of proprietary metals resist corrosion well.

Bond Failure. This has been largely lessened by modern manufacturing techniques. The bond layer or "tinning" is very thin and inclined to be brittle. But metal loss occurs if entire area isn't properly bonded. This condition is easy to tell because the roughened portion of bearing back is visible, either with no tinning or faulty tinning apparent, Fig. 5.

Now, let's roll out the bearings and examine all of them. By checking a complete set, it's often possible to tell if a condition is general or localized. A look at the engine log and any recorded abnormal events may help.

Fatigue. Let's say evidence of a bearing fatigue condition is found in all bearings. Cause may be high peak cylinder pressures, possibly from long periods of overload (found from engine's log) or bad adjustment, (such as early timing).

Fatigue of only one or two bearings might mean unequal fuel delivery to cylinders. Then overworked bearings fail. Check this with pyrometer readings. If readings are OK, check the instruments and gages.

Bearings failing near flywheel are often from special causes. Consider method of power takeoff. Weight attached to already overhanging flywheel, radial loading from belting or chains, or a poorly aligned outboard bearing is straw that breaks the camel's back.

Obviously, all engine parts are interrelated and so failure's 'cause and effect' will be related. With tips given here, your trouble shooting success depends on you.

TROUBLE SHOOTING

Before replacing bearing shells, note all faults and make repairs. For example, suppose there's evidence of misalignment and this is confirmed by crankshaft strain-gauge readings (this chapter). Condition may be from sprung shaft or poor bearing alignment. Re-

move suspected shaft, place between lathe centers or on roller supports at ends. Then check with a dial indicator while turning.

Forged shafts can be straightened but it's a tough job and might need a light regrind for final truing. If bearing mountings aren't right, maybe bedplate wasn't originally level. Or maybe it settled from cracked foundation. Re-leveling bedplate should take out sags or warps.

If fault lies in bearing supports, problem isn't so easy. But this is uncommon. Two courses are open, the simplest is using oversize or semi-finished bearings that are align bored. This has disadvantage of making the bearings non-standard. Next best is to remove some metal from cap faces, replace them and align-bore bearing supports.

This takes good equipment for boring and expert workmanship. Remove only enough metal to clean it up. And metal should come from bore's top and bottom, leaving parting lines untouched. If shaft's axis is slightly changed, check drive gears for meshing.

Before installing bearings, check the crankshaft. Mike journals and crankpins for being out-of-round or taper. If these conditions are found in excess of maker's limits, replace or regrind and use oversize bearing shells.

Another condition, especially with softer shafts, is wear on each side of oil hole. Many bearings have fairly long centrally located oil grooves lining up with oil hole. Area around hole is more heavily loaded, so expect greater wear. This might not effect old bearing since the two surfaces have conformed to each other. But such a ridge may rapidly destroy a new bearing shell, Fig. 6. Remove the ridge by carefully stoning journal.

REPLACING SHELL

Outer diameters of "thick wall" bearings are ground. This insures concentricity and close fits with saddle and caps, both for support and for maximum heat conductivity. Good contact is further insured by providing "crush". The shell ends should extend beyond the faces of saddles and caps, so there is an interference fit. The pressure at the parting line forces the bearing into close contact with the supporting surfaces. The resulting friction also holds the bearing against rotation. Tangs or dowels are also used.

See that this crush exists and that there are no burrs or high spots to cause excessive interference at the parting line. There's usually a relief at this

point and on the flanges' faces. It takes care of any swelling when compressed and acts as an oil reservoir.

You can usually tell shell looseness by carbon deposits on their backs. Such looseness may result in bearing turning. Then bearing will fail from oil holes misaligning.

ASSEMBLING

Take great care during installation to exclude dirt. A sizeable particle behind bearing's back will raise lining. That reduces clearances and creates a highly loaded area. Fatigue and corrosion follow rapidly, as well as actual melting and metal displacement. Dirt on bearing surfaces causes immediate scoring and abrasion.

Make sure bearings are not reversed and that all tangs, dowels, etc are properly located. Then there won't be distortion. When shims separate bearing halves, install them so that they prevent excessive oil escape without galling the shaft.

Clearances. Clearances should never be less than recommended by the manufacturer. At times they must be somewhat larger, depending on mechanical conditions. Journal's surface condition is important. Success of present close tolerances are partly due to improved shaft finishes.

Scratched or roughened surfaces have tiny sharp crests and valleys. Crests protrude through oil film and thus reduce actual effective oil-film thickness. Then more clearance is needed to form a thicker film.

Carefully stoning or polishing journals with fine emery and crocus cloth helps bring good finish. Be very careful to remove all abrasive particles afterwards.

Some large engine bearings can't be taken up for wear. This is justified by long bearing life. In one case bearing lasted 30,000 hr before clearances reached their limit. By then fatigue failure comes in and replacement is best solution anyway.

If bearings can be taken up, restore clearances close to minimum practical recommendations. Clearance of 0.003 in. is equivalent to same amount of wear. Then bearings will have to be overhauled sooner.

When wear approaches halfway mark, decide whether to continue shells in service. Bearings' overall condition, cost of replacement and expected remaining hours of operation must balance cost of early replacement job and engine down time.

With this information, you're all set to do a little trouble-shooting.

Quick Way to Shoot Your

DIRTY?_____ CHECK STRAINERS AND WEAR_____

EMBEDDED foreign matter found in the soft bearing metal may be either metal or foreign particles

MAGNET tells if matter is metallic. If so, check liners, pistons, rings for excessive wear. If not metallic, check air filters for foreign particles being drawn into cylinders. Grit in bearing metal scores shafts, increases bearing clearance, causes overheating

CORRODED?_____ STOP OIL CONTAMINATION_____

FATIGUE and corrosion are often difficult to tell apart. Both are very hard on the bearing lining

ACID COMPONENTS of combustion and condensed water vapor collect in lube oil, cause corrosion. Affect all bearings alike. Etched bearing surface reduces effective bearing area, causes failure from high unit pressure if not corrected soon

LOOSE?_____ HAVE GOOD 'CRUSH' FIT_____

CARBON deposit on back of shell advertises that the bearing shell is loose and in need of correction

CHECK by placing straightedge on bearing, insert feelers at 1, 2, 3. Shell must have "crush" fit at these points. Loose shell may turn, cut off oil supply. Relief at parting edges cares for swelling; they give oil storage for good operation

Diesel Bearing Troubles

MISALIGNED?_____ MAKE STRAIN GAGE TEST_____

MECHANICAL cause is often from misalignment. That deforms bearings, causes high-pressure areas

CRANK BEARINGS' misalignment is found with strain gage (see pp 206–207). Distorted crankshaft may break if not found in time. Bearing-shell distortion is also from foreign particles under thin bearing metal; then heating affects bearing

BOND FAILED?_____TIN AND BABBITT CORRECTLY_____

BOND FAILURE causes the thin pieces of the soft bearing metal to break away from the bearing shell

UNTINNED shell areas before babbitting are chief bond-failure cause. Improved methods make this fault uncommon in modern bearings. When tinning shell, preheat carefully and tin completely. Also preheat mandril and shell to right temperature

OIL PRESSURE LOW? TAKE UP ON CLEARANCES_____

LOW OIL pressure will be hard on the bearings. Keep the pressures that the manufacturer recommends

EXCESSIVE bearing clearance oozes badly needed oil, lowers pressure, makes engine noisy. Heavily loaded bearings may even fail. Shims must seal bearing halves at relief ends, but not touch to scrape oil off shaft. Reversed shells may leak oil

Your Liners Can Crack

SLOW CRACKS

LINERS USUALLY CRACK a few inches from top—run circumferentially around

SLOW CRACK allows water to leak into cylinder. Water washes off cylinder oil

DRY SPOTS rust when engines shut down. Then liner heats fast, seizing the piston

HOW TO PREVENT *Continued*

RUN ENGINE for 15 minutes with no load if the cooling pump is on the engine

ALWAYS COOL SLOWLY by unloading for 15 min; then run pump another 15 min

BUILD UP METAL with welding torch if no metal-to-metal contact at the bottom

CAUSES

UNEVEN TIGHTENING or too much tightening of cylinder head bolts is bad

TOO TIGHT bolts cause bending strain on cylinder head—transmitted to liner

GOOD THREADS help nut tighten more than bad ones—torque wrench reads same

Unless You Stop Them

HOW TO PREVENT

TIGHTEN NUTS in sequence shown. Take up slack on all, then drive them tight

REMOVE BOLTS and nuts with damaged or worn threads and replace with new

RUN COOLING WATER 15 to 30 minutes after shutting down to cool the engine

PRUSSIAN BLUE under collar, and turn the liner slightly to mark the high spots

SCRAPE HIGH SPOTS shown by prussian blue until the bearing surfaces are level

GRIND BEARING smooth with cast-iron plate. Use new copper and rubber gasket

CUTTING OFF cooling water soon as the engines shut down overheats the liners

PISTON RESTS in same spot each shutdown with water off, overheats the liner

NO CONTACT between liner and block at the top or bottom from warped casting

Next Shutdown,

DAMAGES

CRANKCASE wreck ($4771 here) is most common damage when bolts fail

PISTON SKIRT AND ROD break from smack by crank when bolts let go

CRANKSHAFT jamming loose parts when bolts fail can cause this

CAUSES

Scuffing slows piston

Crank drags piston

Bolts stretch

PISTON SEIZING or scuffing stretches crankpin bolts in a multi-cylinder engine as crank drags piston along

Connecting-rod foot

Pull

CONNECTING-ROD foot flexing bends bolts. This also happens from piston scuffing. Check foot with straightedge

PREVENTION

Renew bolts after 30,000 hrs, if —

RENEW BOLTS after 30,000 hours of engine operation if engine has run smoothly with no scuffing or seizing

CHECK LOG for engine's history since last bolt change to see if conditions existed that could cause bolt elongation

ELONGATION LIMIT

DON'T EXCEED bolt's elongation limit from engine instruction book. Reaching limit means bolt reached yield point

Zero

BEFORE TIGHTENING new bolts with wrench, use your stretch gage like this and set the dial so it registers zero

Check Crankpin Bolts

NUT COMING LOOSE strains other bolt. Tap all nuts every few months with light hammer. Sound shows loose nut

NUT NOT TIGHTENED is next. Or one nut pulled up tighter. Both bolts must have exact strain to avoid trouble

STRAINING bolt from over-tightening is common. Torque wrench is NOT answer as thread friction varies on both nuts

BEST CHECK for bolt elongation or stretch is to mike the bolt **before** loosening nuts, with bolts strained

THEN MIKE the bolt **after** loosening nuts. Compare measurement with that of tight bolt for amount of stretch

IF NO LARGE MIKE, make 'stretch gage' from strap iron and dial indicator. Always read at same spots on crank bolt

TIGHTEN BOTH BOLTS until stretch indicated on dial is exactly same for both. Record bolt's stretch in the log

CHROME MOLYBDENUM steel SAE-41-30, ASA 3125 and 3130, with collars, stretch full bolt length, relieve threads

WHITEWASH TEST shows hidden crack. Clean the bolt, rub with kerosene, wipe dry, then whitewash. Crack will show

Deflection-Test Your

DAMAGE

SHAFT BROKE at web from fatigue as a high bearing bent shaft with each rev.

HOW TO TEST

1 Strain gage set measures misalignment without taking crankshaft out

2 Place gage between webs. Spring tension against points holds it in place

6 Set dial to zero. This gage balances itself so face is always to operator

7 Jack the crank to quarter position. Needle hasn't moved from zero mark

8 Jack to top, then quarter position. Needle stays on zero; this crank's OK

12 Lift assembly a few thousandths by screwjack, enough to remove shim

13 Remove 0.015-in. shim. Then take away jack and tighten down bolts

14 Now gage readings in four spots on generator-end webs read zero

Crankshaft Once a Year

3 Spin balanced gage around a few times after setting the dial to zero

4 0.003 in. reading shows that points have imbedded into metal this much

5 Jack the crank over so that it is near the bottom dead-center position

9 Gage on generator end crank shows top dc out 0.003 in. from bottom dc

10 High pedestal bearing is answer. Removing 0.015 in. should correct

11 This pedestal bearing wore less than crank bearing. Loosen nuts

15 Check air gap all around on generator with feelers. Adjust if uneven

HOOKE'S FORMULA

$$\text{Stress} = \frac{43{,}500 \times d \times t}{L^2}$$

43,500 = a constant
d = distortion in mills (not a decimal)
t = crankweb thickness, inches
L = ½ engine stroke in inches

If greatest difference between any two readings is 0.003, the web thickness 3 in., and stroke 14 in. we get: (43,500×3×3÷ (7×7) = 7989 psi. The answer is strain in psi on webs.

16 Calculate stress on webs from distortion to see if limit is exceeded

17 If stress exceeds 20,000 psi, whitewash test to find cracks like this

Loading Effects on

UNDERLOADING

DIESEL ENGINE is usually built for a certain brake-horsepower or load. It does best all-around work when run full load or not less than two-thirds load

ONE-QUARTER-LOAD: Fuel used in lb per brake horsepower hour is much more than at full load. In some cases it's 30 to 50 percent more than at full load

HALF-LOAD: Fuel used in lb per brake hp per hour is only little more than full load consumption. In most cases it's about ten percent more than full load

OVERLOADING

CYLINDER WALLS AND PISTONS become gradually overheated as engine runs overloaded. That means heavy wear and tear on various engine parts, with possible early engine failure. Watch the cylinder lubricators if engine is overloaded

CYLINDER LINERS, working overloaded are liable to fractures and even cracks. Some cylinder heads crack under overload when head bolts are tightened too much. Cylinder lubrication fails. This will cause you early trouble and expensive repairs

OVERLOAD EFFECTS AND CURES

TOTAL TIME engine is overloaded can usually be controlled. Keep down to one hour and not over ten-percent overloaded. During this time, keep checking engine carefully. Pay especially close attention to the engine temperatures and to lubrication. Underloaded engine's only fault is high fuel use. Engine still runs cool

FOUNDATION SETTLING causes slight misalignment with time. It's exaggerted here, but overloaded engine heats up from this condition, causing more wear

Diesel Engines

THREE-QUARTERS-LOAD: Fuel used in lb per brake hp per hour is about same as full-load consumption. In most cases the difference is small enough to ignore

AT TEN PERCENT overload, fuel used in lb per brake hp-hr is only little more than at full-load and about same as at three-quarters load. But engine wears fast

TOTAL FUEL USED at various loads increases gradually with the load. But fuel used in lb per brake horsepower hour goes down as load on engine increases

EXCESSIVE FRICTION and wear between piston and cylinders in overloaded engine is certain if overload continues for long. This causes rings to stick and pistons finally seize. To avoid, try to balance load with standby engines during peak loads

REDUCTION IN SPEED often occurs when engine's badly overloaded. If water circulating pump is attached directly to engine's shaft, cooling is reduced just when it's needed most. Then engine's temperatures shoot up, causing plenty trouble

LUBE-OIL SYSTEMS get less efficient with time and poor maintenance. Out-of-line engines have lower lube-oil pressures. Overloading knocks out engine

UNEQUALLY LOADED CYLINDERS are found by checking exhaust temperatures from each cylinder with pyrometer. Highest temperature shows cylinder carrying most load. Adjust fuel so temperatures are within ten degrees. High exhaust temperature can be from leaky exhaust valve. Over- and underloading cause black smoke

Know your diesel cooling-

Types of closed systems

COOLING TOWER of induced-draft type with heat exchanger

SHELL AND TUBE heat exchanger located inside building

Putting temperature where you need it

MODERATE TEMPERATURES are held by hookup of this kind

LIGHT TEMPERATURE to engine, lower temperature to lube oil

Emergency supply

EMERGENCY water supply takes over when other system fails

What you must know about diesel

▶ AMOUNT OF HEAT absorbed by cooling water varies with type of engine and design of cylinders, exhaust manifold, pistons, lube-oil system and anything cooled by the system. Most engines are designed for minimum temperature difference between cooling-water inlet and outlet (from 15 to 20 F). This temperature difference is ideal.

Don't overlook seasonal temperature variations that may freeze your cooling water and congeal the lube oil. Hook up system so all cooling water flows through engine continuously. Never throttle flow to raise outlet temperature. Use only soft or treated water for uniform heat transfer and high efficiency of cooling through engine jackets. A value of about 8 pH is best to prevent deposits in water spaces. Keep system vented so water spaces are always completely full.

Selecting types of cooling systems shown depends on many factors. One engine maker may prefer the circulating pump with suction from engine, another insists on installing discharge to engine. Good rule to remember: Always install heat exchangers with outlet at top, unless they are vented.

water systems

RADIATOR uses extracted heat for heating building spaces

EVAPORATIVE COOLER has raw-water spray pump in its base

HIGH ENGINE TEMP from lube-oil cooler in separate circuit

HIGHER ENGINE TEMP from some water by-passing exchanger

cooling systems

Open system has water flowing through the engine jackets, then wasted or recirculated through a cooling tower. *Remember:* Continuous evaporation in cooling tower or spray pond increases hardness and impurities. This accumulation doesn't remain in solution, but deposits on engine jackets and retards heat transfer. This is the main trouble with open system for diesel engines.

Closed system is best, may have one or two complete circulating systems. *Example:* If a radiator is used with air as cooling medium, use single system with water flowing through engine jackets, then through radiator. Jacket-water outlet temperature and rate of flow water must be specified by engine builder. Keep in mind that raw-water temperatures, rate of flow and heat load govern the type of cooling apparatus you should use.

Cooling tower may be one of two types—atmospheric or induced-draft. Atmospheric type depends on natural draft or air movement. Induced-draft type has a fan. In some places it's best to install heat exchanger inside the building.

If so, use a shell and tube type exchanger. Then circulate raw water over the tower and through the heat exchanger.

Emergency water connection, sketch, bottom of page 210, assures flow of jacket water if regular system fails. Use this hookup where city water or other source is available at all times. Make sure this water enters system at point where it will be tempered before going to engine jackets.

Alarms warn you when temperatures or pressures go haywire. Automatic shutdown devices stop engine at set temperature or pressure. Use thermostatic controls to automatically adjust flow of jacket or raw water. You can connect them so engine always operates with a fixed outlet temperature from engine or lube-oil cooler, without reducing volume of flow through your engine.

If possible, bring cooling water to normal operating temperature before starting engine. Do this by using heat from another source or by-passing cooling tower, reactor or evaporative cooler until normal temperature is reached.

Courtesy, Diesel Engine Manufacturers Association.

RIGHT ANSWERS are important, whether diesel is large, or small one shown. Here's handy repair rack for small job

DIESEL Q and A

OPERATING DATA are jotted down by cadet midshipmen at Merchant Marine Academy diesel lab for study in classroom

Q 1—Why should the doors of a diesel-engine crankcase never, NEVER be opened after shutdown, until the engine has cooled?

A—In a totally enclosed diesel crankcase, air and oil vapor mixture is very rich. If the engine has a "hot spot," such as an overheated bearing or bushing, piston or any other part, with temperature high enough to form a seizure, there's danger of a crankcase explosion. Only reason there's no explosion before door is removed is because there's not enough air to mix with oil vapors.

But if crankcase door is opened while engine is still hot, fresh air rushing in through doors supplies needed oxygen for explosion. If operator removes door, he stands in direct line of blast and can be fatally hurt.

Q 2—Explain how to find cause if your diesel engine stops suddenly.

A—First check fuel-oil supply. Next check whether fuel system is airbound. Fuel filter might be plugged or fuel lines broken. Then check fuel system from oil-supply tank to strainer.

Check day tank for water in fuel. Do this by opening drain cock. If there is water, and drain valve is not large enough, remove valve bonnet from suction line to drain tank quickly, when engine is needed badly. But take precautions for fire hazard. Next check transfer pump. The valves, springs, gaskets, plunger, roller or guide may be at fault.

Diesel piston may have seized from poor lubrication. Check by trying to jack engine over. Remove piston and replace with new if it's badly scored. In severe cases, liner may have to be removed and piston jacked out. Be sure to change lube oil before starting engine with new piston and rings. Also clean oil pan, lines and filter.

If you find none of these conditions,

engine may have stopped from bearing seizure. If bearing is not badly wiped, scrape enough to clean it up. If it is badly wiped, replace with new unit.

Or the engine's fuel-pump driving chain may be broken (if it has one). Then engine will have to be retimed.

On some engines the fuel pump's adjustable coupling can slip if not tightened properly. Check this.

Q 3—What is best way to start a diesel engine in extremely cold weather?

A—For quick starting, heat circulating water and lube oil to about 110 F. Then circulate through engine until it is warm. Some installations have permanent connection from hot-water line to cooling water for this purpose.

Or place resistor-type electric heater and fan in diesel's air intake. Heat air so it is warm for engine starting. Some engines have air pre-heater, "flame thrower" for cold-weather starting.

If compression on engine is poor from worn piston rings and liners, lift cylinder heads and smear heavy grease around top of pistons. This helps seal compression for those all-important revolutions before engine starts firing.

If engine burns heavy oil, run on kerosene for a few revolutions before shutting down, until heavy oil is cleaned out. After engine starts on kerosene next time, switch back to heavy oil.

Q 4—What is meant by compression ignition, and what pressures are needed for diesels?

A—Ignition in a diesel engine is caused by the heat of compression. This is done by the piston drawing outside air into the cylinder on the downstroke, and compressing it on the upstroke. Oil injected into the hot compressed air ignites without a spark, open flame or a hot bulb.

Pressures of 375 to 400 psi are enough for medium- and large-size cylinders with open combustion chambers. With engines using precombustion chambers, pressures around 450 to 560 psi are needed.

Reason for higher pressures on separate combustion-chamber-type engines is that the ratio of cooling surface to combustion space volume is greater. Then cooling effect of greater area must be overcome with higher pressures. Higher temperatures are also needed in small, high-speed engines where the fuel must be burned in a shorter time.

Q 5—How is fuel oil tested for carbon residue, and what effect does carbon have on the engine?

A—Fuel oil is tested for carbon residue by the Conradson test. This test is made by burning oil with a limited amount of air. Then carbon formed is measured.

Oil forming over 3% carbon causes rapid cylinder wear. A good distillate oil should run below 0.5% carbon content. Some residual oils form as much as 14% carbon.

Q 6—Many engines have valve seat inserts and valve stem guide inserts. Explain how inserts are renewed.

A—Inserts for valve seats are made of nickel, cast iron, stellite, copper-tungsten, or cast aluminum-bronze. These inserts are pressed or shrunk into valve cages or into cylinder heads.

Valve stem guide insert is either a bushing or sleeve made of close-grained cast iron. This is a drive fit into the valve stem guide of the valve cage. Because bushing is soft cast iron, it takes most of the wear caused by valve stem's up-and-down motion.

Valve-seat inserts are removed with a narrow-pointed cape chisel. First the insert is cut, then pried apart over its seat with the chisel. But take care not to harm the seat. Valve-stem inserts are removed by turning the valve cage bottom up. Then remove valve and stem, drive insert or bushing out with brass stock.

Q 7—What effect will a cocked connecting rod have on the engine, and when is a connecting rod said to be in alignment?

A—A cocked rod may make the piston operate in such a way that the piston ring faces are not parallel to the cylinder wall. Under such conditions, upper and lower portions of the ring face are worn away on opposite sides, but not the middle portion. This makes it impossible for one half of the ring to scrape oil on its downward stroke. Also, misalignment places a high load not only on the piston and cylinder wall, but also on the connecting rod bearing.

A connecting rod is in alignment when its piston's sliding surface is exactly at right angles or square with the connecting rod. This is checked by placing this rod vertically on a surface plate, so rod rests on machined bearing cap's surface. Then insert the wrist pin in the upper bearing and measure distance from surface plate to both ends of wrist pin with a surface gage, inside calipers or inside micrometer. If distance is same on both sides, rod is OK.

Q 8—What is a "jerk-pump" fuel-oil system for diesel engines?

A—Pump-injection system, sometimes called "jerk-pump" system, has an individual pump or cylinder pump serving each cylinder, and connected directly to the fuel nozzle.

The pump meters the oil charge and controls injection timing. Nozzle contains a valve, which opens when injection pressure is reached and closes when pump delivery ceases. Most of these systems contain a cam-operated pump, connected by a tube to the fuel nozzle.

Q 9—What would you look for if the backpressure on your engine was excessive?

A—First inspect the exhaust piping, mufflers and waste-heat boilers for flow restrictions, such as carbon deposits. Check for pulsations that might cause backflow of exhaust gas. Pulsations may be from faulty exhaust system.

Q 10—Give some general hints to follow before overhauling an engine.

A—Before overhauling an engine, it's good practice to observe it in operation. Watch cooling water and lube oil pressures, valve mechanism and governor operation.

If practical, indicate the engine and study cards for compression and combustion pressures, also for unusual signs. Listen for abnormal noises and note any condition that's not right.

When tearing down the engine, do it in an orderly manner by marking all mating parts not already marked. Place the large parts on boards to prevent marring them. Place smaller parts in boxes to prevent loss. First clean all parts with Stoddard solvent.

Try to work out repair schedules so only one cylinder is completely overhauled at a time. This has advantage of localizing to one cylinder running faults caused by overhaul. If entire engine is torn down, it's sometimes hard to find minor adjustment faults because everything worked on must be checked on all cylinders.

Q 11—Explain where heat generated in a diesel and in a gas engine goes.

A—*Diesel:* 34% for useful work. This is just another way of saying that 34% is the thermal efficiency. And 30% is jacket loss. It represents heat absorbed from pistons and cylinder walls by water in surrounding jackets. 27% is exhaust and radiation loss. This is amount of heat carried away in hot exhaust gas and heat radiated from engine and exhaust piping. 9% goes to friction loss and includes friction of crankshaft and connecting-rod, pistons, etc. This friction goes to heat and adds to other losses.

Gas engine: 25% is useful work; 30% is jacket loss; 38% exhaust and radiation loss; 7% friction loss.

Heat Burns

First Aid

In case of heat burn, call a physician at once. Burns are always accompanied by shock, and severe shock may cause death

1. Lay the victim down with his head lower than his feet.
2. Carefully loosen tight clothing.
3. Turn his head to one side, in case of vomiting.
4. Cover the burn with a sterile cloth. If a temporary dressing is necessary, use sterile petroleum jelly.
5. After dressing facial burns, apply a clean cloth mask with holes for eyes and nose.
6. If the victim is conscious, give him frequent small drinks of water.

—National Safety Council

5 timely Q&A on diesel engines to help you do a better operating job

1. What is meant by cetane number of fuel oil?

A. Cetane scale is applied to fuel oils, gasoline, diesel fuel, etc, to show oil's ignition quality. This quality means oil's ability to ignite in diesel cylinder —or under similar conditions. Fuel with good ignition quality will auto-ignite at low temperatures and is, therefore, preferred. This quality affects engine starting, smoking and knocking.

Hydrocarbon family ranges from simple combinations, such as methane gas, through various solids, liquids and gases. Some have complicated chemical structures. All have various ignition characteristics. Formula for cetane is $C_{16}H_{34}$.

Cetane is an excellent diesel fuel in its pure form, but isn't easy to separate completely from other hydrocarbons. Cetane content for various crudes differs greatly. $C_{10}H_7CH_3$ is another hydrocarbon, but it has poor ignition qualities. Cetane number is based on proportions of these two hydrocarbons. Cetane number of 100 means ignition quality equal to pure cetane, while zero means one equal to that of alpha-methyl-naphthalene ($C_{10}H_7CH_3$). Ignition number of 60 means mixture equivalent to one composed of 60% cetane and 40% $C_{10}H_7CH_3$.

2. How is cetane number determined?

A. Ignition-delay method on single cylinder test engine is used to determine cetane number. Fuel to be tested is used in engine, and compression ratio is raised until ignition-delay reaches a standard length. Poor-quality fuels' ignition quality needs a high compression to shorten the delay period, while fuels of good quality take lower compression for the standard delay.

Once compression ratio for test fuel is found, various mixtures of cetane and alpha-methyl-napthalene are burned. From this, compression ratio needed for each fuel, to bring delay period to standard figure, is found.

Cetane number of test fuel equals percentage of cetane in mixture that requires same compression ratio as test fuel, to produce standard delay period.

3. What is meant by right- and left-hand engines. Buyers ordering two engines for same plant often specify one each. Is this good practice?

A. Engine is classified as right or left when flywheel is on right or left sides, viewing engine from operating side. Reason one engine of each is specified is because owners believe such an arrangement is easier to attend if controls are on same aisle between engines.

Diesel Engine Manufacturers Association says there may have been reason for such an arrangement years ago before the days of remote governor control, but not with today's common use of such control and the little attention modern engines need.

Such specifications require one engine to be nonstandard. Disadvantage to owner is loss of interchangeability, which is so important to avoid carrying large stocks of spare parts. Then, again, any advantage is lost if a third diesel is installed.

4. If engine fails to start, what would you look for?

A. Engine rarely fails to start, but such difficulty can be divided into two groups: (1) Starting mechanism fails to turn engine through an operating cycle. (2) Engine turns through operating cycle but fails to fire. If starting mechanism doesn't work, check starting air-pressure or look for rundown or weak battery—depending on starting system.

If air-starting pressure is high enough, air-starting valves may not open fully or at right time. Or by sticking, leaky exhaust or inlet valves may prevent engine from quickly raising compression high enough to fire. Leakage from stuck rings or badly grooved cylinders may have same effect. In cold weather, heat of compression may be too low unless cooling water and (in extreme cases) even lube oil are heated.

At times, bearings are too tight (after overhaul) or there is no lubrication on cylinder walls, etc. All this puts extra load on starting mechanism.

5. Should operators spend time listening to diesel engines to become familiar with normal noises so they can tell when unusual sounds develop?

A. Yes. It's good practice to spend a few minutes each day listening to natural beat of engine. Listen at each cylinder to detect knock in crank, wrist pin, etc. Also, listen to gear train or drive chain. Backlash in gears and sprockets can be detected, also loose chains.

Listen for vibration at supercharger. Vibration is caused by dirt on impeller, too much bearing clearance or increased end thrust. To avoid supercharger trouble, run a spin test every six months. Just remove supercharger silencer and spin air rotor by hand. If there's noticeable drag, dismantle supercharger. Usually all that it needs is just a good cleaning.

Change in air manifold pressure at given load shows supercharger needs attention. Trouble may be air leaks in intake connections. Watch air-manifold pressure, or engine will starve for air— causing many headaches for you.

11
Electricity

Moisture Damages

SPLASH from hose while washing floors grounds out motors in plants where floors are washed around electric motors

DRAINS running to floor from stock chests in paper mills. Motors are stuck under tanks in overcrowded room

CURES

CHECK COILS for dampness with megohmmeter. Caution! Even 100v can puncture wet insulation—so crank slowly. Have only qualified electrician make test

WASH INDUCTION MOTOR that's been submerged with warm water to remove oily film. Use 25-psi pressure (not more) and not over 194 F. Oil shaft, bearings

DRY OUT WET MOTOR in temporary oven. If no oven, make one of bricks, asbestos-board and heating elements. Blow heated air through motor like this

PREVENTIONS

FIRST STEP to stop needless damage to electrical equipment is to train personnel. Tack this OPERATOR'S NOTEBOOK on wall and have all persons working around equipment read it

STOP OVERFLOWING TANKS by placing overflow pipe at highest safe level. Another way is float shut-off to cut off incoming water. If not practical, move electric motors

Electric Motors

GLANDS LEAKING from motor-driven pump shoot water into motors in numerous plants. Then motor burns out fast

OVERHEAD LEAKS through ceiling from floor above. Having floors above that are continually wet is common cause

DAMP AIR during idle periods coats expensive motors with moisture. This cuts down the insulation resistance

PASS LOW-VOLTAGE (about half-load value) through windings if no oven. Cover with canvass partly for ventilation. Don't let temperature go over 175 F

TO DRY dc generator, short the armature, reverse series field, weaken shunt field. Use ammeter like this, then check with megohmmeter after a few hours

INFRARED LAMPS speed drying-out process. Position lamps carefully and check winding temperature often; otherwise the insulation may be damaged badly

DRIP-PROOF   SPLASH-PROOF   TOTALLY-ENCLOSED   FAN-COOLED

REPLACE MOTORS that are in trouble spots (if not practical to move them) with one of these four types. Money saved from no outage and repair bills will easily pay for change

NEVER, NEVER start motor after long shut-down, without having electrician check for moisture. That holds true after wet cleaning, long shut-down or suspected moisture

Need Another Outlet?

1 You can tap the new outlet from nearby lighting circuit—if new load is small (extension light, clock etc). First step is to kill circuit and drop fixture

2 Box extension saves chopping concrete. Fasten it with 8-32 machine screws to ears of ceiling box. Set new wall box and take measurements shown

3 To bend **offset** lay conduit on floor and slip bender in place. Can use a 1-in. pipe tee instead of regular conduit hickey. Bend offset before making L bend

▶ FROM TIME TO TIME every small-plant operating engineer juggles the question of installing an occasional electrical outlet in the plant. Many engineers farm out all electrical wiring. But where local laws do not require all electrical wiring be done by a licensed electrician, the operator is often the boy called on to do it.

Keep the National Electrical Code as your bible on all wiring jobs. You can get a copy of the Code by dropping a card to the National Board of Fire Underwriters, 85 John Street in New York City. Many localities have in addition to the NEC, local laws which must be lived up to.

Wiring Methods. Your best bet in running in an extra outlet is to follow the present type of wiring used in your plant provided it is in accord with the code. In other words, if branch circuits are wired with No. 14 wire in ½-in. conduit, follow through using the same materials. Many plants use ¾-in. conduit and No. 12 wire for branch circuits. Also your plant may be wired in electrical metallic tubing which cuts out threading.

Separate Circuits. Tapping your new outlet into the existing lighting circuits should be done only where the new

4 Chalk line guides you in making L bend. Don't make bend in one step; move the hickey forward and take small bites till bend checks with the chalk line

5 Put conduit in place, using locknut and bushing at each box. Use **snake** to pull in wires. Splice black wire to **black** and white to **white**; solder and tape

outlet is to be used for small loads such as a portable extension cord. If load is to be heavy (grinding wheel, heavy soldering outfit, etc), use a separate circuit. You can use the same wiring method for new circuit by looping the

wires through each ceiling outlet to the panelboard.

Where ceiling lighting fixtures are controlled from a distant panelboard, you will have to run your wiring back to the panel to pick up a "hot" leg.

Put Your
Ammeter To Work

► CONSIDER THREE THINGS before using an ammeter: (1) Check whether the current you plan to measure is direct or alternating. (2) Determine about how much current is flowing. (3)

Select the proper ammeter and connect it in the circuit as shown in the sketches below. Always connect an ammeter in series with the load. Select your meter so the reading falls about

mid-scale. Close short-circuiting switch before starting a motor. This prevents damage to the instrument by heavy starting current. Keep shorting switch closed except when reading meter.

1

- Line
- Safety switch
- Switch handle
- Dc motor starter
- Shorting switch
- Shunt
- Dc ammeter

2

- Line
- Ac motor
- Shorting switch
- Ac ammeter

3

- Ac line
- Load
- One turn primary
- Current transformer
- Secondary winding
- Shorting switch
- Ac ammeter

1 The dc ammeter measures voltage drop across a shunt. Select the shunt whose current rating equals or exceeds the maximum current to be measured. Each meter should have a set of calibrated leads. Millivolt rating of the shunt and meter must be equal for the needle to read full scale

2 Some ac ammeters measure to 200 amps directly. Don't use shunts with ac meters. Many ac instruments can be used for dc measurement. When so used take the average of normal and reversed meter readings. Use heavy meter leads. A split-core ammeter is handy for large current measurement

3 Current transformers are used with ammeters on high voltage systems. They are also used when measuring large currents. Normal secondary current is 5 amperes with full load in the primary. Never open-circuit the secondary of a current transformer while current is flowing in the primary

1 Typical departmental meter board in an industrial plant. This installation facilitates keeping accurate records of the electrical consumption in key areas

2 Polyphase watt-hour meter with case removed to show the motive element

First Aid for Plant W-Hour Meters

Here is handy information for the man metering his plant's electrical use: fundamentals of watt-hour meters and practical trouble shooting for five common causes of inaccuracy

▶ BEFORE DIGGING into this matter of watt-hour-meter maintenance, let us get a few fundamentals under our belts. First of all, the polyphase watt-hour meter is a combination of the single-phase meter's elements, driving a shaft at speed proportional to total power. Meter consists of a multi-element motor, means for balancing the torques of all elements, a magnetic braking system, a register, and the necessary compensating rigs. All these components are put together in a meter frame.

Wattmeter measures basic unit of electrical power, the *watt*. Electrical power is the rate at which electrical energy is used. And this unit of measurement, the watt, implies a time factor. Hence, the term watt includes the implied meaning "per hour."

But when we think of watt-hour meters we are dealing with electrical *energy*. We can define that as the total use of electricity over a period of time. It is the product of rate of use (power or watts), multiplied by the number of

time units, usually measured in hours.

Meter Operation. The instantaneous speed of the watt-hour meter's motor is proportional to the power passing through it. On top of that, total revolutions in a given time are proportional to total energy, or watt-hours, drawn during that time. Besides the motor, this meter has a magnetic braking system and a register.

Calibration. Now for a thumbnail sketch of calibration and adjustment. You have seen men from the local utility run checks on watt-hour meters in your plant. They use a standard indicating wattmeter to measure the average power over a stated period. And, at the same time, they count revolutions of the watt-hour-meter disk with the aid of a stop watch.

Then they check the average meter watts by taking the watt-hour-meter constant K, and multiplying it by the number of disk revolutions. This quantity is multiplied by 3600, and the product is divided by the time in seconds, ticked off on the stop watch. Value of watts they get through these mental gymnastics is compared with the watts read on the standard wattmeter. That constant K, mentioned above, is the number of watt-hours represented by a single revolution of the aluminum disk.

Meter is adjusted to read properly at full load and unity power factor simply by varying the position of retarding magnets. It is adjusted at light load by the light-load adjustment lever

that you can readily spot once the glass disk is removed.

Keep in mind that the above facts are set down in a very general way. The actual job requires more specific information and a better knowledge of the meters. But plant men should give careful thought to the trouble-shooting side of the picture, since watt-hour meters are widely used to measure electrical energy generated or consumed in various parts of a plant.

Trouble Shooting. The modern watt-hour meter is an inherently accurate and a stable device. When properly calibrated it retains its accuracy for a long time unless subject to influences outside normal operating conditions. So, in brief, the purpose of your maintenance of a trouble-shooting program is to (1) identify meters that have become inaccurate (2) find out what has caused the error (3) restore the meters to normal accuracy.

A watt-hour meter may be found inaccurate because of improper calibration or adjustment, incorrect installation, or the adverse effects of foreign influence.

Actually, when you boil it down there are few meters that are found inaccurate as a result of the first two causes. Possibility of mistakes from incorrect installation should not be ruled out when a maintenance check is being made, but outside influence is a much more common cause.

For instance, when the meter runs slow, the things to look for are: pos-

3 Testing boards like the one shown here are sometimes used to carry out meter trouble shooting. This method centralizes all checking and repairs in one area

Potential and current transformers bring line voltage and current within range of meter

Permanent magnets act on disk to establish shaft speed for a given load. Disk speed in turn determines watthour constant, K

4 Meter may be connected directly in a low-voltage, low-current circuit

sible dirt; a magnetic short; bearings that are dry, worn or out of line; shaft bent; gears too tight; shorted current coil; open potential coil; meter not level; incorrect adjustment.

On the other hand, if the meter runs fast, think in terms of a possible weakened braking magnet or incorrect adjustment. If meter runs backward, you have an incorrect connection.

A creeping meter may result from vibration, stray fields, excessive voltage, shorted current coils, incorrect adjustment, voltage leakage or ground in wiring.

Dust or Dirt. Dust, dirt or other foreign particles, like tiny pieces of metal, are the most common cause of incorrect operation after watt-hour meter has been in service. They may tend to close the air gap between the disk and the electromagnetic polepieces, and thus cause friction. This is especially likely if the dirt includes particles that cling magnetically to the magnet, and align themselves with the flux so they rub on the disk. Dirt in any of the gearing will increase gear loading, and may cause the meter to slow down. The answer is simple. Just give it a thorough cleaning.

Bearing Troubles. Worn or poorly aligned jewel-and-pivot or ball-type bearings increase value of the friction load, and thus make the meter run too slowly. Wear from this source is greater with light loads than with heavy, and becomes progressively greater as long as the worn bearings are left in service. Bearing friction may also be increased to an objectionable degree if the meter is out of level. Lack of lubrication of the pivot bearing is another source of increased bearing friction.

Misalignment or improper adjustment

of bearings is generally an indication that the meter has been subjected to severe shock, either mechanical or electrical. Best solution is simply to replace the bearings. Of course, there may be cases where it is best to oil the pivot bearings or level a tilted meter. On the other hand, it may be best to have a repairman completely overhaul the meter.

Bent Shaft. When you come across a bent shaft or a warped disk, you invariably meet up with measurable errors in watt-hour-meter registration. A bent shaft or a warped disk generally indicates that the meter has been roughly handled or subjected to excessive surges. Your best bet is to return the meter to the shop for a thorough and complete overhaul.

Defective Coils. A shorted turn in a current coil of a watt-hour meter lowers the torque near or at full load. At light load, however, meter may run fast or slow depending on which of the two coils is shorted. A short circuit in the potential coils changes the meter's torque and also its lag adjustment. Shorted turns may indicate that surges have been applied to the meter. Once again, your best bet is to return the meter to a competent shop for an overhaul job.

Weak Magnet. A braking magnet, whose strength has been reduced by exposure to excessive stray fields or surges, allows the meter to run faster than it should. In this case, the answer is simply to replace the weakened magnet with a new one.

Creeping can mean a headache. By this term we mean a continuous, slow rotation of the disk when there is no load on the meter. The disk may creep either forward or backward. Although

the creeping has to be rather fast to have any effect on meter registration, it is standard policy among meter-test men to permit no creeping of meters in service.

Watt-hour meters are intended for use where they are not subjected to vibration. And any unusual amount of vibration may cause creeping.

But if stray fields are strong enough to produce eddy currents in the meter disk, these currents may react with the potential coil flux to develop a torque that causes creeping. Once again, the only answer is to relocate the meter, or shield it magnetically so the stray fields will not affect it.

Or if the line voltage is considerably above normal, the compensating torque due to a light-load adjustment is proportionately higher, with the result that the meter may creep. The solution is simply to recalibrate the meter to operate at the higher voltage.

In short-circuited turns in the current coil, flux of the potential coils induces voltage. This causes eddy currents in the disk, and results in creeping. You will find it best to return the meter to the shop for complete overhaul and replacement of the defective coil.

Creeping may also be traced to improper connection. For example, when the meter is installed with potential coil connected on load side of the meter, creeping results.

Sometimes creeping is caused by leakage or grounds in the metered circuit. This false creep disappears when meter is tested with load wiring disconnected from meter. You can make a quick check on this by disconnecting meter and running a megohmmeter test on the load wiring.

To Clean Motors Safely . . .

. . . use common sense, compressed air at proper pressure, avoid all highly flammable solvents while inside plant*

Don't let air pressure exceed 40 psi

Goggles

Respirator

1 Compressed air does good cleaning job on assembled motors. But first use a vacuum rig to pick up light surface dirt before it's scattered by air pressure

▶ SAY WHAT YOU MAY, one of the most important jobs in maintaining electric motors is keeping them clean. And generally speaking there are only two ways to clean motors. One is with compressed air as sketched up in Fig. 1 and the other is by use of solvents, Fig. 2.

Hazards. Before we say much more about either method of cleaning, let's consider the safety angle. First off, no cleaning job should be attempted unless the motor is shut down and the safety switch open. On top of that, it's wise to padlock this switch in the open position.

When using compressed air for cleaning, protect yourself with safety goggles *and* a respirator. With both flamable and non flamable solvents, there's the everpresent danger of breathing the vapors. Then we have the fire and explosion hazard.

Compressed Air. Wherever possible, a brush or vacuum system should be used instead of compressed air for cleaning light dust, chips and oily dirt. On the other hand, compressed air may be the only practical method unless motor is taken apart.

Where you use compressed air, make sure there's a separator in the line to remove all impurities and moisture. Pressure should be regulated, and *not* over 40 psi. Where you go above that pressure, there's danger of damaging motor's insulation.

In the case of grain elevators, candy factories and cotton mills, dust can become an explosive mixture if blown into suspension by an air hose. Hence, never use an air hose for cleaning motors in these locations.

Now here are a few tips when using an air hose: Before turning on pressure, check all air line connections for tightness. Then while opening or closing valve hold the air hose nozzle. Otherwise the hose may whip about and hurt somebody.

Solvents. Unfortunately, its a common but unsafe practice to use solvents with low *flashpoints*. Flashpoint is lowest temperature at which solvent will vaporize into an inflamable mixture. And low means anything below 100 F. This includes gasoline, −45 F, naphtha, 20 F, and benzol 12 F. Such solvents create a serious fire hazard and may produce an explosive mixture with air at room temperature. *They should never be used for cleaning purposes inside a building.*

In a small enclosed space such as an engine room, only a small amount of vapor from these solvents is necessary to set up an explosive mixture. In the case of gasoline vapors only 1.3 to 6 percent by volume in air is needed. A spark or exposed flame coming in contact with this mixture could cause a violent explosion. Then again, small amount of naphtha or benzol left on a motor's windings could form an explosive mixture that would explode when motor is started.

On top of that, some low flashpoint solvents are highly toxic. You're running the chance of being overcome by fumes when used in an enclosed place.

For instance, benzol in concentrations greater than 100 parts per million in air is unsafe for an 8-hr exposure.

High flashpoint solvents—those having a flashpoint of 100 F or above—are often used for cleaning motors. Kerosene and the Stoddard solvents are in this family. Although the explosion hazards of these solvents are not as great as the low flashpoint babies, they do present a fire hazard when not used safely. Keep a carbon-dioxide fire extinguisher handy when using these solvents.

If high flashpoint solvents are used where the temperature may reach 100 F, they will vaporize rapidly and may form explosive vapors. And at high temperatures these vapors may have the same toxic effect as vapors of low flash point solvents.

Other Solvents. Vapors given off by petroleum solvents (both low and high flash points) are heavier than air. They travel rapidly and in confined spaces tend to settle and remain in low areas. They will enter pits and flow through drains and sewer lines. A spark may easily ignite them and cause a serious explosion.

Carbon tetrachloride and trichloroethylene won't burn. They do vaporize rapidly, and their vapors are poisonous. Carbon tetrachloride is unsafe for continuous exposure if it is present in quantities greater than 50 parts per

KEEP WORK AREA WELL VENTILATED . . .

Keep carbon dioxide fire extinguisher handy

Use approved one-gallon safety can

Elbow grease helps too!

2 Solvents do trick when cleaning frames. Good mixture is 25% methylene chloride, 70% Stoddard solution and 5% perchloroethylene (use it, don't spell it!)

SAFE STORAGE . . .

Fireproof cabinet

Dont store more than 10 cans in one cabinet

1-gal safety can

3 Storage of all solvents should **agree** with Fire Underwriters and local **laws**

HERE'S HOW TO GET RID OF USED FLAMMABLE SOLVENTS . . .

DON'T....

Flammable solvent

Drain

4 Disposal in sewer drains is taboo. Solvent may give off flammable vapors that can be ignited at some distant point

DO...

Firefighting equipment

Wind in this direction

5 Disposal of used solvents is best handled outside **plant.** Dump flammable solvents in small quantities, then burn

---

million in air. So where carbon tet has to be used in confined areas, provide either forced ventilation or have the job done under an exhaust hood with a downdraft flow of air. Make sure the exhaust fan is big enough to hold a negative pressure under the hood at all times. Another out is using an air-supplied gas mask.

If carbon tet is mixed with low flashpoint solvents, will the explosion hazard be eliminated completely? Nope, the fire and explosion hazard may still be present if one of the fluids evaporates

faster than the other fluid. Then when the mixture is being used in a confined space the hazard of toxic vapors still remains. A reasonably safe cleaning mixture contains 25% methylene chloride, 70% Stoddard solution and 5% perchloroethylene.

Storage. When it comes to storing any of these solvents, your first step is checking with the Board of Fire Underwriters and your local fire-ordinance body. Then follow whichever regulations are most strict.

It's poor practice to store any type

cleaning solvents in the work area. If you must keep them in the building, they should be stored in one-gallon safety cans in a fireproof cabinet, Fig. 3. If you find it necessary to store more than ten such cans, use a spearate fireproof storage vault away from the building.

As practical men, we know there are many times when such solvents must **be** stored in the plant. But in such **cases** make provision for adequate **ventilation.** In confined areas use **down-draft** forced ventilation.

1 Multiply the hp rating by service factor to find maximum load motor can handle continuously. Use 50-C temperature rise rather than 40 C when using the service factor rating

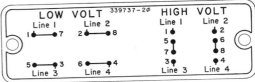

2 Dc-motor nameplates tell whether winding is shunt, series or compound. Approximate resistance of rheostat used in the field circuit and max field current may also be included

5 Some manufacturers use a separate tag like above to show connections for a two-voltage job. Actual voltages along with other standard info are listed on the regular nameplate

6 Case of an explosion-proof motor is designed to withstand internal explosion and at the same time prevent ignition by explosion within casing of gas or vapor surrounding motor

7 Type letters pin the motor down as far as the manufacturer is concerned, but no type markings used by two companies are alike. Design letters now give dope on motor performance

What Your Motor Nameplate

Your electric motor nameplate is chock full of info. Here's a quickie telling you how to get the most from it along with dope on recent changes

▶ MOST MOTORS you will come across have nameplate markings that follow a standard established by the National Electrical Manufacturers Association. To agree with this standard, most motor nameplates are stamped with the manufacturer's type and frame number, motor rating in hp, time rating, temperature rise, rpm at full load, voltage and full-load amperes.

In addition to this list you'll find the nameplates on dc machines also tell whether the winding is shunt, series or compound. On top of that ac motors also include frequency and number of phases the motor is designed for in addition to design and code letters. The list is pretty much rounded out by mentioning power-factor and dc excitation information on synchronous-motor nameplates.

Types. For many years motor manufacturers have included type markings on the nameplate. It is a mystical

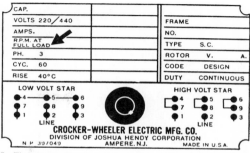

3 Hp rating is the full-load output of motor keeping within rated temperature rise shown. When handling this load, motor draws current marked after amp listing on the nameplate

4 The **slip** is difference between speed listed on nameplate and synchronous speed of rotating magnetic field. Slip is usually expressed as percent of the motor's synchronous speed

8 Power-factor marking on synchronous motor nameplate may be either unity or leading depending on specific design

Means

combination of numbers or letters that help each motor manufacturer keep his records in order but mean little to the repairman or operating engineer. Rather than change this established system, the NEMA added a series of *design* letters for polyphase squirrel-cage motors to give the motor user additional specific information about the motor's performance characteristics.

The design letters used are *A, B, C, D* and *F*. We won't go into all the specifications covered by each design letter in this short article. The following data will probably be sufficient for most operating men. For the lads who want it, all the dope is available in tabular form from the NEMA at 155 East 44th St, New York City. The point we want to get across is that motors with design letter *A* are normal-torque, normal-starting-current motors while design *B* refers to normal-torque, low-starting-current jobs.

When you meet a *C* design, it's a high-torque, low-starting-current motor while design *D* is a high-slip machine. The picture is rounded off with the *F* design which is a low starting-torque, low starting-current motor. Design *B* is by far the most popular of all types.

Code Letter. The code letter on an ac-motor nameplate (ac motors rated 1/20 hp and larger except polyphase wound rotor) is simply an indication of the kilovolt amperes per hp the motor will draw if the rotor is locked to keep it from rotating. There are a string of code letters running from *A* to *V* with a few left out in between. This series boils down to a graduated scale with code *A* motors pulling between zero and 3.15 kva per hp and code *V* motors at the other end drawing 22.4 kva per hp and up. This little nugget of knowledge comes in handy in determining the setting of motor branch-circuit protective devices.

Horsepower and Service Factor. The hp marking on any motor is its mechanical output at full load. Of course the input will be greater because of losses in the motor. Incidentally, it's smart to select motors to operate at or near their hp rating. Your motor will not only be running at a higher efficiency but at better power factor to boot.

Service factor ties in with the hp marking since it tells you the permis-sible overload the motor can carry continuously. For example if a motor has an allowable service factor of 1.15 (most 40 C rated motors have), that motor can carry 1.15 times its rated hp continuously when operated at rated voltage and frequency. Keep in mind that a motor marked for a 40-C rise above room temperature at full load will not exceed a 50-C rise when carrying its service-factor load.

Another marking to keep in mind is the *time rating*. This gem of information tells the length of time the motor will carry it's full load. Most motors you'll run across will be stamped "Continuous," meaning they will carry rated load continuously without overheating. Still others will be marked 15 min or ½ hr which is the maximum time the motor will supply full load output without overheating. After that period, the motor must be allowed to cool to room temperature.

By the way, the figure 40 C stamped on most nameplates means the manufacturer guarantees temperature rise in the motor will not exceed 40 C at rated load. This holds true provided there is nothing to block the free flow of air around the motor and the ambient (surrounding air) temperature is not more than 40 C. The word "Open" on your nameplate refers to the motor housing. It means the motor frame is open, permitting free passage of cooling air through the motor. Enclosed motors have a higher temperature rating.

That figure following the ampere marking is the current the motor draws when the baby is pulling it's full load, hence delivering rated hp output. In a two-voltage job there will be two values of current listed, Fig. 1. The larger of these will be the current drawn at the lower voltage; this is usually twice the current pulled at the higher voltage.

1 Remove heavy surface dust with vacuum hose. Finish off the job with compressed air, keeping the pressure to 50 psi

2 Make sure the metal protecting hood is either large enough or properly ventilated to prevent the motor overheating

Meaty Tips on Maintaining

Bearings and insulation are two most common reasons for motor failure that can be licked by smart maintenance. Here are facts that should be woven into any good maintenance plan*

▶ IT'S REALLY A TOUGH JOB to sit down and set up rules about how often electrical equipment should be inspected and maintained. Every operating man you meet has his own ideas and probably some pretty good reasons to back up those ideas. We're going out on a limb with the following suggestions, based on conditions you are liable to run into in the average plant. Our guess is you can take most of the following time schedules and adopt them in your own plant. As the years roll by, you may find the schedule should be changed. If so, go right ahead and custom-tailor it.

Weekly. Make a weekly check on bearing-oil level, and see that the oil rings turn freely. At the same time, use your hand to check temperature of motor bearings and motor frame. Sniff the warm air coming from the motor; you can't miss the smell of overheated insulation.

Monthly. Make a thorough check on brush holders, pigtails and brushes. While you're at it, blow out the motor with compressed air. After the blowing-out job, check that the brushes ride free in their holders. See that the brush springs are doing the job they were designed for. Last but not least, check the insulation resistance with a megohmmeter. And keep a record of that reading, so you can compare with previous and future readings.

Yearly. Just about once a year check the motor air gap with a feeler gage. After that, go ahead and check line voltage with a voltmeter, and motor load with an ammeter. During this checkup, replace grease in ball and roller bearings.

Every Two Years. Figure on a general overhaul about every two years. Take the motor apart and do a class A-1

*Based on information supplied by the Westinghouse Electric Corp

cleaning and inspection job. See that all windings are tight. Replace any loose wedges and bands before dipping in varnish and baking. Inspect commutator and risers.

In addition to the many things that common sense will tell you to take care of during the general overhaul, wash out the bearing housing while you have the end bell off. A good way is with hot kerosene followed by compressed air.

Bearings. After sounding off about setting up the time schedule, let's spend a little time talking about bearings. You know that bearings rank right up near the top when it comes to causes of shutdowns, delay and expense. This isn't surprising when we remember that they are the victims of poor foundations, misalignment, vibration, dirt mixed in with lubrication and frequently the wrong lubricant. All this is in addition to their real job of supporting a rapidly revolving part.

Lubrication is the first requirement for successful bearing operation. And this means more than just squirting in oil. That all-essential oil film between bearing and shaft prevents metal-to-metal contact. Properly designed bearings have adequate area for the load and proper oil grooving. In a sleeve bearing, oil sticks to the shaft and is dragged around by rotation of the shaft. This action forms an oil film between shaft and bearing. It is this film that carries the load and prevents metal-to-metal contact.

There are two things to keep in mind for proper maintenance of sleeve bearings. First, use the right oil to be sure you'll have an oil film once the shaft has started rotating. The second is to reduce damage of metal-to-metal contact by using the right babbitt. Remember, you'll have metal-to-metal contact when the oil film is lost either by accident or during the starting period.

Oiling Sleeve Bearings. Keep your oil well filled to the proper level, and see that the oil rings turn freely. Add new oil only when the motor is at rest, to prevent flooding. When you run into a sleeve bearing that needs too frequent refilling, it is probably leaking oil into the stator windings. A smart thing to do is replace it with a sealed sleeve bearing.

Now for a few tips about that felt seal at the bearing end:

3 That fellow with the oil can plays a big part in bearing maintenance. Overlubrication is just as bad as not enough

4 Thanks to intelligent maintenance, this bearing wore less than 0.0015 in. after eleven years of continuous service

Your Motors

The real reason for using this felt seal is to keep out air and dirt. It does little in preventing oil leakage, once it becomes oil soaked. Hence, whenever you replace a sleeve bearing, put in a new felt seal. Your best bet is to order these seals from the motor manufacturer. If you have to make them in an emergency, use a high-grade felt not less than a ¼-in. thick before compression. Make the seal's inside diameter equal to the shaft diameter or slightly less.

Under normal operating conditions, the bearing temperature shouldn't rise more than 40 C above that of the surrounding air. A bearing will feel comfortably warm to the hand at this temperature. Look into the matter if your bearing rise is more than 40 C. Keep in mind that oil leakage is generally much worse at high temperatures.

Anti-Friction Bearings. When the shaft is moving in a sleeve bearing, shaft and bearing are separated by an oil film. But in a ball bearing, the balls themselves do the separating. This is true both when the shaft is turning and standing still.

Most ball bearings used in horizontal motors are grease lubricated. Follow the advice of the motor manufacturer when it comes to selecting a suitable grease. Soda-base greases are usually preferred because of their high melting point. When a ball bearing is in trouble, you can usually spot it, since it will overheat or give off an unusual noise. Remember that broken or nicked balls will spell a quick finish to any bearing. You can usually pick out these cases by the *clicks* as the shaft turns.

Check for overgreasing when the bearing temperature runs high. The first thing that will show up after overgreasing is excess heating caused by churning of the grease. A general rule to follow is that the bearing should not be over half full. Leave the drain plug out for a few minutes after regreasing. This will let the excess grease drain off as the motor spins.

Insulation. Poor insulation runs hand in hand with bearings as a main cause of motor failure. In addition to the regular insulation around the individual wires, you have insulation between commutator bars and individual coils.

5 To seat new brush, fit a strip of No. 00 sandpaper around commutator. Hug commutator, pull in direction of rotation

The first rule for maintaining good insulation is: Keep it clean; keep it dry.

Dust and dirt not only contribute to insulation breakdown, but they tend to increase motor temperature by preventing proper ventilation. The best way to remove dust and dirt is with compressed air at about 50 psi. Don't blow compressed air against the insulation until you are sure the compressed air is free of any moisture that may have gathered in the air line. And remember that too great a pressure may loosen the binding tape or injure the insulation by sand-blasting with the abrasive dirt that is nearly always present.

Where you find that compressed air doesn't remove all the dirt, try using a solvent such as cleaner's naphtha or carbon tetrachloride, or a mixture of the two. The cleaner's naphtha is not the safest thing in the world, so take precautions to prevent a fire or explosion. A mixture of 50% carbon tetrachloride and 50% cleaner's naphtha won't burn. But their vapors, mixed with the right proportion of air, are explosive. In extreme cases it may be necessary to use straight carbon tetrachloride, but remember that the vapors are poisonous. Apply your varnish treatment after the motor has been thoroughly cleaned. Two dips and bakes are enough for most normal jobs.

CAPACITOR-START MOTORS just won't start if capacitor is bad. Note how condenser connects with auxiliary winding

SIMPLE TESTS to help you spot a faulty condenser shown above. If ohmmeter is handy, use it to check for shorts and grounds

Spotting Faulty Capacitors

You can save a repair bill next time one of your capacitor motors gives trouble

▶ SMALL MOTORS seldom get the care and attention that large ones do. As a result many small motors are unnecessarily sent to the shop for repair or replacement, when the plant operator with a little "know-how" could put the motor back in service.

Most small motors are single-phase, alternating current, using either capacitor or repulsion starting. Unless the service requires high starting torque, capacitor-start motors are satisfactory.

The capacitor-start motor may be readily recognized by the capacitor unit mounted on top or side of the motor frame, Fig. 1. Most troubles with capacitor-start motors involve the capacitor, therefore, it is mounted so that testing and renewal is simple.

If your capacitor motor blows fuses, has poor starting torque or hums without turning over, the trouble is usually in the capacitor. To check it, remove its cover, disconnect it and make the following tests:

Shorted Capacitor. A test for short circuit in a capacitor must be made with direct current since even a capacitor in perfect condition will pass alternating current. If a 110-v dc source is handy a test light connected as in Fig. 2 will light when the capacitor is shorted. Otherwise, use a bell and battery test set. If the capacitor is shorted, you probably have located the cause of

blown fuses. Replace the shorted capacitor with one of equal voltage and current rating.

If the capacitor is not shorted, it should be checked for grounds as in Fig. 3. Either ac or dc may be used to make this test. Be sure to make the test on each of the capacitor terminals. If the lamp lights, it indicates a grounded capacitor and a possible cause of blown fuses.

Loss of Capacity. A simple though not accurate test for possible deterioration of a capacitor is checking its spark. The capacitor should be charged by connecting it across a dc supply (properly fused) for a few seconds, as in Fig. 4. The capacitor is now "charged" and the terminals should not be touched, due to the possibility of a severe shock.

With the capacitor disconnected from the line, the terminals are shorted with a screwdriver having an insulated handle, Fig. 5. If the spark is snappy, the capacitor is probably in good condition; if the spark is weak or cannot be seen, the capacitor has either lost capacity or is open. In such a case, the motor starting torque will be affected. This may prevent the motor from coming up to speed and cause overheating or damage to the starting winding.

Checking Capacity. A capacitor of the *electrolytic* type may deteriorate with age causing a gradual decrease in

capacity. An electrolytic condenser is formed by winding two sheets of aluminum foil into a cylindrical shape. The dielectric is usually gauze treated with an electrolyte. If such a condenser loses more than $\frac{1}{4}$ its rated capacity, it should be replaced. The previous test of charging and noting the spark of a capacitor on discharge is simple but not accurate.

Capacity may be accurately measured by connecting the capacitor in an alternating-current circuit with a voltmeter and ammeter. Connect capacitor across the line with an ammeter in series. The voltage rating of capacitor should equal that of line. Close the switch and take voltmeter and ammeter readings. If no current flows, you know the capacitor is open circuited.

If you do get a flow of current, take the current reading and divide it by the voltage across the condenser; then multiply by 2650. The result will be the capacity in micro-farads. The value 2650 holds true only when the frequency is 60 cycles. As mentioned above, if the capacitor has lost more than $\frac{1}{4}$ its rated capacity, a new one should be installed.

Although the capacitor is not always responsible for troubles in capacitor motors, it is a likely suspect. For this reason the above tests will solve the majority of your capacitor-motor faults.

12
Shopwork, maintenance

NONRETURN valve on boiler is opened by maintenance man at Eastman Kodak Co's Hawkeye plant. Experience shows that it pays to instruct all personnel in the operation and maintenance of valves because their upkeep can run into big money

'KNOW COMMON VALVES,'

CLASS of power-plant youngsters at Dow Chemical Co soak up valve info to help pass upgrading exams. Testing relief valve *(right)* assures operation when installed

► METALS for valves are usually brass, bronze, iron or steel. When selecting valves for steam, cold water, oil or gas, almost any of these metals is good enough. But each has certain limits. Never use brass above 550 F. Bronze can stand 350-psi pressure and temperature up to 700 F. Bronze and brass valves usually come in smaller sizes, up to and including 3 in. Above this size, use iron up to 250 psi and 500 F. Steel is for temperatures above 550 F, and is also used when valve must withstand high internal or external pressures or shock.

Markings. Look at the valve body. A

Each Valve Is Designed to Do a Specific Job. Your Operators and . . .

GATE
Outside screw & yoke (os&y) has a rising spindle and double disk taper-seat for sealing

GLOBE
One-piece bonnet, renewable composition disk helps this valve stay tight a long time

ANGLE
Union bonnet and regrinding bevel-disk. Save fittings by putting these at elbow spots

NEEDLE
Union bonnet with a stainless steel disk assures a tight closing and long wearing seat

"2," "4," etc. on the body gives smallest pipe size the valve should be used on. Body marked 200 WOG means valve is safe for no more than 200 psi, whether it's cold water, oil or gas.

Instead of 200 WOG, valve may be marked 125 or 125S. "S" means that this valve is for 125-psi saturated-steam pressure at its equivalent temperature of 344 F. But for "cold" service same valve may be used up to 300 psi. So temperature and pressure are important. Of course, it's dangerous to use a "cold" valve for a "hot" application.

Type and Use. Gate valve is about the most common type. It is named for wedge (or gate) that's lifted for full, free flow. But it's not designed for throttling, close-control, or quick-closing operation. One reason is disk must be opened full diameter of valve, making operation slow. Throttling also causes severe wiredrawing or wear, thus damaging seating surfaces.

Globe valve is so named because of its body shape. This valve is very popular for flow control. Reason is disk travels only a short distance from full-open to closed position, saving operator's time. Besides, wear caused by throttling is easily remedied by replacing or regrinding disk and seat. But trouble is that it restricts or offers more resistance to flow.

Angle valve is same as globe, but body is built on an angle. Use for a 90-

deg turn in the line, thus reducing number of fittings needed. Then there is also less resistance to flow in line.

Y or blowoff valve has all the good features of a globe valve, but it offers less resistance to flow. It is very popular for blowing down boilers. If this or any other valve has a replaceable composition disk, it gives a little. That allows disk to adjust itself to scale and other particles. So tightness is maintained and seat protected against damage.

Needle valve resembles globe valve but has a tapered disk. Use where extremely close regulation is required. Taper fits accurately in its seat, giving drop-tight closure with very little effort.

Check valve is needed for many services. Use as a safety precaution to prevent any back flow. Or use to keep liquid in a line when pump is shut down —then it is called a foot valve. Check valve operates automatically, flow keeps it open, and gravity closes it. Two basic types are swing and lift. Use swing check in series with gate valve and operate it either vertically or horizontally. Lift check valve teams up with a globe or angle valve. It is made in three types, horizontal and vertical.

Special type of lift check is called a stop check. Use on parallel boiler installations to prevent back flow, and so cut out boiler. Stem is not connected to disk and acts only as a wedge to hold

disk down. Or stem allows pressure to raise the disk.

Plug valve or cock has a tapered plug with hole for liquid to run through. Plug is turned to line up with pipe openings. Plug can be opened or closed quickly. It handles thick or dirty fluids easily. But keeping it tight may be a problem.

Plug valve comes in various designs. But basically all types are the same, except for variations in stem, bonnet or disk.

Operation of all valves is based on the principle of a nut and bolt.

Rising stem is the simplest and most common type, used for many globe and angle valves. Also for gate valves with split wedge, allowing stem to turn.

Nonrising Stem. Here threaded disk travels up and down on stem. Because of this feature, this design is used only on gate valves.

If a standard right-hand thread were used, turning stem clockwise would allow disk to "run" up, making it come off the seat. To have conventional operation, a left-hand thread is used. Then turning the stem clockwise causes disk to go down, closing the opening. Thread makes the difference between a right-and left-hand valve.

Outside screw & yoke. Here nut is turned, but it can't move up or down so stem moves without turning. Used on solid-wedge gate valves. In this type,

231

STOP CHECK
With stem open, disk opens by itself to equalize pressure, closes as check

PLUG
Stick-proof plug valve is loosened before opening or closing, then tightened

Y TYPE
This has all good features of a globe valve but offers less resistance to flow

CHECK
One-piece cap and renewable composition disk prevent back flow in the pipeline

SWING CHECK
Bolted cover, requiring disk in larger type, is used in a horizontal pipeline

BYPASS
Prevent wiredrawing the large valves by running small equalizing line, as shown

threaded section of stem is entirely out of valve body. That makes it easy to lubricate, and it is free from exposure to line conditions. Besides, you can see if valve is open or closed.

Plug valve often sticks because of very tight fit between body and plug. To eliminate this, some plugs can be raised, giving more clearance and making it easy to turn. Cone checks or very large plug valves have this feature. Besides manual operation, valves can be operated with levers, motors, gears or pistons.

Bonnet Types. Screwed-in is the simplest of the bonnet types, being of one-piece construction. Used on small, low-pressure valves, it is not good when frequent opening of the body is necessary, or where shock or vibration is common.

Union-ring bonnet is a type with internal parts that are easy to inspect, repair or service. This construction gives added strength to body against internal pressure and distortion; it's practical only on valves up to a certain size.

Bolted bonnet is most rugged type,

and so is used on almost all large valves. Gasket between body and yoke enables it to withstand high pressures and temperatures.

Type of Ends. Brass, iron and stainless-steel valves in smaller sizes usually have screwed ends. Larger sizes have flanged ends. Solder ends are used when valve is to be connected to copper pipe and tubing. Steel valves may have screwed, flanged or welded ends.

Bypasses are used on steam lines to prevent water hammer by equalizing pressure and temperature on both sides of valve. Water in steam or in "dead" line causes damage if a valve is opened too quickly. Opening valve slowly to prevent this injury may damage seat and disk from wiredrawing. So open bypass valve slowly while main valve is still closed. Because bypass valve has only a small area compared to a large valve, don't worry about wiredrawing.

Maintenance. Preventive maintenance for valves simply means lubricating stem threads, if exposed, and bearing, if there is one; also, tightening down packing nut or gland flange to prevent

needless leakage around the valve stem.

Extensive maintenance depends on service. Repacking may be needed while valve is under pressure. Some valves may be repacked under pressure with valve wide open.

If a valve leaks at the seat, repair it at once. If neglected, seating surfaces may be so badly wiredrawn that valve must be scrapped. Gate valve is repaired by lapping in disk and seats with fine emery cloth.

Metal disks of globe and angle valves are repaired by grinding in with an abrasive compound. Use slug, coin or pin to take up play between disk and stem, and thus keep disk from turning on stem. Disk must turn with stem to allow oscillating motion needed for grinding.

Valves with composition disks may be serviced by replacing disk insert. Right disk is needed for each job. Asbestos composition is often used for saturated steam, air, gas, hot water or condensate. Leather or rubber disks do good job for cold water or other liquids. Special soft metal is for air, gasoline or oil. You must know service for each.

How to Store Portable Lamps

1 Pick up two ends together

2 Coil the double wire

3 Wrap loop around coil

4 Pull loop end through

5 Test before unwinding

Yours for Better Filing

1 File racks like these get more mileage out of files. NEVER THROW FILES IN A DRAWER. You wouldn't throw your razor blades around and expect a smooth shave. Files have numerous sharp edges. Save them by handling as a delicate tool

2 Filing is an art. Have work at elbow level. Take long even strokes, keeping file level. Ease up on return stroke. NEVER RUB FINGERS OVER FILE. That makes file slide. Tap file handle on bench every few strokes to knock out chips

3 Lathe filing is dangerous. A good mechanic learns to file with either hand to get away from revolving chuck. Move file forward and sideways over turning piece—never hold in one place. Moving helps file clear chips and avoids ridges

4 Draw-filing makes flat surfaces like valve faces, valve chests, smoother than "straight filing." Hold file in both hands and push, then pull across work. Use mill file for most draw-filing jobs. Clean file with file card often

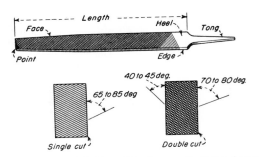

5 Machinists' files: A, FLAT, cuts all around; B, HAND, for finishing flat surfaces; C, HALF ROUND, for curved surfaces; D, ROUND for openings; E, SQUARE for keyways, slots; F, PILLAR for slots; G, THREE SQUARE for angles

6 Single-cut and double-cut files have angles shown. Larger file cuts are designated as: rough, course, bastard, second-cut, smooth, and dead smooth. Small files are numbered, as 00, 0, 1, 2, etc. Courtesy, Nicholson File Co

Grind Drills Right

1 All operating engineers use drills. For best results, certain basic facts should be known. Lip-relief angle, chisel point, cutting-edge angle, body-diameter clearance and web of drills are important. Drills are only as good as the user

2 Lip relief is important. With no clearance, cutting lip can't cut into metal. It only rubs. By having right clearance drop, cutting edge digs into material. Use less relief at margin than at center. Edges chip from too much relief

3 For normal all-around jobs, a 59 degree angle is best. Skilled machinists grind drills to about this angle without measuring. More angle makes drill "walk" and hard to center. Less angle gives long cutting edges, strains drill

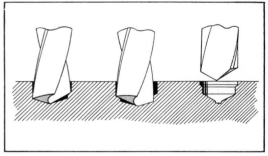

4 Grinding cutting edges right is important. If angles are different, one lip cuts more than next. If lips are off-center, they will cut larger hole than drill diameter. If off-center and off-angle, hole is larger and drill is strained

5 Chisel point of drill widens as drill is ground shorter because web widens toward shank for strength. This wider web should be thinned so drill's cutting edges can dig into metal without being held back by wide chisel-point rubbing

6 To grind drills right (A) learn to hold at right angle when starting to grind. Then swing drill downward with a slight twist to lessen relief angle near margin. To thin web (B) hold near wheel edge. Never grind on side of wheel, it may break

Practical Drilling Hints

1 Always check drill's cutting edges before using. If corners are rounded, grind down to dotted lines shown. Rounded corners wedge into metal, causing friction. Then heat anneals edge. Grind away roundness, no matter how small

2 Drills break like this when work springs against it. The blocks should have been nearer drill. When the drills point cuts through metal, spring action forces metal against drill, causing greater "bite." Drill can't take it and snaps

3 Drilling thin metal is extremely dangerous if not clamped down right. When drill's point cuts through, cutting edges dig in. Drill either spins metal, hurting operator, or drill breaks. Grind drills with slightest lip relief for thin metal

4 It's OK to hold work being drilled with stop bolts placed so they keep work from turning. Nuts on bolts keep work from climbing up when drill cuts through, hurting operator or breaking drill. For safety, clamping work down is better

5 Quick way to find nearest drill size for tap is using new nut like this as drill gage. Largest drill just slipping through nut is right size. Of course best method is looking up size on chart. In hurry, nut takes place of chart or gage

6 Oiling drill is important. Oil above work near shank so oil feeds onto work (for slower speeds). Use oil when drilling steel, wrought iron, monel metal or malleable iron. No oil is needed on cast iron, copper or most bronze and brass

7 Rope is handy when using portable drill like this. That's why OE should know knots. Any knot that's easy to untie when tight is OK here. Pushing heavy portable drill overhead is hard work unless easily adjusted fulcrum gives leverage

8 Another way to have rope do hardest part of drilling job is by using like this. Fasten to drill and then throw ends over any convenient pipe. Two half hitches are simple knots anyone can make. They don't get tight, untie easily

LAYOUT AND DRILL FLANGE

1 Laying out and drilling flange is important because holes must line up. This is right way to layout and drill holes. Measure old bolt circle from the edge of opposite holes like above. This is easier than trying to guess at center of holes

2 Saw off piece of wood to plug flange center and rub chalk over surface. Hermaphrodite is right tool to find center. Scribe four arcs on chalked wood 90 deg. apart like this. Then prick punch center for dividers to find the true center

3 With center punched in wood, it's easy to scribe bolt circle with dividers. Set dividers to one half of circle diameter found in Fig. 1. Chalk flange face and use sharp-pointed dividers to scribe a clear circle. Good tools save time

4 Circle must be divided in two exact halves. Place machinists' scale on the flange so it splits center mark. Then scribe line on flange like this. Machinists' scriber is best tool. Learn to use the right tool for each operation

5 Center line is shown dotted here. Since this is a six-hole flange, distance between holes is equal to radius of bolt circle (distance from center to bolt circle). Prick punch circle at line and scribe four arcs shown on bolt circle

6 After prick punching each arc on circle, set dividers to half of drill diameter. Then scribe six circles around flange. Next, punch four marks around drill circles. These marks are important. All machinists lay out holes this way

7 In clamping flange to drill press, make sure drill lines up with a slot in table or place wood under flange to protect table. C-clamps, used like this, are OK. Making table and flange tight helps drill do better job and prevents accident

8 Drills seldom follow prick punch. They have tendency to "walk." Then distances **X** and **Y** aren't equal. As soon as drill point "spots" flange as shown, check for being centered. If not centered use prick punch as chisel and cut groove from spots top to bottom shown by dotted line, center sketch. Drill this groove out and check again. Size of groove depends on distance drill is off center. Be sure to bring drill back so the four center punch marks are split and distances **X** and **Y** are equal

Centering Stock for Lathe

1 There are many ways to center round stock for turning between lathe centers. First, chalk stock ends. Use center-head and scribe marks at right angles, **A**. Hermaphrodite is handy, **B**. Scratch four marks at right angles. Punch center

4 To center stock in lathe, chuck it and start machine. Steady hand on compound rest, **A**, and hold chalk so revolving stock barely touches. Stop lathe, **B**, loosen jaw opposite chalk mark and tighten jaw nearest mark. Repeat until stock is true

2 Surface plate, surface gage and V blocks are a third way to center round stock. Place stock to be centered on V block or blocks, depending on stock's length. Scribe lines as with hermaphrodite. Hammer deep centerpunch mark with lines

5 Another way, **A**, to center stock is by placing tail end of tool holder against stock, with set screw tightened finger tight. Don't run lathe but turn chuck by hand, tightening or loosening jaws. The quickest way is with a dial indicator, **B**

3 With centers punched at both ends, place round stock against lathe center in tail stock and support other end with hand to take weight off center drill. Feed slowly against drill. Center drills are delicate. Oil drill, feed carefully

6 In Fig. 5 **B**, set dial indicator in tool holder and move compound rest against work until dial reads zero. Turn chuck by hand and move jaws with key until reading is the same all around. When stock's true, oil drill, run it slowly into stock

Measuring Instruments

1 To fit replacement parts accurately, operating engineers must know how to use machinists' measuring instruments. An outside caliper is a precision instrument in hands of trained machinist. Set it right by hooking to end of scale like this

2 Never set inside caliper from scale, but from outside caliper like this. First set outside caliper as in Fig. 1. Then rest inside caliper on outside caliper and swing back and forth. Adjust until the contact can just barely be felt

3 There is only one way to hold an outside micrometer. Rest frame in palm of hand and press with tip of ring finger. Adjust thimble or ratchet with thumb and index finger of same hand. Now other hand is free to hold piece being measured

4 Easiest and most accurate way to set inside caliper is with outside micrometer. Rest one leg on micrometer anvil and swing other leg against spindle like this. For right setting, you should feel for slightest contact with mike

5 In measuring inside diameter of large cylinder, hold caliper's bottom leg against surface to keep it from moving. Then find exact opposite point by swinging caliper slightly sideways as well as back and forth. Set caliper by tapping leg

6 Eccentrics wear out-of-round. Wear in thousandths of inch can be found by caliper. First caliper largest diameter. Then hold feeler gage under caliper leg to find smallest diameter. Try different feeler gages until exact wear is known

Finding Condenser Leaks

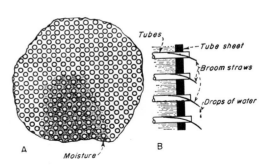

1 Finding leaks in tubes of surface condensers, oil coolers and other heat-transfer apparatus isn't easy. First fill condenser's steam side with hot water from feed pump through boiling-out connection, until it reaches exhaust-trunk vent

2 Tiny pinholes in tubes will keep tube sheets moist over large area, even with 160-F water inside condenser, evaporating moisture, **A.** Stick broom straws, **B,** into tubes suspected leaking. Water dripping from straws will show leaky tubes

3 One way to find leaks inside tubes is to look through tubes against light at other end, **A.** Larger leaks show up easily. To be sure about smaller leaks, stick soft pine tapered plug into both ends of all tubes appearing to leak, **B**

4 After 15 or 20 minutes, remove one plug, **A,** from one end of a tube. If it leaks, water that filled tube will flow out, proving tube has hole. If compressed air is handy, another way to find leaks is with air, **B,** then look as in **3A**

5 On large condensers connected directly to steam turbines, connect air hose to vent valve or to turbine-casing drain. Build air pressure in exhaust trunk and turbine until compound gage shows 7 psi. This pressure helps show up leaks

6 After leaks are found, plug both ends with bronze plugs of size shown. Never use wood because wood often leaks before tubes can be renewed. White lead the plugs and tap into tubes. After 10% of tubes are plugged, install new ones

Making Condenser Plugs

▶ Because operating engineers are responsible for machinery repairs, knowing how to run a lathe is important. A small repair shop in any plant saves time and money. But engineers who haven't learned the machinist trade pick up many *wrong habits* without expert advice. Then they never learn the correct way and do faulty work.

This series on machine-shop practice is intended to point out important details by starting readers on the right track. Instead of showing how to do various things on a lathe we're using an old trick we used in teaching trade school. That's showing how to do jobs that come up in most plants. A good machinist uses either hand so he can file and saw facing chuck as in Fig. 6, to keep from leaning over revolving chuck.

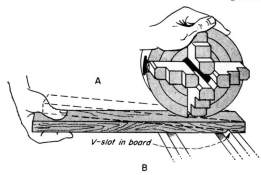

V-slot in board

About 2½ in. from chuck

1 The facing page shows in detail how to plug condensers. Here's way to make plugs, cutting taper with compound rest. Use board to rest chuck on, **A.** Place ¾-in. brass stock through headstock with end near chuck to prevent chatter, **B**

Lathe chuck

Chalk

Chalk marks

2 To center stock in chuck, rest hand on lathe and touch chalk to revolving stock, **A.** Move stock in chuck by loosening and tightening jaws until chalk makes even ring around stock. Mark two jaws, **B,** to use for next pieces

90°

Set for taper

Wrench

Compound rest

3 Loosen compound-rest nuts, set rest to 4 degrees and tighten nuts. Grind tool for brass at 90 deg so top surface has no rake, **A.** Set tool with center gage against stock after compound rest is set for taper. Oil all the lathe bearings.

Center

4 Set tool tip to center of stock, **A.** Set lathe speed for about 200 ft per minute. That's surface speed of stock, or about 1000 rpm for ¾-in. stock. Now move lathe carriage so tool tip is 1¾ in. from stock end, **B,** length of the taper

5 No oil is used on work when turning brass. With tool set for plug's length, run in crossfeed to make deep groove, **A.** Now run compound rest out by hand and knock sharp edge off plug with tool, **B.** This does neat job and prevents cut hands

6 File tool marks from plug, holding 12-in. smooth-cut mill file as shown. Push file towards work and towards chuck lightly, **A.** Use hacksaw in groove with lathe turning same speed as before. Saw shown askew here but keep in line

Don't Shut Down for A Lost Spring
Wind One Like This

Start expansion spring like this

Start compression spring like this

1 Because springs get lost easily when taking machinery apart, operating engineers should know how to wind them, especially when there's no time to order a new one. Use a mandril of slightly smaller diameter than inside of spring. Drill small hole through mandril for wire. Wind first few coils by turning lathe slowly. Wind expansion spring turns close together. Start compression spring same way but spread coils proper distance apart

Clamp can be held while lathe is turning

Copper or fiber jaws

Clamp

2 Make clamp of hard wood or flat metal stock with jaws of soft wood or fiber. Tighten clamp enough to wind compression coils tight against mandril. Hold clamp with one hand and control the mandril speed with the other by having it on the starting button or belt shift. Make sure coil of wire is fed into clamp without binding. The safest way is to cut off approximate length of wire for spring and have assistant hold other end

Lathe feed set for threads

Brass screw

Crossfeed

Tool post

Clamp resting against tool holder for heavy wire

3 Winding heavy compression springs like this assures a good spring. Rest clamp on offset tool holder. Replace hardened toolholder set screw with brass one and tighten lightly against wire. Then set lathe for same number of threads as turns per inch in compression spring. As lathe feeds wire along revolving mandril, oil wire to keep it from scoring. Loosen clamp and tool holder slowly when spring is wound to prevent violent recoil

A drill press will do for winding very light springs. Use a clamp like this on a short rod in chuck.

4 Small springs of ¼-in. outside diameter or less can be wound by placing fiber jaws in a parallel clamp and holding clamp by hand. The faster the mandril turns the better "life" your spring will have.

If you don't have a lathe, a drill press can be used for turning the mandril. Any good grade of spring wire stock of brass, monel or steel will do, depending on the service the spring is made for. Wear goggles when handling spring wire

Small springs can be wound on fast-turning mandril and a machinist's parallel clamp with jaws

Dry Ice Makes Expansion Fit

1 Use drum having at least 2-in. of space around liner. Place drum in box and insulate with sawdust. Add industrial alcohol to cover liner

2 Dry ice is sold in 10-in cubes. Place a cube in a burlap bag and break it into small pieces by pounding with a wood mallet or a hammer

3 A metal disk and hook are handy but not essential. A disk like this one might be needed if the liner has to be tapped in with a hammer

4 Place liner in the alcohol bath and pour in the dry ice. The drum should be at least 6-in. above the liner because the alcohol will "boil"

5 Dry ice and alcohol cools liner to about 80 F below zero in about 15 minutes. This 6-in. diameter liner shrinks about 0.006 of an inch. Alcohol lubricates liner and prevents frost forming on it

6 Work fast after removing liner from bath. Delay warms liner quickly. It might seize when part way in cylinder. If the liner won't slide in by hand, cool again and force it into cylinder like this

▶ DRY ICE IS a quick practical chilling medium for making expansion fits, because its temperature is 109 F below zero. It works better with alcohol because alcohol prevents frost and acts as lubricant. Dry ice is solid carbon dioxide, (CO_2) and it changes to gas when it absorbs heat. It's sold in 10-in. cubes, weighing fifty pounds each. While 97 F below zero is possible, the liner is cooled between −50 F and −90 F. That's because it warms rapidly when taken from bath. Count on about 0.001 in. per inch shrinkage for each inch of liner diameter. This holds for most metals. A water-tight container, same shape as liner, is best. Alcohol should cover liner or parts to be shrunk and container should be higher than liner to allow for alcohol boiling. Putting bag of dry ice in liner while fitting keeps it from expanding too fast. Caution: CO_2 displaces air. *Ventilate working area.*

Shrink and Expansion Fits

Lube oil
at 180 F

Steam

Copper coil

1 Gear is lowered into lube oil, heated to about 180F (do not boil) till hole expands. Steam coil or electric heater does trick. Oil should just cover gear

10-in. cube of dry ice
in burlap sack

2 Break dry ice by placing cube in burlap bag and hitting with mallet

Alcohol

Crankshaft

Tank

Sawdust

Casing

Dry ice

3 Fill sawdust-insulated tank with industrial alcohol to cover crankpin. Tank should be 6 in. higher than alcohol level. Keep adding dry ice to keep "boiling"

4 To fit expanded gear hub over shrunk crankshaft, have tools ready and work fast. Tap gear lightly using wooden block and move crane to help slide gear on shaft

▶ COMBINING SHRINK WITH expansion fit does tightest possible job on two pieces to be held together. This method has been used where machinery works under severe conditions.

Lubricating oil heated to about 180 F expands female part about 0.001 in. per inch of diameter. Dry ice shrinks male part about the same amount. When parts, machined to allow for diameter change, are expanded and shrunk and quickly fitted together, fit is twice as tight as "one-method" job.

Dry ice is a quick practical chilling medium for making expansion fits, because its temperature is 109 F below zero. It works better with alcohol because alcohol prevents frost and acts as a lubricant. Dry ice is solid carbon dioxide and it changes to gas when it absorbs heat. It's sold in 10-in. cubes, weighing fifty pounds each. While 97 F below zero is possible, piece is cooled between —50 F and —90 F. That's because it warms rapidly when taken from bath. Count on about 0.001-in. shrinkage for each inch of diameter. This holds for most metals. A watertight container, slightly larger than piece being cooled, is best. Alcohol should cover part to be shrunk and container should have high sides to allow for alcohol 'boiling'. Ventilate working area.

Removing Cylinder Shoulders

1 Internal-combustion-engine cylinders wear from pistons and rings. Dotted line shows exaggerated wear, with shoulder at top and bottom of liner. Shoulders form even when counterbored

2 Effect of liner wear on piston rings is shown by ring and clearances. New ring at top is closed but new rings at worn section are opened against walls. Excessive openings make new rings leak

3 To remove shoulders before installing new rings use grinding rig like this made by threading round stock and securing small grinding wheel. Brass pipe is for holding wheel against cylinder wall

4 Bottom shoulders can be removed by using electric drill with long rod hung above cylinder. Place oil on rod so it slides easily within brass pipe while you are holding wheel against shoulder

5 For grinding top shoulder a shorter rod can be used if there is little head room. Pull wheel against shoulder as shown. Caution: Don't grind worn cylinder wearing surface. **Always wear goggles**

6 This is the cylinder liner after top and bottom shoulders have been ground and counterbores (shown exaggerated) ground. Larger rings can now be used with minimum gap openings. This saves fuel

▶ NEVER fit new rings into diesel-engine cylinder liner without removing shoulders at top and bottom. New rings fitted over shoulders aren't much better than old rings because they open in worn part, Fig. 2. Shoulders don't have to be removed by reboring liner; they can be ground out with home-made rig if necessary. Caution: Keep grit out of crankcase by spreading canvas.

Hand Boring Small Cylinders

1 Small cast-iron cylinders are easy to bore with hand boring rig. Make boring bar 2½ times cylinder length of same diameter stock as piston rod. Make brass-bushing guide to fit stuffing box and rod snugly. Cut 40 threads to inch on bar and in flat disk to fit. Drill bolt holes in strongback and tool and screw holes in disk. Tap holes and file flats on bar.

¾" disk

File flat for wrench

Neck bushing of brass or white metal

½" flat plate strongback

2 Replace old packing and neck bushing with new neck bushing. Hold it in place with gland. Oil rod and bolt boring rig in place after centering bar with inside caliper against counterbore. Don't start boring until bar is perfectly centered. Adjust pointed tool. If tool chatters, insert second tool on opposite side. Take finish cut with round-nosed tool. Complete each cut to end of cylinder.

Washer

40 threads to inch

Counter bore

Neck bushing acts as guide

Open end wrench

3 Make wooden plug for polishing tool marks from cylinder. Cut plug from 3-in. block in lathe with parting tool. Tack on No. "0" grinding cloth as shown. Bore plug for piece of pipe to be pressed into center. Pipe bushing can be loose fit on bar. Turn outside diameter of plug so it is snug fit when felt or canvass pad and grinding cloth are tacked to rim.

Grinding cloth

Felt or canvas

Set screw

Pipe bushing

Tacks

Wooden plug

2"

4 Remove tool disk from bar and attach polishing plug. Bolts can be left loose against strongback, as plug will center itself. If grinding cloth doesn't fit snugly enough to remove tool marks, wrap another strip of felt or canvass cloth underneath. After cylinder is polished bright, remove rig, clean and coat cylinder with beeswax. This gives cylinder a hard glazed surface for low friction.

Boring bar with polishing plug

Removing Broken Nipple

1 Neatest job is to saw broken nipple with a hack saw blade like this. Be sure to stop sawing when you reach threads

2 If saw can't be used, sharpen a diamond point chisel and split nipple all way through like this. Don't cut threads

3 Split nipple is easy to collapse with chisel if it's split right. Caution: Don't knock nipple through hole to inside

4 If broken nipple is in a close place and you have no "easy-out", hollow grind hard stock and work with lever and wrench

5 Don't let nipple that's rusted solid stop you. Heat with blowtorch after it's split. Then it will collapse easily

6 This "easyout" won't always work. If the nipple is badly frozen and won't budge, don't spoil threads by wedging

▶ A BROKEN NIPPLE can lead to a major shutdown and a big maintenance job. Whether it's made of cast iron, steel or brass, there is a right way and a wrong way to go about removing it. If it's a cast-iron nipple, the easiest way is sawing to the threads with a hacksaw blade. Then chip pieces out of it by collapsing near the cut. With a little patience, the nipple can be removed so there are no marks and it won't even be necessary to clean the threads with a pipe tap. The nipple can be split with a chisel, but it's easier to botch up the job. Then again hammering may loosen fittings unless they're properly braced near the nipple. If an "easyout" won't remove the nipple on the first try, don't spoil the threads by putting your beef on it and wedging this tool deeper into the nipple. Heat it with a blowtorch. Heat expansion usually breaks rust or dried thread sealing compound so that "easyout" will remove nipple. A hollow ground square file end—turned with a wrench and held in place with a lever often does the trick in tight spots. Using the *right* method for *each* job makes it quick and easy. Caution: Don't knock chips into line.

Repair Your Valves

MACHINING, RESEATING and GRINDING

TAKE light cut on valve disk after truing up with a dial indicator at *A* and *B*. Set the compound rest to valve's angle

RESEAT valve body with reseating tool, or by truing in a lathe chuck. Remove only enough metal to clean up the pits

CAUTION! Do not use reseating tool if valve body has no machined finish. Only way then is to machine seat in lathe

NO LATHE or RESEATER? USE HAND TOOLS

BADLY GROOVED and wire-drawn seat and disk look like this. The sealing surfaces can be restored with hand tools

REMOVE DISK from stem and place washer or small piece of gasket under stem to lock parts solidly so disk won't turn

REMOVE metal on both sides of groove with file (after giving light grind to find bearing surface), keep turning disk

TESTING

LIGHT PENCIL-MARK the disk. Place stem in bonnet, and turn disk few degrees against seat to see that all marks rub

LEAK-TEST for normal work pressure with small hydraulic pump like this, after assembling valve. If not tight, grind again

NO HYDRAULIC PUMP? Use outlet from feedwater discharge for test. C-clamps handle all size valves on the test flange

Before They Give Trouble

GRIND valve and disk by locking disk to shaft loosely with split pin. Use bonnet as guide. No nuts; no packing

IN-PLACE grinding of large valves. Use wooden guide if upside down. Then you won't have to support the heavy bonnet

FINISH by cleaning compound from disk and valve. "Grind" with oil, clean, and again grind with oil until seat shines

GRIND with coarse compound each time you file around once. Don't touch ground surface with a file—don't file deep pits

FILE EVENLY all around by turning disk so you don't file flat in disk. Idea is to remove metal that would be machined

FINISH FILING carefully with fine mill file. Give grind with fine compound and grind out all pits. Then give oil rub

Don't Forget These . . .

OVERHAUL bonnet assembly before finishing job. Shine up valve stem with 00 emery cloth. Replace old packing in bonnet. Replace old gasket with new. Clean threads in Stoddard's Solvent, oil. Work stem full thread length

Select <u>Right</u> Lock Nut

SELF-TIGHTENING

FIRST THREE NUTS are square; they deflect when wrenched down. That causes friction lock between bolt, nut threads. Use these nuts either against a washer or against a bottom nut. Reason: The unit pressure at bearing corners is high

FOURTH NUT has recessed bottom-face at outside thread pitchline. Tightening against seat forces upper slotted threads. Last nut also flexes when wrenched down. No slots are needed because nut is thinner, presses upward against threads

INSERT TYPE

FIRST AND FOURTH NUTS have resilient insert of nylon or some chemically treated material to resist moisture, oil and drying. Because hole in insert is smaller than thread's diameter, it sets up constant pressure against bolt. This friction grip

stays tight during vibration. Second nut has oval spring steel insert that hugs the bolt. Third nut has nylon plug that squeezes against bolt when nut is wrenched down over the threads. All of these nuts can be used for locking a number of times

SPLIT GRIP

ALL THESE NUTS have same principle of hugging bolt threads near top because upper slotted section is slightly squeezed in after nut is tapped. First nut is designed for temperatures ranging from 250 F to 1200 F without bolt seizure or galling.

Second nut is for many degrees below zero to at least 550 F. Third and fourth nut are not built for unusual temperatures. Fifth nut has twelve serrations for standard socket wrench; it's for below zero to at least 550 F; handy for tight spots

For Each Problem

THESE NUTS have a constant spring-tension grip against bolt or screw threads. They are made of pressed sheet material for various services. First nut spins freely down over the thread. Second nut can be "zipped" on, then tightened. Third nut spins freely until outside edges contact at bottom. Then threads grip at bottom. Top part also partly presses against bolt. Fourth nut also locks at bottom and top when tightened down. Nuts shown are handy for holding light parts of sheet iron

GRIP TYPE

FIRST NUT has three depressions at top, forcing threads inward to cause friction against bolt. Second nut's top three threads are deflected slightly to hug bolt. Depressions in top of third nut also distort threads enough to give locking action against bolt threads, as in first nut. Fourth nut has torque locking notches in thin section around its top. All these nuts spin freely until the top threads come near the bolt threads. Then they must be tightened down rest of way with a wrench

WEDGE TYPE

BOTTOM slotted male-nut wedges against bolt when top female nut is wrenched down against it. Can be reused many times—popular for moving parts. Second nut also has same principle but wedging action is caused by collar nut tightens against

Because of vibration, temperature and severe service around most power plants, the common nut used as a locking device comes in for work that it's not designed for. But there is a nut to fit most needs and the power engineer and maintenance mechanic should know his lock nuts. The twenty-five shown here are stock items, designed by various nut and bolt firms. If these do not solve your problem, write to the Industrial Fasteners Institute, Cleveland 44115, OH, for a catalogue giving detailed information on thirty-six different locknuts.

Aligning Couplings and

Why You Must Check Couplings

FOUNDATION SAG may not be evident to the naked eye, but vibration, building settling, etc, lower or change level of machinery, wearing the flexible couplings

BEARING WEAR may be uneven. Then most accurately aligned equipment gradually gets out of line. That wears contact surfaces of even best-designed coupling

EXPANSION or contraction from hot, then cold piping often pushes equipment out of line. Lining up cold machinery does not mean it will be aligned when hot

Handy Uses for Indicators

THRUST CLEARANCE or end play in most rotary equipment is highly important. If this clearance is excessive, vital parts, like rotors of turbines or pumps, rub on casing, causing costly damage. To check end-play without taking equipment apart, break coupling, then move shaft both ways while the indicator is set up as shown here

BEARING CLEARANCE (also wear) is easy to check. Break coupling, clamp indicator to coupling and rest the indicator on bearing. Not necessary to take leads

Simple Checks

SCALE or straightedge placed across two coupling halves at four places around coupling is visual check. But indicator is much more accurate, saves costly wear

FEELERS between coupling halves at four places around circle show if shafts are aligned. But this naturally cannot be as accurate as using a dial indicator

How to Align Couplings

Angular check

1 Place rubber between shaft ends (not near rim) to hold halves apart. Mark halves, attach indicator as shown, turn the halves together for angular check

Other Handy Indicator Checks

New Machine Has to Be Level

HORIZONTALLY ALIGN equipment with machinist's precision level by placing level on shaft or on machined surface. This helps parts to wear as intended

HORIZONTAL alignment also means that machine must be level in all directions. This helps in right setup of the piping and other equipment hooked to machine

INDICATE machine base by setting gage to zero. Tighten down most rigid unit first (pump). If base springs, loosen and place shims until there's no spring

BENT SHAFT or coupling half not square is found by turning one half, while indicator is on the other. Don't assume couplings are true on shaft. Check them

CROOKED SHAFT shows up with indicator anchored to bearing while pin touches shaft. If shaft is true, check coupling half for being true on shaft (right). If not true, hole may be eccentric to rim, or coupling may not be snug fit on shaft. Soon as you know that shaft and coupling's half are true, you can line up coupling (below)

Rim check

Figuring rim misalignment

2 For rim check, change indicator as here. Turn two halves together, watch dial. Raise machine legs same amount with shims if angular check is perfect

3 With 0 reading at top and 0.020 in. at bottom; 0.020 in. divided by 2 equals 0.010 in. that one machine is out. Raise low machine with 0.010 in. shims

4 End clearance between the coupling halves must be right. If not, the worn thrust bearing or heavy rotor may load lighter motors, causing serious damage

Tips On Making

A   B   C

MANUFACTURED RINGS ARE BEST—but sometimes you have to "roll your own". Start by chucking cast-iron ring pot in lathe as at **B**. Use dead center on wood block to prevent "chattering" of long pots. Rough-turn outside until diameter equals cylinder diameter plus 12 thousandths in. per in. cylinder dia. (to allow for cut in making joint) plus ⅛ in. (to allow for final finishing). Then rough-bore inside so thickness is 1/30 cylinder diameter plus ¼ in. for finishing. Now lock carriage to lathe bed, face edge of pot with facing tool and cut off ring **C**. Width of ring **W** is determined by width of most-worn part of piston

G   H   I   J

FOR ALL JOINTS the critical dimension is the end clearance **X**, sketch **G**. This allows for expansion and should be 3 thousandths in. of cylinder diameter when ring is closed. (Make cut a little larger). We allowed for this by adding to the diameter when rough-turning ring pot. To make step cut, take scriber and square and lay off as in **G** and **H**. Then make four cuts with hacksaw as shown at **I**. Note that middle cuts are made off-center. Now spread gap with a distance piece **J** to allow room for saw to finish cuts. Cut along heavy black lines indicated by arrows. Note: these two cuts are also off center.

N   O   P

MAKING THE SECOND CUT. With ring bound, clamp it to face plate as in **N**. Any number of rings may be trued up at the same time. Sketch shows two. Use distance piece under washers on the inside for uniform purchase. Note step joints are directly under washer. Overhang gives washer enough purchase to hold ring closed. Remove wire and turn ring down to exact cylinder diameter less desired clearance. Then clamp ring from outside as at **O**. Don't let ring open while clamping. Inner bolts can't be removed, so tie them to center with string to keep out of way. Turn inside of ring until thickness is 1/30

And Fitting Piston Rings

Angle Joints

Step Joints

D

E

Left hand   Right hand

F

Compression

groove **A** as groove will have to be turned to this dimension to prevent rounding of new ring. After cutoff, finish other side of ring with grinding cloth on face plate. Now we are ready for the joint. What kind of joint? Angle or step. An angle joint is easier to make, but a step joint makes a better ring. Some

people say combine left-hand and right-hand angle joints as in **D** because it makes for turning action to keep cylinder round. Actually it encourages blowby. Other people claim a step joint can not leak, yet gas can leak around the back as at **F** if ring is loose. Still **step joint is best**. Let's make one to find why

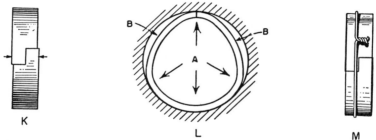

K

L

B   B

A

M

If cuts are made right, the joint will look like **K**. The indicated over-hang is very important. Now comes a step a lot of people overlook. The ring is not yet a true circle. Put it in the cylinder and it will leak at **B** and bear at **A**, sketch **L**. Why? Well, cut a piece out of a circle and put the ends together. The result

will have an egg shape—not a true circle. To make the piston ring a true circle it is necessary to make a second cut. The second cut can be made only with a step joint. Without the overhang in **K** the ring can not be held closed while turning it to a true circle on the lathe. Next, tie ring closed as at **M**

Q

R   Grinding compound

S   Piston

T

of cylinder diameter, more if piston groove is deep. Note: Always leave tool marks on ring as at **P**. They help ring wear to cylinder quickly and prevent blowby. Next, slightly round ring as at **Q** to let it slide over lubricant without scraping it off. Then remove overhang at step joint by grinding inside

surfaces as at **R**. Never file the outer edges. **S** and **T** show quick checks for fitting to piston. If ring is too tight, turn piston groove, not ring. Groove has to be turned down to heavy dotted line in sketch **A** anyway before new ring will fit. These tips aren't the whole story. Practice will tell you the rest

MAKING EMERGENCY RINGS

1—Why Is 2-Cut Ring Right; One-Cut Wrong?

Sooner or later, most engineers get into a spot where the only answer is a homemade ring. And all too often the "quickie" ring spoils what was a good cylinder. That's because rings are turned and bored to finished dimensions, split and put on the piston. Such "one-cut" rings wear cylinder egg-shaped. Even worse, there's so much leakage that blowby removes cylinder lubrication, causing scuffing and pressure loss.

If you must make rings in an emergency, make them in two cuts as shown here. After casting is rough turned and bored, cut gap and wire ring shut. Then finish turn, while ring is collapsed. That makes a "2-cut" ring.

Slide ring into smallest part of

TWO-CUT RING (black) forms perfect circle, while one-cut ring (gray) goes to egg shape

ONE-CUT RING leaks badly on both sides of gap, wears the cylinder out of round

cylinder; check the gap clearance with feelers. Remove and file until clearance is correct, to keep ring ends from butting when heated to working temperature. Butting causes ring to wipe oil from the cylinder, piston to seize.

3—Preparing Piston Groove and Ring

OLD GROOVE in piston wears wedge-shaped, has filled-in corners. Remove carbon, make sides parallel for the ring

CARBON from the large piston groove is cleaned with an old dull file or a scraper. Don't remove any of the metal

TRY RING in groove by rolling it around piston. This shows tight spots in groove or if ring needs more facing

TURN A SEAT in casting from which ring was cut. Tap ring into it snugly. It serves to hold ring while facing side

5—Finish Turning and Boring (Second Cut)

PLACE FACEPLATE on workbench. Slide four bolts in slots and tie them with string for later use. Put heavy paper on plate and pile collapsed rings, centered as well as possible. Have gaps over slot. Clamp rings down from outside with another

set of bolts and washers. Attach plate to head stock, loosen bolts and true-up rings. Take finish boring cuts. One at a time, remove outside bolts and clamp with inside bolts. With outside clear, remove wires and take light cuts to finish OD

2—Roughing-Out the Ring (First Cut)

FIRST STEP: Know cylinder diameter. Figure oversize for tension and gap clearance, then add enough for finish turning. Sketch shows taking rough cut after truing-up casting

ROUGH BORE casting to size such that there is enough metal left for finish boring after ring is collapsed. Take boring cut deep enough to get under mill scale, or tool dulls fast

FACE END of casting after rough turning and rough boring. Turn cast iron slowly. *Caution!* Beware of fine cast-iron dust blowing into eyes. It's painful to get out. Wear your goggles

CUT OFF roughed-out ring with cutting-off tool. On long casting you might have to run tailstock against a block of wood resting against casting. That prevents chattering on this job

4—Sawing, Filing and Collapsing

SCRIBE guide marks for sawing out gap. Figure on removing enough metal to give tension needed in the ring

CLAMP in vise between copper jaws and cut gap with hacksaw, if you've no miller. If ring has flaw, split at flaw

FILE saw ends parallel. This is not finished job as ring is only rough turned, bored. File only to scribed marks

COLLAPSE ring by twisting a soft iron wire around. This does job of expensive cylinder and clamp makers use

6—Leave Tool Marks on Ring, Handle Carefully

GROOVES worn in cylinder allow blowby, removing badly needed oil on walls of newly ringed cylinder. This may cause scuffing and eventual failure of ring

TOOL MARKS left on finished ring wear quickly into slight grooves and irregularities of most cylinders. Smoothly finished rings take longer time to seat

CAREFUL HANDLING of any new ring pays. Nicks and burrs can score cylinders, or keep ring from seating. Remove them before putting any ring into its groove

You can spot breakdowns

Some plant uses for this handy tool—the reflectoscope

Crankshaft may have flaws

Ultrasonic testing of this new diesel crankshaft shows up any hidden flaws. This is good insurance before placing new parts in machine. Here, old crank cracked through web. Operator is placing crystal (of reflectoscope) on crank web. Crystal projects a mechanical vibration or wave of extremely high frequency into the metal. Wave is projected through length of web, in path directly under the crystal, tests only segment of steel directly in its path. Keep moving crystal.

Crankpin tested in place

Here instrument is set up near machine, test made to find flaws in engine crankpin. If a defect is present, wave is reflected from it. If this defect happens to extend in a direction parallel to the wave, it presents practically no obstruction to the wave and won't be detected unless crystal is changed to new location. To detect such a defect, move crystal to another surface so wave travels at angle to defect, thus being partially or completely blocked in wave's path.

Flywheel shaft

Fatigue cracks in axles or shafts, which are hidden from view by flywheels, bearings or crank disks, show up distinctly under reflectoscope inspection. Besides, size and location of such cracks may be found. That helps operators to use the shaft as long as possible, yet alerts them to remove it from machine before failure from any defect growth.

Metal thickness

Lost metal is detected on location. Extent of both internal and external storage-tank corrosion is found without emptying—or entering tank. Pitting plates has same effect on sound waves as defects or cracks. Thickness of plates can also be measured with reflectoscope. You learn if plates have become dangerously thin with time.

Pipe for flaws

Critical-high-temperature, high-pressure pipe is inspected for dangerous defects. Here the angle search unit is used. Sound ricochets off inner and outer walls, circling the pipe until it returns to the search unit. Defects in the path of sound reflect back and appear as an intermediate signal on screen, which operator watches.

quickly <u>before</u> they happen

Here's how the reflectoscope works

▶ IF YOU CAN'T AFFORD costly breakdowns, answer is preventive maintenance. One way to prevent work stoppage is to "look" inside metal for flaws, replace such parts before they fail. Instrument coming into common use for this job is the reflectoscope. This is a portable electronic instrument producing pulses of ultrahigh sound. These pulses are induced into an object under test by a crystal search unit.

Nondestructive tests can be quickly made for fatigue cracks, internal flaws, corrosion, wear and imperfect welds. You not only don't have to drill or cut open part being tested (for loss of metal) but you often don't even have to remove part from machine, as these photos show. That saves time and costly labor.

Two beams. You can use two search beams: a straight beam and an angle beam. With a straight beam, sound waves travel straight through the object. They bounce off the opposite side, and are picked up again by the search unit. This gives you thickness measurements. This beam also detects any defects in the material that are in the wave's path, doing a twofold job.

With an angle-beam search unit, sound waves don't bounce straight back to the search unit, but ricochet off all sides and through all areas of the object. This search unit is especially useful for checking welds, pipe and tubing—also, other objects where a straight unit might prove impractical or difficult. Instrument penetrates and indicates reflections through 30 ft of aluminum or steel. In porous metallic and nonmetallic materials, maximum penetration varies—depending on wave length to within fraction of an inch. *Courtesy, Sperry Products, Inc, Danbury, Conn.*

Let's test your pipe lines for flaws and thickness

1. Sound bounces off inside walls

2. Flaw shows on viewing screen

3. Crack sends the sound back

4. Screen shows crack like this

5. Thickness is measured here

Typical maintenance test is of high-pressure steam piping. Here sound is beamed into pipe wall at an angle with search unit. Sound ricochets off inner and outer pipe-wall boundaries in its travel around pipe, and returns to search unit, Fig. 1. Uninterrupted flow and beam's return are indicated on reviewing screen, Fig. 2. Any discontinuity in path of sound beam reflects a portion of energy back to search unit, Fig. 3. This shows on screen, Fig. 4.

Thickness measurement. Here the sound is beamed into pipe wall with a straight beam and reflects from inner face, Fig. 5. Length of time needed for sound to reach inner face and return to search unit varies with thickness of material. Thickness is indicated on screen. You can thus measure degree to which bent pipe (for example) thins on outside and thickens on inside of bend, assuring full specified thickness in high-pressure work. It has many other plant uses.

MANUAL IMPACT WRENCH builds up spring power; number of blows indicates bolt tension

Stress bolts right to make up tight flanges

Today more than ever before, greater speeds, heavier loads, higher temperatures and pressures demand accurate flange tightening. In selecting the best fastener for each unit and specifying torque-wrench settings, designers consider and balance factors such as (1) tension needed to keep bolt tight under maximum stress (2) smallest adequate bolt to reduce weight and cost (3) type of bolt material (4) class and type of thread (5) safe maximum tension and clamping to prevent distortion and breaking.

All this adds up to a truism: after such careful engineering analysis went into selecting the component, it's only good sense to follow proper procedures when tightening the assembly during manufacture or maintenance.

When a flanged joint is assembled, each component comes under varying tensile or compressive stresses. During service, especially at high temperatures, the flanges transmit heat to the stud or bolt, causing it in turn to expand at the rate of its coefficient of linear expansion. But normally this is such an infinitesimal difference that it can only be calculated on paper or measured under laboratory conditions. You can ignore it in the field.

Initial bolt stresses for flanged joints must be high enough to insure a tight joint. But at the same time you'll want a reasonable safety factor, to be sure stress doesn't come too close to the yield point of the bolt material. Usual practice in field-erected flanged joints is shown in table at left. According to Crane Co's survey, these initial stresses are customary when alloy-steel bolt-studs are pulled up by hand. Experience has proved these stresses OK for American Standard steel flanges. Values in this table are graduated by size—if you can't apply higher torques, it's best to keep initial bolt stresses at about one-third of the material's yield point.

Stress can be applied by one of three methods: (1) figuring proper torque and pulling up the well-lubricated bolt-stud by that amount (2) figuring elongation needed for the desired stress and pulling the bolt to that length (3) prestressing bolt with a pneumatic tensioner. If you're

Initial stresses for American Standard steel flanges

Size of alloy-steel bolt-stud (note 1)	Average stress applied manually, lb per sq in. (note 2)	Approximate torque to obtain stress, ft-lb (note 3)	Elongation, in. per in. of effective length (note 4)
¾	52,000	175	.00173
⅞	48,000	255	.00160
1	45,000	370	.00150
1⅛	42,500	500	.00142
1¼	40,000	665	.00133
1⅜	38,000	860	.00127
1½	36,500	975	.00122
1⅝	35,000	1285	.00117
1¾	34,000	1700	.00113
1⅞	33,000	2200	.00110
2	32,000	2350	.00107

Notes: (1) Coarse-thread series, 1 in. and smaller; 8-pitch-thread series, 1⅛-in. and larger. (2) Average stress applied by maintenance man in assembly, using a lever and wrench or sledging. (3) Based on well-lubricated threads. (4) Based on modulus of elasticity of 30,000,000
Courtesy, Crane Co, Chicago, Ill.

BEAM TYPE TORQUE WRENCH flexes as torque increases, changing relationship between the pointer and the dial near the handle

PNEUMATIC IMPACT WRENCH has integral torque control. Its setting can be adjusted to any value between 500 and 1000 ft-lb

using the second method, you will need to calculate total elongation of the bolt. Do this by multiplying the value given in the table for elongation (in. per in.) by the effective bolt length (distance from center of nut at one end of the stud to center of nut at the other end).

Most operating engineers work with seals, know the effects of gasket flow, bolt pressure and flange-surface finish on sealing action, and the consequences of improperly tightened bolts. In this connection, remember that properly specified new gaskets and flanges *should not leak*. If they do, it's a signal that they were tightened improperly.

Torque is a twisting, the effort we apply in turning a nut, bolt or cap screw. It is measured in units such as ft-lb or in.-lb—in other words, length multiplied by weight. Let's say you have a ft-long socket wrench extending horizontally from a nut. If you hang your own 200-lb weight on the wrench's end, you're applying 200 ft-lb or 2400 in.-lb of torque to the nut.

Last month we said a bolt stretches about 0.001-in. per in. of length for each 30,000-psi tensile stress. Torquing is the common way of getting this stretch. How much torque do we need? A torque chart supplied by the wrench or equipment manufacturer—table at left—will give the approximate answer.

Three basic torquing tools are the (1) spanner (2) impact wrench (3) torque wrench. Bolts can also be expanded by heat or with tensioners, but in the strict sense neither of these methods is torquing.

(1) Hand-tightening with a spanner or a box, socket or monkey wrench is an old hit-or-miss, unscientific way. But it's still used. To make matters worse, someone is likely to add a long piece of pipe to the wrench for leverage, or try a hammer as a "convincer". Results are over-stretched bolts or stripped threads, making the flange assembly useless and sometimes dangerous.

(2) Impact wrenches are pneumatic- or electric-power driven, or spring-actuated. There are many models on the market, each with its own distinct features. Photo, above right, shows a heavy-duty pneumatic-operated impact wrench with integral torque control. This wrench shuts itself off once it has reached the proper torque. Torque

setting can be adjusted to any value between 500 and 1000 ft-lb, simply by turning the adjustment screw with an Allen wrench. A lockout button lets tool do the work of conventional impact wrench with full 1½-in. bolt capacity.

Calibrators are used to determine proper settings for many types of air-pressure-operated and torque-control impact wrenches. They are especially necessary when you're installing high-strength structural bolts. Calibrator gage reads directly in lb of tension. Actual bolts, nuts and washers from the job are tightened in the calibrator —a new set of these should be used each time the wrench is applied. Adjust wrench setting until a series of three bolts can be tightened to more than the needed minimum tension, and no bolts break.

Spring actuated wrench, photo at left, is self-contained, hand-operated—a rachet type impact wrench. It uses principles of the hammer, lever and spring; automatically delivered rotary hammer blows or impacts produce the torsional force. This wrench will apply brute force to loosen a frozen nut or it can be switched to the precise task of tightening a bolt to desired tension with little manual effort. Torquing is based on the relationship between number of impacts and bolt tension—a click clearly indicates each blow, and mechanic estimates the tension by counting the clicks.

Sensory torque wrench sounds a distinct click, then imparts a strong impulse to the operator's hand the instant a preset torque has been applied to the nut. Operator can also read torque on the dial.

Center photo shows a torque wrench which releases automatically when proper torque is reached. To set it, operator turns the handle until the micrometer adjustment indicates desired torque. Then he locks the handle, attaches the socket, applies his wrench to the nut or bolt and pulls smoothly until the wrench releases.

Torque readings, do not always mean that the load has been placed in the stud. With torquing devices, the true load can only be determined by actually measuring elongation of the bolt in question. You can do this with an extensometer, measuring over the full length of each bolt. Or you can use a measuring device.

Stretch

Most stretch here

Effective diameter

Stretch

Removed metal

Stretch

Removed metal

A   B   C   D   E

Let's talk about bolt stretch and measurement

Tightening flange bolts uniformly and to the right tension is a vital chore. Experience shows that when you're working with high-temperature flanges, uneven bolt loading causes real trouble. Not only do gaskets deform, but flanges bend, bolts relax and finally the flanges start leaking. More loading may twist bolts or strip threads.

Many articles have been written that discussed prestressing bolts with hydraulic bolt stretchers instead of tightening them with wrenches. Now let's see how bolts act when they're pretensioned or torqued. We'll look at some typical and special-purpose bolts, see how to get the most life out of a bolt and how much it can be loaded to do most efficiently the job for which it was designed.

Elongation of the bolt is an effective measure of the stress it's carrying. Lab tests show that any steel is elastic, will stretch about 0.001 in. per in. of its length under a load of 30,000 psi *if* load is kept below the yield point. Soon as load is removed, steel returns to its original length. This happens regardless of the steel's hardness, whether it is stainless, carbon or alloy.

External forces acting upon a material produce tension, compression, bending, shearing or torsional stresses. And these stresses cause deformation. But as we mentioned before, if the stress isn't too great the material's elasticity returns it to its original shape soon as the stress is removed. If the metal doesn't take its original shape and dimensions, the *elastic limit* is exceeded. Then the bolt is *dead;* dangerous to use, it should be scrapped.

You can easily find the elastic limit or *modulus of elasticity* by dividing load in psi by elongation, in. per in. of unit length. Bolts made of steel average 30,000,000-psi modulus of elasticity, a figure to remember.

What bolt load should each stud carry? According to Edward Valves, Inc, bolting used with steel valves for pressures over 300 psi is usually alloy steel, having a yield strength of 105,000 psi. For best initial tightness in flanged joints, they recommend bolting loads under half this value, around 45,000 psi.

Ideal bolt would be tightened by upsetting the end like a rivet, sketch **A.** If it could be ideally loaded, entire bolt between the two heads would stretch evenly because the shank is minimum diameter over its entire length. But

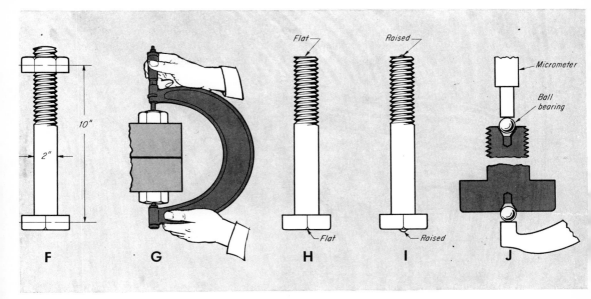

most bolts are constructed like **B.** Their weakest part is the threaded section from the bottom of the nut to the end of the threaded shank. Effective diameter is measured from the thread roots as shown, so most stretch of the loaded bolt takes place in only a part of the length. Of course, parts above and below nut ends are not stressed uniformly, and threaded portion carries a varying stress. But rule of thumb for any bolt says total stress is divided equally along portion of bolt between nut centers.

Reducing shank is one way to increase the strength of a bolt, especially to withstand shock loads. Besides the remarkable increase in impact strength, modified bolts with reduced shank allow greater stretching or elongation if overtightened. That gives the nut less chance to loosen because of vibration. Resistance to fatigue failure also goes up and the bolt lasts longer.

Most of bolt **C**'s shank is turned down to the same diameter as the smallest diameter through its threads. Shank is reduced to a given optimum size, which depends on type, method of manufacture, etc. Interesting thing about this bolt: although it is modified to weigh less, it can absorb several times more impact energy than conventional bolt **B,** which may surprise you.

Another way to reduce shank area or theoretically make it equal to effective diameter through threaded section would be to drill metal from shank's center, bolt **D.** This would have same strengthening effect as removing metal from outside the shank. Bolt **E** is a special-purpose model, used for the connecting rod of large diesel engines, or for standard railroad-car carriages, large rock crushers, etc.

How to figure loadings? F is a 2-in.-dia bolt 10-in. long between nut centers. How much should this bolt be loaded? Answer is no mystery.

Let's assume we want stress of 45,000 psi. Divide 45,000 by the modulus of elasticity, 30,000,000. That gives us 0.0015 in. Since the bolt is 10-in. long between nut centers, multiply 0.0015 x 10 in. This means we'll stretch the bolt 0.015 in., which is to be added to the bolt length over ends. If all bolts in the flange are elongated or stretched exactly this amount, and taken up in sequence by tightening bolts opposite each other, flange and gasket should give maximum service for life of installation. This method is important and should always be followed.

Method of measuring bolt stretch with a micrometer is shown in **G.** But if both ends of bolt are not parallel, readings will be off. You could grind or face off ends at right angles to the bolt shank by chucking in a lathe, **H.** Another way is to machine a small raised tit or bump on both ends as shown in **I.** Still another way: use a center drill and countersink both ends, **J.** Then you can set in ball bearings, measure with the micrometer.

Any measuring device that reads in thousandths of an in. will do. This might be an outside micrometer, vernier caliper or a dial indicator as used with tensioners.

If you're using a torque wrench, remember that friction between the threads and bottom nut will vary. So there will be a difference of elongation between bolts even though torque indicator reads the same for each. More-accurate method: predetermine the elongation, then, through sequence tightening, set all bolts to this value.

Test your
fastener know-how

Russel, Burdsall & Ward Bolt and Nut Co

Q 1—In a gasketed joint, protection against leakage increases proportionately as the strength of the bolt is increased.   T F

Q 2—Rated clamping force indicates the strength of a bolt only if it is tightened properly.   T F

Q 3—Undertightening a bolt is as serious a mistake as overtightening. In other words, a minimum torque rating is as important as a maximum one.   T F

Q 4—A jam nut should always be instal!ed above the full size nut.   T F

Q 5—Severe overtightening of either a fine- or a coarse-thread bolt will lead to the same results; the bolt will break.   T F

Q 6—Mathematically, a fine-thread bolt has a greater stress area than a coarse-thread one. It is, therefore, safer.   T F

Q 7—A bolt with no head marking has an ultimate tensile strength of about 70,000 psi.   T F

Q 8—Head markings should always include the manufacturer's identifying symbol.   T F

Q 9—A bolt should always be replaced with one of an identical head marking, or with one known to be stronger.   T F

HEAD MARKINGS on hex head bolts, indicate their strength level. For safety, you must know the identification symbols

Energy-systems engineers know that you can't replace one bolt and nut with another simply because it fits. The safety factor built into threaded fasteners works only if the right class of bolt is selected and the assembly is tightened properly. Industry standards have therefore been established to help users with fastener selection and proper installation procedures. Unfortunately these standards are often misunderstood or misapplied.

How about you? Do you know all that you should know about fasteners? Let's review your answers to the True or

False questions above. If you work with fasteners and miss more than two answers, you had better read carefully the discussions following each of the answers.

A 1—FALSE. Bolt-users who experience joint failures often go to larger bolts. But these do not solve their problems, because the bolts are not torqued to their proper loads. Besides, using larger bolts for greater tightening often results in bowing of the flange. This was the problem of a manufacturer of sealed compressors designed to contain a toxic gas. The larger bolts resulted in leakage. Why? Because larger, stronger bolts caused more bowing of the flange and still greater leakage. We solved the problem by recommending harder gaskets to minimize compression, and slightly smaller bolts tightened with less torque. The bolts were also spaced closer together to help achieve a more uniform seal area.

A 2—TRUE. One of the most misunderstood points about threaded fasteners is their rated clamping force. Rated clamping force indicates the strength of a fastener *only* when the unit is properly torqued. The more a fastener is torqued, the stronger it gets—within, of course, the limit of the fastener material's inherent strength. Thus, clamping force of a threaded fastener depends on two factors: (1) inherent strength of the material, and (2) tightening procedure. If a fastener has a rated clamping force of 20,000 lb but is tightened to develop only 5000 lb, the extra inherent strength does nothing but resist shear.

A 3—TRUE. Recently I noticed stenciled NAS (National Aerospace Standards) numbers alongside some screws fastening the skin of a jet plane. They read "maximum torque 20 in.-lb." This worry about overtightening is typical. But what some people don't realize is that undertightening can be as damaging as overtightening. In what way? Undertightening can cause external premature failure caused by a joint loosening under applied load. On the other hand, severe overtightening—which can cause the bolt to stretch beyond its elastic limit—will prevent its safe reuse after the joint is disassembled. The point to remember is that listing a minimum torque is as important as listing a maximum one. So don't overlook the importance of torque.

A 4—FALSE. Jam nuts help prevent loosening of nut and bolt combinations. It seems to be traditional to place the thinner jam nut on top of a standard hex nut. Yet from a technical point of view, this is incorrect. The jam nut, when placed on top, merely adds frictional resistance at the interface with the standard nut. But if the position is reversed, not only are the same things accomplished, but frictional resistance is also introduced in the thread flank. The standard hex nut, being stronger than the jam nut, stretches the bolt thread off lead. (Lead is the distance a threaded part moves axially in one complete rotation with respect to a fixed mating part.)

A 5—FALSE. Ten years ago, the automotive industry used 80% fine-thread fasteners. Today, the industry uses 80% coarse threads. Why the switch? Because, mathematically, a fine-thread fastener has a greater stress area than a coarse-thread one. But when a nut is tightened on a fine-thread bolt, full strength cannot be developed in the stress area. If a nut is overtightened on a coarse-thread fastener, the bolt fractures. With a fine-thread unit, overtightening will cause the thread to fail in progressive shear. This results in a difficult-to-detect stripped connection. For this reason, one automotive company now uses only coarse-thread fasteners in all critical connections. If a coarse-thread bolt is overtightened (torque wrench out of adjustment), it breaks. In effect, breaking becomes an automatic inspection procedure, and the bolt is removed and replaced. If fine-thread bolts are used, the torque wrench could kick out as usual and the machine might have a de-

ASTM and SAE grade markings for steel bolts		
Grade marking	Specification	Material
	SAE – Grade 1 ASTM – A 307	Low carbon steel
	SAE – Grade 2	Low carbon steel
	SAE – Grade 5	Medium carbon steel, quenched and tempered
	ASTM – A 449	
A325	ASTM – A 325	Medium carbon steel, quenched and tempered
A490	ASTM – A 490	Alloy steel, quenched and tempered
BB	ASTM – A 354 Grade BB	Low alloy steel, quenched and tempered (medium carbon steel, quenched and tempered may be substituted)
BC	ASTM – A 354 Grade BC	Low alloy steel, quenched and tempered (medium carbon steel, quenched and tempered may be substituted)
	SAE – Grade 8	Medium carbon alloy steel, quenched and tempered
	ASTM – A 354 Grade BD	

*Manufacturers identification symbol required

fective stripped-thread connection. This of course could be critical and lead to a major breakdown and costly repairs.

A 6—FALSE. The reason for increased use of coarse-thread fasteners in the automotive industry is safety. Don't be misled by the fact that fine-thread fasteners have a greater stress area. This is true theoretically, but in practice coarse-thread fasteners are safer, especially in installations in which fasteners are tightened automatically.

A 7—TRUE. Identification markings on fastener heads must be understood for safe and proper use. No grade identification marking on a bolt head means that the bolt is an SAE Grade-2, or its ASTM equivalent, and has an ultimate tensile rating of 69,000 psi. This classification is now being raised to 74,000 psi.

A 8—TRUE. One of the head markings (SAE Grade-5) of the type shown in the table above indicates an ultimate tensile strength level of 120,000 psi in sizes of ¼ through 1 in. A bolt with such a head marking cannot be replaced by an unmarked bolt because the latter would not meet the required strength level.

A 9—TRUE. Many markings are used to indicate strength levels of bolts. The main ones are shown in the table above. The manufacturer's identifying symbols are also part of head marking requirements. That is why it is important that bolt users check these markings carefully and that nothing should be done to alter the physical properties of the bolts. Posting these markings pays dividends. ●

20 ways smart maintenance management can get their money's worth out of epoxy resins

Almost across the line in the energy systems, epoxy resins are stepping in as versatile maintenance tools. Epoxies are thermosetting materials. Converted by a hardener and cured by heat, the resins become hard, infusible plastics; some harden at room temperatures.

Epoxies are naturals for coatings, castings, laminates—some are tough enough to cast drop-hammer dies. They are superior adhesives for most metals, unaffected by a wide range of acids, alkalies and solvents. A filled-epoxy-resin system has virtually zero shrinkage.

But on the other side of the coin, epoxies aren't elastic. They won't bond to everything. To help you skirt the pitfalls, here are answers to 20 questions on best use of epoxies for industrial repairs.

1 I know epoxy compounds do a good job, but I've heard they are toxic. How about that?

A True, epoxy compounds can cause skin irritation, and they should be well ventilated. But if your men are properly instructed, resins can be used without fear. A pamphlet about safety is available from most manufacturers.

2 Why do epoxy-resin repair compounds come in two parts? Mixing is a problem. Why can't they be packaged in one part like the cold solders?

A Epoxies harden chemically, rather than drying through solvent evaporation. The best of them contain no solvents, so they are 100% reactive. Result: very little shrinkage; however thick you apply them, they harden throughout the mass, not just on the surface. Mixing is certainly a nuisance. But when you want maximum performance, it's worth the trouble.

3 I notice that the makers of epoxy-resin cements want us to wait, sometimes up to 24 hours, before using a component that's been repaired. But often we have to get a machine back to work in a hurry. What can be done about this?

A Some of the so-called *room-temperature-setting* epoxy cements harden in a few hours, some need a day. In general terms, the cements with the best chemical or heat resistance require a longer time. But any of them can be hardened in minutes with a little warmth. Heat from any source will do the trick: the sun, lamps or a heat gun.

4 According to some claims, epoxy resin bonds everything to anything. Is that 100% true?

A No. Nothing we know of does everything equally well except maybe money. Epoxy-resin cements adhere best to cold-rolled steel, aluminum, magnesium; to glass (but not Pyrex); to concrete and plaster (if they are dry); to wood (if it's roughened); to other epoxy materials, phenolics and polyesters. They do fairly well with copper, brass and low-nickel stainless steels. But they don't adhere well to nickel, tin or zinc, nor to thermoplastics like polyethylene, vinyl, styrene.

5 Is it necessary to clamp parts while bonding them?

A No. Contact pressure is all that's needed.

6 The manufacturers of epoxy cements stress proper surface preparation. Sometimes this is easy, sometimes not. Coming down to brass tacks, what preparation is really basic?

A The area *must* be dry, oil-free, clean, and that's basic. Beyond this point, adhesion is greatly improved by sandblasting or hand sanding; this not only assures that the area is clean and bright, but increases the area of bond. Epoxy repairs should be made in a dry atmosphere.

7 Should I weld, or use epoxy instead?

A Depends on your need. Welding is stronger and takes heat, which epoxy won't. But epoxy is often good enough.

8 **How permanent is an epoxy repair?**

A This question is asked frequently, but it's impossible to answer; there are too many variables. Epoxy resin is a plastic. Whatever you put in it—steel, iron, aluminum, sand—the epoxy is doing the work. You can repair steel with epoxy, but epoxy doesn't *replace* steel. Why? Because steel has many times the tensile strength of any epoxy-resin formulation.

9 **How do I know what epoxy compound to use where? Aren't they all more or less the same?**

A Epoxy-resin compounds are formulated for a given job. A "general-purpose" resin is a compromise of the possible properties: speed of set, heat resistance, chemical stability. Any epoxy cement on the market, if it's any good at all, will do run-of-the-mill repair jobs. But if you have one particular property in mind you have to buy a formulation which emphasizes that property. Heat resistance and chemical stability are gained at the expense of speed, for instance. Rapid hardening is obtained at the expense of mixing life. Putty-like qualities are achieved by sacrificing wetting ability.

10 **I used an epoxy compound to fill seams in an old wooden tank on our roof. It stood up well for a while; then the epoxy fell out. Why?**

A Even so-called *flexible* epoxies are not elastic. A polysulfide-rubber caulking compound should have been used here, and for similar repairs.

11 **I tried to bond some galvanized metal with epoxy and got a very poor bond. Why?**

A That's because epoxy bonds poorly to zinc, tin, nickel. Trick here is to remove the zinc coating by sanding. Then you'll have a good bond.

12 **We repaired a steam line in our plant with epoxy, and it didn't take long for steam to start leaking again. Are epoxies overrated?**

A Ordinary epoxy cements, the kind you buy at the hardware store, can be ruined by live steam. You'll have to try again with a special epoxy compound formulated to stand up to steam. Spread it on glass cloth and wind the cloth about three times around the (dry) hole. Let the repair harden 24 hours, then turn on your steam. You'll be happy. Of course superheated steam is another story.

13 **I understand you want me to turn off the water in the system before I make an epoxy repair. But I can't turn it off. What can I do?**

A There are various ultraquick-setting compounds. Some people insert wooden slivers—the wood swells, holds up long enough. Have glass tape cut and ready to apply. Spread your epoxy cement on the tape and wrap it around on itself three or four times. Set the epoxy up rapidly by applying 175-F heat.

14 **Should we use epoxy or grouting compounds for expansion joints in our cold-climate plant?**

A Epoxies, even the flexibilized epoxy compounds, are too rigid—not suitable for applications where there is a lot of movement. Use polysulfide sealants instead.

15 **Will epoxy resin grout heavy machinery?**

A Yes, but choose a very heavily filled compound. Or you can fill it yourself. Start with epoxy resin and a flexibilized hardener. Mix these and add dry aggregate, about five parts of aggregate to one of liquid material by volume.

16 **I have an old eroded condensate tank made of concrete. The water is somewhat acid and around 200 F. Can I repair this tank with epoxy materials?**

A No. That's a mighty rough environment for an organic compound. Instead, cut out all the weak concrete, dry the tank thoroughly, paint on an epoxy-Thiokol tie coat, and, soon as this is tacky-hard, trowel or pour a thick coat of portland-cement concrete.

17 **We are changing the process in our chemical plant, exposing our many electric motors to highly corrosive acid fumes. Will coating the windings with epoxy resins protect them?**

A Epoxy resins are noted not only for adhesion and insulating value, but also for chemical resistance. This includes most solvents, alkalies, and acids except for the strong oxidizers (for example, nitric acid). But to be on the safe side, the windings should be thoroughly impregnated as well as coated on their exterior.

18 **We may change the hydrocarbon lubricating oil in our large steam-turbine generating units to a newer synthetic phosphate-ester fire-resistant lubricant. Would an epoxy coating prevent insulation on the electric-generator end from softening —if some fluid gets by the shaft seals?**

A Epoxy-resin insulation will not be adversely affected by the phosphate-ester type lubricant. But whether an external application of epoxy will succeed here depends on the type of existing insulation. Ideally, generator windings should be completely impregnated, then coated.

19 **Piping, valves and other equipment around the cooling towers in our petrochemical plant are corroding badly because of the humid atmosphere. Could we spray them with an epoxy-base paint?**

A Yes, epoxy coatings provide excellent resistance to corrosion. Despite a tendency to fade and chalk, they are dependable and longer-lasting. But epoxy compounds are just that—compounds. Even the 1-component high-temperature epoxies contain a latent hardener. Choice of the proper hardener is at least as important as choice of the proper epoxy resin.

20 **Can we use epoxy to seal a multiple-wire electric connector?**

A Epoxy will work, but an elastic polysulfide or silicone rubber would probably do a better job for you.

Good Maintenance Cuts Fluorescent Lamp Costs

These pointers on fluorescent lamp maintenance will help you reduce lighting troubles and cut costs. Good service means good light which is an important factor in any plant

1 Board for testing lamps and starters before putting them in the fixture, should also be used to test them before discarding

▶ IN PRACTICALLY EVERY TYPE of power plant and industry today, fluorescent lights are used more and more. Although most fluorescent lights give little trouble, a certain amount of attention will pay dividends in low-maintenance costs.

Here are some practical reasons why prompt and careful attention should be given to the small amount of maintenance required by fluorescent lamps. In most cases, the fixture and its installation represent a considerable investment of time and money. Prompt attention, and proper care guard this investment, insure better light, and reduce the number of replacement parts required.

Whenever something goes wrong with the light, auxiliaries or fixture, turn the light off immediately. This requirement should be thoroughly advertised and preached throughout the plant so that maintenance can be started before troubles get out of hand. Some defects put abnormal loads on the lamps themselves or the ballasts and starters. Turning the light off when trouble starts quite often saves burned-out starters, lamps, ballasts, and other ruined equipment.

Trouble Symptoms. Common symptoms of trouble include blinking or flickering lamps and smoke. For other troubles see check list opposite. The fixtures quite often hum and smoke, and sometimes even smell.

Some of these conditions are normal developments, while others result from one or more simple causes. First, let us look at certain indications, plus a few experimental tests that tell where the trouble lies.

When testing defective or damaged equipment practice maintenance economy by trying the simple, least expensive remedies first. For example, if a lamp flickers for some time, it may be the fault of the voltage, ambient temperature, tube, starter or ballast. So check voltage and ambient temperature and look for a cold draft striking the lamp.

Sometimes a lamp fails to light because the starter has worked loose from its socket. Making such a checkup first quite often saves the cost of a lamp or starter.

If at all possible, keep a few spare lamps, ballasts and starters on hand for emergencies. They often help to make quick checks right in the fixture and

thus determine which element is at fault.

Tests. Here are some trouble-shooting tests that can be used to advantage on most fluorescent lighting fixtures:

Sometimes the cathodes are broken or air leaks into the lamp. In this case test each end of the lamp separately by connecting the two base pins in series with a 60-watt light bulb on a 115-v circuit. If cathodes are in good shape, the lamp filament will light and the fluorescent bulb will glow. If the fluorescent lamp does not glow, it indicates a burned-out cathode or that air has leaked into the tube.

A 60-watt light bulb is all right for testing fluorescent lamps from 14 to 40 watts. For larger fluorescent tubes of from 65 to 100 watts, use a 200-watt, 115-v test lamp. Commercial cathode testers are also available, and can be purchased for this test if desired.

Test Board. A simple test board, as shown in Fig. 1, will also help in checking starters and lamps right on the job. This test board, of course, checks only the tubes and starter and does not show wiring defects or ballast troubles in the fixture itself. When testing the various parts, be sure that test conditions such as correct voltage and ambient temperature are ideal. Be sure to test each lamp and starter.

When using the test board, place the fluorescent tubes in the holder so that they make good contact and then close switch A. Press pushbutton B and hold it down several seconds. If the lamp is good, both ends glow and it lights immediately upon releasing button B.

To test a starter, place a good lamp in the holder of the test board. With switch open, insert the starter in the

2 Made from an old starter switch, this manual starter helps in testing lamps, starters and auxiliaries without taking the lamp fixture apart

FLUORESCENT LAMP RECORD

	LAMPS						STARTERS						BALLASTS					
	Originally Installed			Replaced			Originally Installed			Replaced			Originally Installed			Replaced		
Date	Number	Cost	Date	Number	Cost	Date	Number	Cost	Date	Number	Cost	Date	Number	Cost	Date	Number	Cost	
Total																		

Use this record sheet to keep track of number and cost of equipment replacements. It also serves as a check on lamp life

Fluorescent lamp troubles listed with possible causes and suggested remedies. List serves as a maintenance check chart

socket and close the switch. If the starter is in good condition, the lamp will light and burn steadily.

If lamp ends glow continually, but the lamp does not light, it shows that starter contacts are stuck. If the lamp does not glow, this shows that starter contacts have not closed. This defect generally indicates that breakdown voltage is too high. For satisfactory testing results, be sure that voltage at the test board is never less than 110 v.

Portable Starter. The manual starter shown in Fig. 2 checks starters and lamps while they are mounted in their own fixture. When trouble symptoms appear, remove the starter from the lamp fixture and insert the manual starter. Move the switch on the end of the cord to the "on" position. The ends of the bulb should glow if the lamp is good and the wiring correct. When no glow appears, insert a new test tube in the fixture. If the ends of the new unit do not glow with switch in the "on" position, it indicates that lamp holders, wiring or ballast is defective.

On a good lamp the ends glow with the switch "on," and it starts immediately when the switch is moved to the off position. If the lamp does not light, snap the switch several times from on to off before condemning the tube. When the light comes on properly with the manual switch, but does not operate with the old starter, this indicates that the trouble is in the starter.

Clean Lights. In addition to testing and replacing lamp parts when they are defective, it is also a good idea to keep reflectors and glass light bulbs clean. This alone can increase light output as much as 30 to 50 percent.

To clean porcelain or synthetic enameled reflectors, remove lamps, then wash reflectors with soap and water and rinse with clean water and dry with a cloth. Dry-wiping leaves much of the dirt on the reflector and does very little

good. Wash the lamp tubes thoroughly, especially if the fixtures are mounted in a dirty, smoky atmosphere. Caution: Don't use either a strong alkaline or acid solution on aluminum reflectors.

Hum produced by the magnetic forces within the transformer or ballast may be objectionable in certain office or shop locations. If the lamps are in a room with noisy machinery, obviously this small hum does not matter.

Manufacturers design and assemble ballasts to minimize hum, but sometimes the fixture is noisy in spite of this. If the fixture is noisy, first check ballast mounting, and, if necessary, attach it to a rigid part of the fixture. Second, fasten all metallic parts securely because loose screws or bolts amplify the noise. If excessive hum persists, and cannot be traced to a defective ballast, replace it with a new ballast unless the noise can be eliminated upon disassembling.

269

NET, woven to protect steam turbines, is inspected by Art Cannon. Ropes are strong enough to support workmen, materials. This is also good insurance against outage

SAFETY NET with dropcloth is spread above machinery before painters work on 70-ft-high ceiling in power plant. Paint splashes and dropped tools kept from machinery

▶ "SAVE THE SURFACE and you save all" is good advice, but just *how* do you save the surface? For an answer to this and other plant-painting questions, I visited Art Cannon of Oliver B Cannon & Sons, in Philadelphia. With 32 painting crews about the country, Art Cannon is our leading industrial painting contractor.

Regardless of who does your painting, you should know how this topnotch industrial specialist goes about a job. Then you will see that painting takes plenty of know-how.

Staff Meetings. Art Cannon has staff meetings each month. His key men fly in from the field. Sales, estimating, production and testing personnel attend. An expert from a paint manufacturer. one from a chemical firm, or a specialist on corrosion, etc, is usually present.

Current jobs are discussed and solutions arrived at. Art's own paint chemist gives lab findings on new products. He also advises on tough surface-protection problems and licking some unusual local condition.

Everyone has a chance to put in his oar. So supervisors go back into the field ready to tackle their problems from a new slant. Do these meetings pay off? They helped put this firm at the head of the list, competing with local contractors everywhere.

Another reason Cannon does so much work in power plants is that his experts do critical areas, without a shutdown. The plant's regular painting crew does routine safe areas that almost any painter can do.

Every Cannon supervisor has ten years

SAFETY SIGNS tell plant employes that men are working above them here

PROTECT running machinery with dropcloth over frames so no shutdown's needed

IDENTIFY pipes with removable plastic strips, either lettered or plain colors

...Information Before Painting Your Plant

of previous training with the firm. Safety is a big item, especially when working high above and around machinery, as in power plants. A green painter always works with a trained man until he knows the ropes.

Estimating is important. Condition of each building or section is listed as good, fair, poor or very poor. When to paint, next scheduled repainting, quotation and comments are listed. Here are some typical remarks I copied from an $85,000 estimate:

"(1) Buildings can wait until rehabilitation program is scheduled, then interior and exterior painting can be done. (2) Buildings viewed by public need painting for public relations and appearance. (3) Can wait until more essential buildings are finished. (4) Building important to production should be done before more replacements are needed."

Scheduling. All preparations are completed *before* plant is shut down. That means putting up protection, having enough men for shutdown equipment so work isn't held up, etc.

His average power-plant team of painters is about 40 men. Idea is to put as many men as possible on a job at one time as overhead costs the same for 10 men as for 100.

Safety. Each painting super checks scaffolds, ladders, rope and other equipment before sending them to a job. Safety nets are homemade because Art isn't satisfied with ones on market. Every rope, scaffold, ladder, etc, is numbered and recorded on an office card. Age, repairs and all other data are

jotted down. All rope is scrapped after one year's service, regardless of condition. "CAUTION, men working overhead" signs are placed below painters. Because of an excellent safety record, this firm has very low insurance rates.

Doing the Job. Here are a few rules on painting: Wire-brush, scrape or chip all rust and corrosion spots. Vacuum-clean to prevent loose dust, flyash, etc, getting into machinery.

Spraying can be more expensive than a brush job as it needs more protection. But, usually, spraying gives a thinner coat, does not protect as well.

Never red-lead new metal exposed to weather. When building outdoor metal tanks and steel structure, let them rust for a year or two until all mill-scale rusts off. Coat of primer and paint lasts on such a job. If red lead is painted over new metal, mill scale peels off with time, taking new paint with it.

In our electric-generating plants, 35% burn coal, 35% burn oil or gas, 30% are hydroelectric. Newest plants can switch from coal to oil, whichever is cheaper. These dual-design boilers present temperature problems.

Ducts get hotter when oil-fired, need a heat-resistant paint to cover changing temperatures. Aluminum is often used —one coat. Aluminum's varnish-vehicle stacks up with other enamels. If not aluminum, use a light pearl gray, a semi-alkyd oil-base paint, with a semigloss finish. There is no good heat-resistant paint in green or blue. Here are a few color schemes for power plants.

Turbine Room, ceiling—oyster white for right light reflection. Ceilings are

usually 60 to 70 feet high. Light green or cream adds good character to this room. Walls—light gray. Machinery —soft green enamel.

Boiler room, ceiling—oyster white. If walls are brick, well struck and pointed, or of good monolithic concrete, leave unpainted. Don't paint unnecessarily if surface is attractive and needs no protection. Once painted, a surface must always be painted. Use heat-resistant aluminum or a gray paint for boilers and adjacent piping, boiler doors, buck stays and ducts.

Piping. Paint the same color as boiler. Paint all insulated piping connected with boiler feed system a contrasting enamel of dark gray or green. On other machinery, such as pumps, drivers, coal pulverizers, coal feeders, use a gray hard-drying gloss surface. Colored plastic bands are best for piping. If over 12 colors are used, have labeled bands. Adopt U.S. Bureau of Ships color scheme.

Crane. Paint high-visibility yellow for safety.

Railing. Paint top rail black, so soiling from hands won't show. Paint middle rail safety yellow, and the rest gray. Yellow middle-rail makes stairway "stand out" for quick access to trouble spot.

Fire Lines. Paint red or use red bands to identify them.

Stacks. Paint with special coating to withstand temperature. Paint aluminum stacks at top with broad black band to camouflage smoke discoloration.

"Remember." Art Cannon reminded me as I left, "It costs more not to paint."

CORROSION is tough on steel sash if putty fails between glass and the sash

CALK steel sash with gun after digging out old putty—before painting sash

SPRAYING: not usually as good as brush work, calls for mask in close quarters

Making a Glass Face For Gage

1 Broken glass gage faces are often neglected until needle is damaged and dirt discolors face so graduations can't be read

2 Circular glass face can be made easily. First step is cutting template to exact size from thin piece of sheet metal

3 Next, place template on piece of glass and hold against two edges. Template is needed because glass is hard to mark

4 Hold glass and template in a bucket of water. Now cut glass with tin snips under water. Water prevents glass flying out

5 Dotted lines show approximate shape of glass after first cutting operation. Caution: Don't try to cut circular line

6 Glass will have rough edges like this when corners are cut off with tin snips, cutting close as possible to template

7 Glass can be made into perfect circle by filing rough edges with bastard file like this. Use gloves to protect hands

8 Finished glass like this can be made by any operator from scrap glass found around most plants. Protect your gages

13
Bearings, oil grooving

ABC's of Oil

WHY GROOVE IS NEEDED

AT REST, with oil cut off, oil leaks out of bearing. Weight of shaft squeezes oil film until shaft and bearing come into direct contact as shown

DRY SHAFT starts climbing up on bearing's side when shaft starts turning. Always make sure oil is flowing to bearing before turning shaft

PUMP-ACTION of shaft starts as speed increases. Then oil is forced between shaft and bearing. Oil separates shaft so it slides to bottom

HYDRAULIC PRESSURE creates oil wedge when shaft turns rapidly. This pressure is so great that oil wedge forces shaft up against opposite side

WHERE TO GROOVE

SPLIT BEARINGS must be chamfered at edges. That holds true for both halves. Shims must be cut back to keep from scraping oil off shaft and should almost touch shaft near ends, to keep oil in

HEAVY-DUTY split bearing (especially if bearing is long or uses heavy-bodied oil) needs extra attention. Besides chamfering split edges, longitudinal groove at top assures even oil supply all around

LONG BEARINGS must have more than one source of oil supply to feed long distributing groove at top. For every 8-in. length of bearing, add another oil supply or bearing will starve for lack of oil

HEAVY-DUTY bearings turning at slow speeds need an extra oil groove in lower half just before pressure area. This gives heavy oil supply just before pressure area, where support is most needed

PRESS-ROLL bearings have thrust **up** on top bearing and **down** on bottom bearing. Oil distributing groove must be placed on side before pressure area. Then there's good oil-wedge from hydraulic action

BELT PULL fixes oil-groove location. Oil-distribution groove is on bottom so pressure area on top is supplied with even oil wedge, not broken up with grooves. Watch this bearing for leakage at top

Grooving Bearings

NO OIL GROOVE in bearing is bad. It causes oil to be carried around shaft directly under oil-supply opening. Then because shaft gets no oil near ends, bearing runs hot and finally burns

OIL FOLLOWS least resistance. If bearing has no oil groove, but is out of line as shown, oil flows along clearance between left side of shaft and bearing. Right side runs dry because oil doesn't spread

RIGHT WAY is to have oil-distributing groove shown here along top of bearing. Oil from supply flows into groove to within a half inch or so from bearing's end. As shaft turns, oil is carried down

FOUR-PART BEARINGS must have distributing groove in top. All quarter-box edges must be chamfered to keep from scraping oil off shaft with sharp edges. Cut shims to fit along chamfered edges

RING-OILED one-piece bearing needs groove at top only. Ring carries oil from sump to this distribution groove. Hydraulic action of oil carried by shaft forces oil-wedge between shaft and bearing

. . . **CIRCLE-GROOVE** pressure area of any bearing. Oil escapes through groove into low-pressure area—then won't support shaft

. . . **CROSS-GROOVE** pressure area of any bearing. Grooves cut down bearing area besides helping oil film to leave the area of wear

VERTICAL BEARINGS need groove around top to feed oil down by gravity. If leakage is bad, spiraled groove against shaft rotation helps shaft force oil up. Thrust bearing at bottom takes load

THRUST-BEARING needs oil fed to shaft and not to collars at top. Centrifugal force throws oil away from shaft so that it oils the collars' thrust sides. Oil fed to collars' top would not feed shaft

. . . **CHAMFER** split-bearing to end of bearing. Oil leaks before it can be forced under shaft by hydraulic action of oil wedge

HOT BEARINGS—

WHY BEARINGS GET HOT

1 **NO OIL** or not enough oil is common cause of hot bearings. Check lubricators and bearing at set periods to be safe

2 **DIRTY OIL** causes undue friction and bearing soon heats. Keep oil supply clean and stored away from dusty air

3 **OUT-OF-LINE** shafting and bearings concentrate load on small area. Then bearing runs hot. Line up the bearings

FIRST STEP TO COOL

1 **FLOOD WITH OIL** by screwing up on cup adjusting nut. Or unscrew oil cup and pour oil into the bearing

2 **REMOVE WICKS** if wick fed. Then pour oil into box to flood bearing. Idea is to have oil wash through bearing

3 **FEEL SHAFT** near bearing with back of hand for heat. If temperature goes up after flooding, try to find the cause

EMERGENCY STEPS TO COOL

1 **DON'T GET EXCITED,** that won't help. If engine bearing is sizzling hot, throw the load off engine quickly

2 **GET ASSISTANT** to throttle engine to slow speed, **but don't stop.** If engine is stopped bearing may seize the pin

3 **REMOVE CENTER-OILER** (if this type). Flood bearing with oil can while engine turns slowly, until bearing cools

Causes & Cures

4 WRONG-GRADE OIL for bearing won't support shaft. Oil film collapses —shaft and bearing metals make contact

5 UNEVEN BEARING or journal surface has same effect as out-of-line shaft. Reduced bearing area causes heating

6 TIGHTLY KEYED bearings allow no space for oil between bearing and shaft. Loosen bearing bolts until shutdown

4 EXAMINE OIL on back of fingers like this. If oil is gritty and dirty keep flooding bearing to wash out grit

5 INCREASE WATER-FLOW if water-cooled bearing. Feel discharge line to make sure water is flowing and not plugged

6 POUR OIL on shaft if temperature keeps climbing. This will help carry away some heat so the bearing can cool

4 NEVER POUR COLD WATER on bearing or shaft. Bearing will contract and seize shaft. Shaft may even crack.

5 SLACKEN SETSCREWS or bearing cap nuts and flood bearing with cylinder oil. Heavy oil stands heat much better

6 SPEED ENGINE SLOWLY when bearing cools and keep checking temperature. If okay, throw on load.

1 Here are the basic parts of a ball bearing. Going from left to right we have the outer ring or race, retainers to position the balls and then the inner race

4 This is most popular single-row ball bearing; carries a fair thrust load

2 Working surfaces of a tapered-roller bearing's cup, cone and rollers are tapered for true rolling action. This type has high thrust capacity in one direction

3 Straight-roller bearings are suitable for loads at right angles to the shaft (called radial loads). Needle and hollow-spiral roller bearings are in same boat

Pointers

Every operating man needs a little dope from time to time on anti-friction bearings. Here's a concentrated dose* telling of popular types with a dash of maintenance info

▶ STARTING FROM SCRATCH. let's define an *anti-friction* bearing. It's simply a group name given to ball and roller bearings. The fundamental idea behind them is to change sliding to rolling friction. This is handled by simply putting a ball or roller between the stationary and rotating parts of the bearing.

Advantages. Low friction, high reliability, little maintenance and close clearances that remain throughout the bearing's life are the chief advantages of anti-friction bearings. Keep in mind that there is but little difference between the "running friction" of anti-friction and plain bearings when maintained properly. Remember that the starting friction of anti-friction bearings is much lower than that of plain bearings. Why? Well, picture the shaft in a plain bearing when it is at rest. The lubricant is squeezed out between the point of contact at the load zone. Now as soon as the shaft starts to rev up, you have metal-to-metal contact between shaft and bearing. It's at this time that you get greatest wear.

Likewise a ball or roller bearing will

5 Filling-notch bearing is suitable for high speeds

6 Double-row type takes big radial and thrust loads

7 If shaft is not in line, use self-aligning bearing

8 Felt-seal type holds the lube in and keeps dirt out

on Anti-Friction Bearings

show but little wear during many years of service if properly selected, installed and maintained. Remember that bearing housings can be sealed snug to hold in oil or grease. This explains why anti-friction bearings don't need relubricating very often.

Bearing Parts. Most all anti-friction bearings you will come across have four main parts: an outer ring, inner ring, balls or rollers, and the retainer or cage, Fig. 1, 2 and 3.

The retainer or cage has the simple job of spacing the rolling elements. If a retainer were not used, the balls or rollers would not stay equally spaced as they enter and leave the point of maximum bearing load. Then again, the retainer reduces any friction that would be present if the balls or rollers were allowed to touch one another. Some types, however, mainly needle bearings, are designed to be used without a retainer for certain applications.

Type Bearings. While there are many types of bearings on the market, we'll just talk about those most generally found in the small plant. The most popular single-row bearing is shown in Fig. 4. It's sometimes called the deep-groove, non-filling slot, no-notch, or continuous-raceway bearing. This bearing will carry high thrust loads in either direction since the grooves are deep.

The four general classifications of roller bearings are tapered, square, journal and needle rollers. Each is available in designs suited to specific applications. Probably the most common tapered-roller design is the single-row bearing shown in Fig. 9. This particular design is suitable for heavy

radial loads, heavy thrust loads in one direction, and for use at moderate speed. When you come across heavy thrust loads in either direction you will probably find the double-cup tapered-roller bearing used, Fig. 10.

A worthwhile point to remember in connection with straight-roller bearings, Fig. 11, is careful ordering of replacements. To make sure identification is correct, get the following information: (1) manufacturer's name (2) bearing number if possible (3) bore (4) outside diameter, width, number and diameter of rolls.

Maintenance. No, we're not going to go ahead and describe the features of all bearings shown on this page. They're to let you know the popular types. Your job as an operating engineer is to keep them in top-notch shape. In a nutshell, this means keeping them clean and well lubricated. Follow the advice of the manufacturer regarding

lubrication. He'll specify the kind of grease or oil to use, how often to renew it and how much to use.

While talking about lubrication, remember that more than one operating engineer has caused more harm than good by over-lubricating. This can easily happen if you are applying grease with a hand gun. Avoid this trouble by opening the plug located on the bottom of most bearing housings. Keep this out for a short period after starting up machine; this gives any excess grease a chance to drain off. When you're sure that all excess grease has been removed, replace plug.

As a parting shot, remember: Dirt is a bearing's worst enemy. So work with clean tools in a clean place.

When wiping bearings, use only clean lint-free rags. Then, before putting new bearing in place, wipe housing to prevent any dirt present from slipping into the clean bearing.

9 Tapered-roller style usually mounted in sets to take heavy thrust

10 The double-cup tapered job handles thrust in either direction

11 Here's straight-roller bearing without inner race; runs direct on shaft

HOW TO HANDLE BALL BEARINGS

Courtesy of Marlin-Rockwell Corp.

DON'T LEAVE BEARINGS EXPOSED OVERNIGHT

PLACE UNWRAPPED BEARINGS on a clean newspaper, never on dirty bench. Oily bearing surface collects dust particles from air if left open overnight. Dust shortens bearing life. Always cover open bearings with cloth during maintenance

CLEAN BALL BEARINGS LIKE THIS

WASH IN CLEAR KEROSENE while moving inner ring back and forth to loosen grease. Blow with air, but don't let bearing turn or it may scratch. Flush again, then blow. Oil to keep from rusting. A clean-oiled bearing runs very quietly

REMEMBER THESE TEN POINTS

1 Don't take bearings from box until ready to install
2 Keep hands and tools clean
3 Don't wash packed grease out of new bearing
4 Keep grease can covered
5 Use clean rags; they are cheaper than bearings
6 Keep assembly bench clean
7 Lay bearings on clean newspapers
8 Get shaft and housing seats perfectly clean
9 Paint housings inside to seal core-sand, chips
10 Cover exposed bearings if left overnight

14
Gaskets, packing, mechanical seals

Gasket material must flow

To seal tightly, gasket must conform to joint's mating surfaces. Most gaskets are made relatively soft, so compressing them causes the material to squeeze into surface irregularities. With some materials, heat does the same thing. Another form of "flow" results from combined compression and dragging action, called attrition.

Bolt pressure, surface finish affect sealing action

For any given gasket material, a certain pressure produces just so much flow. That may be enough to fill irregularities of a fairly smooth surface, say an ordinary machined flange, but not those of a rough surface, like a serrated flange (*top sketch*). If joint can take more bolt pressure, tighter squeeze will fill both sides (*bottom*).

If flange serrations are "phonograph" type (continuous spiral) they must be completely filled or there's a path for fluid pressure to escape. With concentric serrations, partial filling of grooves may be enough. Parallel shaper or planer grooves make life tough for gasket—there are paths for pressure escape on two sides of the flange.

Gaskets supply the <u>give</u> that makes a tight

In piping, and all kinds of machinery, we've got the problem of making pressure-tight joints between two rigid elements —flanges, say. We can do it, but only if flanges are machined to mate perfectly, and can be counted on to stay that way. Such a joint is practical only in rare cases, so we turn to gaskets. Because they're designed to give, they make up for imperfections of the average joint—surface roughness, small misalignments, etc.

At relatively small cost, gaskets do away with need for precision machining. And it's easier to replace a gasket than to re-machine surfaces when something goes wrong.

How Gaskets Seal. To stop leakage, flat gasket in a flanged joint must be squeezed between joint faces, tight enough so it exerts more pressure against faces than fluid pressure tending to leak past it. Since most joint surfaces aren't perfectly smooth, gasket material must "flow" into hills and valleys of joint faces to get good mating and a tight seal with no leakage pathways.

Usually, gasket materials are made to flow by compressing them between flanges. In some cases, heat may do the job, as when a bell-and-spigot joint is sealed with molten lead. In other cases, copper ring on screw plugs, for example, flow is caused by combined compression and dragging action— called attrition.

Surface Condition. In ideal contact, gasket material fills every depression in joint surfaces completely. This depends on gasket material and condition of joint surfaces (*sketches above*). Whether partial filling of tool marks, etc, is good

enough depends on nature of grooves. Ordinary spiral tool marks (so-called phonograph finish) leave a continuous path for leakage unless filled. But concentric serrations leave no such path and complete filling isn't as important. Really tough problem comes on joints finished with planer or shaper tools—parallel grooves give paths for pressure leakage on two sides.

What Finish? Smooth surface is usually best. Serrations —V-grooves 1/32 in. apart—are OK with soft gaskets. Never use phonograph finish with metallic gaskets, or any gasket for high pressures.

If joint surfaces are smooth, problem of leakage can't be solved by merely refinishing with rougher tool marks. If you do that, bolts must be strong enough to stand extra stress needed to flow gasket into deeper grooves.

Forces on Joint. Pressure in pipe tends to push flanges apart. Bolts resist this. In addition, they must be loaded enough to squeeze gasket and keep pressure from leaking past. Since pressure pushing joint apart (*hydrostatic end force*) works against exposed portion of flange faces as well as internal diameter of pipe, always try to use gasket with about same ID as flange. This keeps hydrostatic end force to minimum, reduces needed bolt load.

Seating Force. The force needed to make gasket material flow into surface imperfections is known as *seating force*. It's usually expressed as unit stress in gasket, psi, and is considered independent of internal pressure.

For most gasket materials, it's important to apply a high

These forces act on joint

In a working joint, with pressure in the pipe, hydrostatic end force tries to push flange faces apart. Bolt load resists this action, and must be enough greater to put sufficient squeeze on gasket to maintain a tight seal against internal pressure.

Carbon print of pressure distribution shows most pressure, and hence most

Bolt layout and tightening make a big difference

flow, is around bolts (*upper sketch*). Joint of this type will give trouble because sealing effect midway between bolts is poor. Solution lies in better bolting layout, with more uniform pressure distribution.

If you tighten one bolt full, flanges cock out of parallel. Taking up on another bolt won't bring them back

even. Resulting poor pressure distribution encourages leaks, speeds failure. Right way is to take up all bolts hand tight first. Next tighten two bolts diametrically apart, then two as near 90 deg from first two as possible, and so on as indicated in sketch. Tighten bolts again after joint comes up to operating condition.

---

seal in a rigid joint

stress initially, to get full flow and good seating. Then load on gasket can be reduced without hurting seal. Such reduction of load on gasket usually occurs naturally as a result of bolt expansion and creep, gasket "settling."

Bolting Layout. Most gaskets are used in bolted joints. Carbon prints, *sketch above*, show pressure is greatest close to bolts, weakest midway between. This puts a premium on good bolting layout—adjusting number and spacing of bolts to distribute total load uniformly around gasket. It also means knowing the operating conditions to be met and the kind of gasket best suited for them. In other words, the gasket should be designed into the joint, not selected by cut-and-try after the joint gives trouble.

Tightening Bolts. Follow right sequence in bolt tightening (*see sketch above*). Equally important: Don't overstress bolts by whanging away at the end of a long hickey or oversize wrench. Overstressed bolts take a permanent set, won't keep the squeeze on gaskets.

Gaskets at Work. With joint at working conditions, hydrostatic end force tries to separate flanges. Bolts, gasket and flanges expand. Net effect is less contact pressure on gasket. That's why bolts are easy to take up after they've been in service. At high temperatures, don't overlook effect of bolt creep, slow flange deformation. And sudden temperature and pressure changes give joints a real wallop, hence often cause serious leakage problems.

To see how designers build resiliency into gaskets to meet operating conditions, turn to following pages.

Four types of gasketed joints

Confined

Unconfined

Partly confined

Self-energized

Gaskets may be broadly classified by the type of joint they seal:

Confined gaskets have no freedom to flow except for slight clearance between mating faces.

Unconfined gaskets are free to flow away from and toward fluid pressure.

Partly confined gaskets are free to move on either inside or outside rim.

Which Joint? In unconfined joint, gasket can be blown out if it isn't loaded enough. With confined or partly confined joint, worst that can happen is a slow pressure leak.

Because unconfined joint is most practical it's commonly used for piping and many other joints. Most metallic gaskets can be considered as self confining. Some gasket makers design semi-metallic gaskets for unconfined joints by putting a retaining ring, or similar device, outside.

Self-energized joints are used mostly on boiler handholes and manholes. These may have only one bolt; don't depend on bolt pressure to keep them tight. Fluid pressure working on entire plate area provides gasket load.

Critical period is while pressure is being brought up—initial bolt pressure drops to zero as fluid pressure climbs. So you have to tighten bolt from time to time until final pressure is reached. In this type of joint, make sure surfaces are even, gasket area is big enough for load.

Right Way to Remove, Lay Out,

First: We remove old gasket

1 Open drain to relieve pressure before breaking any pressure-vessel joint

2 Remove all bolts if flanges are sealed with a full gasket (shown at right)

3 If gasket is ring type, remove only enough bolts to slip old out, new in

Second: We lay out new gasket . . .

1 Bolt circle is found from bottom (or top) of holes

2 Radius of bolt circle is equal to half of diameter

3 Draw line through circle's center for opposite holes

4 On 6-hole flange, radius is distance of bolt holes

Third: Now the new one is cut out

1 Gasket cutters turn out clean holes fast. Top unit needs only one hand

2 Tin snip is next best way. Never hammer gasket as that causes lumps

3 Punch holes after placing material on hard wood to protect the punch edge

Cut and Install Your Gaskets

4 Flange spreader: handy to open joint. You might require two on some jobs

5 Steel wedges, driven in at opposite edges, take place of flange spreader

6 Scrape out old gasket. Be sure every particle is removed. Don't harm face

5 To check distance, walk dividers around the circle. Lay out holes a little larger than their actual size

Or trace it out

A   B   C

TRACING a gasket's form (if you can get at flanges) is often faster. *A*, graphiting the flange; *B*, making impression; *C*, finished

And last: We install it

1 Leave one corner on gasket so you can hold it for positioning in flanges

2 Slip one bolt through flanges and gasket. Position gasket, place bolts

3 When line is under full pressure, tighten again in the sequence shown

PUMP GASKET MUST BE MADE RIGHT

1 Coat machine surface with graphite and oil so form impresses on gasket sheet

2 Hold gasket sheet firmly against machine and make impressions by rubbing sheet against corners of openings for clear markings

3 **B** Chisel slot for tinsnips or scissors like this

A Punch holes same size as studs and against wood to protect punch

4 With tinsnips or scissors cut gasket hole ⅛ in. or so smaller in diameter than hole in machine

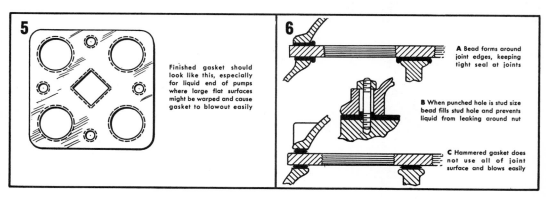

5 Finished gasket should look like this, especially for liquid end of pumps where large flat surfaces might be warped and cause gasket to blowout easily

6 **A** Bead forms around joint edges, keeping tight seal at joints

B When punched hole is stud size bead fills stud hole and prevents liquid from leaking around nut

C Hammered gasket does not use all of joint surface and blows easily

13 ways you can stretch the life of your O-rings

Hydraulic-seal application seems to generate plenty of questions. Of course, most of the time I'm solving specific problems. With so many variables in each situation there can be no one formula for success: as in any engineering equation, the solution is usually the compromise that works best.

A lot of design data has been published and is available to anyone. Manufacturers offer excellent manuals. But some basic answers aren't spelled out, and nevertheless we must cope with the problems. From my experience I've distilled 13 fairly general questions about O-ring design and application that crop up again and again. Now I'll try to answer them to help with your stubborn sealing problems.

1 **Some static O-ring seals show the same wear you'd expect to find in a dynamic seal. Why?**

A This happens in a pulsing-pressure system: the O-ring moves with the pressure pulses. As a field remedy, add a backup ring to prevent this motion. If you're tackling the problem in the design stage, change the groove width to equal O-ring's cross section plus an allowance for squeeze.

2 **What is the limit of pressure an O-ring can seal?**

A There is no limit—theoretically. If the O-ring doesn't get a chance to extrude into a clearance gap, it's unaffected by pressure. But from the practical viewpoint that's rarely attainable at very high pressures because of bolt stretch, diametral growth, etc.

A single backup ring with normal dimensional toler-ances seals up to 3000 psi. By putting two backup rings downstream you can extend this to about 4500 psi. But beyond that point you must resort to more drastic measures. Using an O-ring plus a fairly heavy Teflon ring, backed with a bronze ring, it's possible to seal 100,000 psi statically and 50,000 psi dynamically. In this design the O-ring expands the Teflon ring, which fits into lips of the bronze ring, in turn expanding them to produce the required zero clearance gaps.

3 **Why do O-rings sometimes extrude in situations where system pressure and downstream clearance gaps are within accepted limits?**

A There are two likely reasons. In systems where the pressure pulses rapidly, short-duration transient pressure peaks may shoot up to two or three times the indicated system pressure. Gages won't show those peaks, but suitable pressure pickups feeding an oscilloscope will tag them. So you see, your O-ring *isn't* working within accepted limits—it's subjected to shock loads beyond its ability to resist.

A less-frequent cause stems from industry's longtime reliance on Durometer hardness as a measure of an O-ring's limits. From the extrusion standpoint, modulus of elasticity is the value that really sets the limits of O-ring application, and the value may vary considerably from the Durometer reading.

In any given class of compounds, Durometer and elasticity values are close to parallel. But if we shift from, say, Buna N to silicone, we find that a 70-Durometer silicone will have about the same modulus of elasticity as a 50-Durometer Buna N. This is also true to a lesser degree of other types of compounds. Since it's actually the modulus of elasticity which resists stress caused by pressure, there can be varying degrees of extrusion resistance in compounds of the same Durometer value.

4 What can we do about spiral failure of O-rings?

A Spiral failure, which looks like a long-pitch screw thread cut around the O-ring, is caused by eccentricity between the groove root and the mating surface. That produces a varying squeeze. This in turn causes varying amounts of roll around the periphery of the O-ring on the pumping stroke. The return stroke usually will not have exactly the same cycle of eccentricity, so the ring may not completely recover. Repeated again and again, this cycle eventually overstresses the ring to the point of failure.

That's the cause. The cure isn't that simple. Since eccentricity may result from excessive clearances, high side loading, breathing cylinder tubes, too-short bearings, etc, you'll have to review the entire mechanical picture so you'll know what measures to take to correct it. In many cases it is better and more economical to select a different type of seal, such as a V-ring.

5 How coarse a surface finish can we use for static O-ring seals?

A This is always a good topic for a hot discussion between design engineers and the machine shop. Break the answer down into three categories: (1) In relatively steady-state pressure systems or where pulsations are not violent, O-rings in 3/16- and 1/4-in. cross sections can tolerate 125-micro-in. finishes. (2) If pressure pulsations are violent these rings should have a 63-micro-in. finish at the groove root and opposite face but can still live with 125 micro-in. for the groove walls. (3) Because of more limited sealing-contact area, the smaller 1/16-, 3/32- and 1/8-in. sections should have a finish at least as smooth as 63 micro-in. on all four sides. And if pressure pulsations are violent, they will be better off with 32 micro-in. at the groove root and opposite face.

6 What are the upper and lower temperature limits of present-day O-ring materials?

A That's a simple question, but it won't get a simple answer. It's impossible to be general in certain areas.

Fluid or gas environment, type of equipment, duty cycle, life requirement and other factors all enter the equation.

In normal design, where liquid or gaseous environments are compatible with use of silicones, −130 F is the accepted low limit. But in some situations, using squeeze in the 80% range, you can get good static O-ring seals in the cryogenic region of −350 F.

At the other end of the thermometer, in the 450-500 F range, silicones offer good life sealing gas environments, and fluoroelastomers like Viton and Fluorel can handle many fluids. As an extreme, successful tests of 1-minute duration have been run at 1000 F.

Here are two vital rules of thumb: (1) Don't exaggerate temperature problems. That will probably cost money and may result in inferior performance. (2) Do give all—and I mean all—details to your O-ring manufacturer or supplier. Take advantage of his know-how.

7 What about rotary O-ring seals?

A Rotary O-ring applications must be approached with caution! I've seen successful applications under conditions which seem impossible. For example, I've seen an O-ring working as a shaft seal in fairly high-pressure pumps. It's not really a full-fledged seal, but allows a controlled leak with leakage bleeding back to the intake side. What happens here is that the O-ring wears enough to leak a little; leakage then provides a heat-removal and lubricating film.

A rotary O-ring seal always contacts the same small area, so it retains its own frictional heat. Also the system fluid tends to squeeze out, so the seal must run dry. These two factors can become a vicious circle, each reinforcing the other. So unless rubbing speeds and pressures are low, say 250 rpm and 25 psi, or required life is short, better consult with your sales engineer before making a decision.

8 An O-ring rod seal is running leakfree and dry to the touch. How does it get lubricated?

A Looks like a paradox, doesn't it? But there's a fairly simple answer: the rod is *not* dry. Viscosity and film strength of the fluid actually carry a microscopically thin film of fluid past the O-ring. Since there is no accumulation of fluid on the rod, the same forces are carrying the film back past the O-ring against pressure on the retraction stroke.

This can be easily demonstrated in tests. When the rod is solvent-cleaned after extension and before retraction, wear rate increases very markedly.

9 What is the ideal surface finish for dynamic O-ring seals? By the way, can it be *too* good?

A Whether a surface finish can be too good for O-ring seals has been a moot point for years. I believe that your surface can be refined to the point where the fluid film's adhesion to the surface is reduced, so that friction and wear are actually increased! But this is only an opinion. I have never seen a surface this highly refined and I'm sure no one would want to pay for it.

Barring this questionable situation, we can say that the better the finish, the longer O-ring life will be. However, as pages in this chapter explain by illustrations and text, one should always keep in mind that it is the character of the surface profile rather than the RMS micro-in. value which is important. Cylinder tubing, with a maximum roughness of 35 or 40 micro-in., is an excellent surface from the seal's standpoint—its profile is like rolling waves. But the same RMS value produced on some materials by a hard grinding wheel could have sharply defined peaks and valleys, 100 micro-in. or more in total deviation. This would be like the same waves approaching the beach: with the same volumetric deviation, they've assumed the sharper, more violent surface of breakers.

To demonstrate the effect of surface profile on the seal there's a small experiment requiring only simple equipment that you *could* make, although I don't particularly recommend it. First, sit on a $1/4$-in. bearing ball, then on a $1/4$-in. thumbtack. One might be very uncomfortable, but the other is intolerable!

10 Does pressure affect life of dynamic O-ring seals?

A Of course it does, to an extent which depends on surface finish, fluid, O-ring material and clearance gap.

Better surface finishes, as we've said, mean longer life; in addition, pressure increases won't chop as big a percentage off the ring's life. But the O-ring material is usually determined by the seal's environment, so you'll seldom get the chance to improve life this way.

Since size of the clearance gap together with O-ring's modulus of elasticity affects extrusion, keeping the gap to the lowest practicable value will improve the ring's life under increased pressure. All else being equal, if you can prevent extrusion, lower-Durometer compounds will resist wear better. This indicates that if extrusion failures do occur, backup rings would be a better solution than increasing the O-ring's Durometer.

Take two hypothetical cases. In one case extrusion can occur, in the other it's prevented. Let's tabulate the relative effect of pressure on life, assuming both rings have a 1-million-cycle life at 1000 psi.

Pressure, psi	Cycles of life with extrusion allowed	Cycles of life with no extrusion
1000	1,000,000	1,000,000
2000	300,000	800,000
3000	150,000	600,000
4000	70,000	400,000
5000	20,000	200,000

These hypothetical figures show that the decrease in life is exponential when extrusion can occur, practically straight-line when it's prevented.

11 How can we increase friction in O-ring seals?

A Frictional resistance of an O-ring is made up of two components. One is a function of internal energy of the ring itself, generated by the squeeze. The other comes from energy generated by the applied pressure.

If the ring is not under pressure, the second component is zero. At about 250 to 300 psi the two components become equal; beyond this point the component caused by squeeze becomes relatively unimportant. Hence, reducing the squeeze helps in low-pressure applications.

In higher-pressure situations where the squeeze component is unimportant, we can resort to a mechanical device in addition to refining the surface finish. We can use an *undercut* or *dovetail* groove to hold the ring. In this type of groove, pressure will force part of the O-ring down the undercut sidewall, allowing less area of contact and therefore reducing friction. Best results come when the sidewall angle is adjusted so contact length equals about 25% of O-ring section diameter at operating pressure. It's also necessary to drill bleed passages to the bottom of the undercut wall, so fluid can't get trapped to prevent movement.

Sometimes friction can be cut by using a dry lubricant. A number of preparations on the market consist of Teflon or molybdenum disulfide dispersed in resin. In some types of applications they'll be very useful, but only trial will determine which ones.

12 Will it help the situation if I can identify the cause of an O-ring failure? If so, how should I go about it?

A Knowing the mode of failure can be extremely helpful in correcting the equipment. Barring faulty O-rings, which do sometimes get installed, the basic causes of failure are normal wear, extrusion, nibbling, spiraling, pulsation cutting, installation damage, damaged equipment such as scored bores or rods. Appearance of the ring helps you decide which of these is shortening its life:

Normal wear will appear as an more or less evenly worn flat area at the rubbing surface.

Pure extrusion first appears as an undercut groove around the periphery of the ring near the ID or OD on the side towards the pressure. By the time the ring fails an entire section of it may be sheared away.

Nibbling is a form of extrusion which occurs when side loading or some other cause results in shaky travel of the piston or rod. Chunks are bitten out around the periphery. Preventing nibbling, like preventing extrusion, is a question of clearances or anti-extrusion devices.

Spiral failure appears like a very long-pitch screw thread cut around all or a part of the O-ring. We talked about spiraling in answer 4.

Pulsation cutting looks somewhat like extrusion damage. But instead of being undercut, the ring is gouged toward the center of the section.

Installation damage usually appears as a single chunk sheared off the periphery of the ring. It's usually caused by lack of a suitable entrance chamfer or an attempt to install an oversized ring, or perhaps the ring was pushed into place across a threaded shaft. A second rubbing surface will produce fine grooves across the face of an otherwise usable ring. Grooves in a badly worn ring are caused by high-pressure jets of fluid through leakage areas.

13 How can I be sure that a new O-ring-seal design will work out?

A This question is the easiest of all to answer. Your O-ring manufacturer has had long experience dealing with the various combinations of pressure, temperature, fluids, pulsations, equipment design. When in doubt, consult him. Take advantage of his experience with other people's problems. But let me stress that optimum O-ring design, like any other, is usually the compromise that best suits a variety of factors. So give your expert all the details.

1 Remove *all* old packing. Aim packing hook at bore of the box to keep from scratching the shaft. Clean box thoroughly so the new packing won't hang up

2 Check for bent rod, grooves or shoulders. If the neck bushing clearance in bottom of box is great, use stiffer bottom ring or replace the neck bushing

3 Revolve rotary shaft. If the indicator runs out over 0.003-in., straighten shaft, or check bearings, or balance rotor. Gyrating shaft beats out packing

6 Cutting off rings while packing is wrapped around shaft will give you rings with parallel ends. This is very important if packing is to do job

7 If you cut packing while stretched out straight, the ends will be at an angle. With gap at angle, packing on either side squeezes into top of gap and ring, cannot close. This brings up the question about gap for expansion. Most packings need none. Channel-type packing with lead core may need slight gap for expansion

Now let's learn how to install packing correctly and cut down costly outage

11 Open ring joint sidewise, especially lead-filled and metallic types. This prevents distorting molded circumference—breaking the ring opposite gap

12 Use split wooden bushing. Install first turn of packing, then force into bottom of box by tightening gland against bushing. Seat each turn this way

Cross expansion Sectional Diagonal

15 Always install cross-expansion packing so plies slope toward the fluid pressure from housing. Place sectional rings so slope between inside and outside ring is toward the pressure. Diagonal rings must also have slope toward the fluid pressure. Watch these details for best results when installing new packing in a box

4 To find the right size of packing to install, measure stuffing-box bore and subtract rod diameter, divide by 2. Packing is too critical for guesswork.

5 Wind packing, needed for filling stuffing box, snugly around rod (for **same size** shaft held in vise) and cut through each turn while coiled, as shown. If the packing is slightly too large, never flatten with a hammer. Place each turn on a clean newspaper and then roll out with pipe as you would with a rolling pin

8 Install foil-wrapped packing so edges on inside will face direction of shaft rotation. This is a must; otherwise, thin edges flake off, reduce packing life

9 Neck bushing slides into stuffing box. Quick way to make it is to pour soft bearing metal into tin can, turn and bore for sliding fit into **place**

10 Swabbing new metallic packings with lubricant supplied by packing maker is OK. These include foil types, lead-core, etc. If the rod is oily, don't swab it

13 Stagger joints 180 degrees if only two rings are in stuffing box. Space at 120 degrees for three rings, or 90 degrees if four rings or more are in set

14 Install packing so lantern ring lines up with cooling-liquid opening. Also, remember that this ring moves back into box as packing is compressed. Leave space for gland to enter as shown. Tighten gland with wrench—back off finger-tight. Allow the packing to leak until it seats itself, then allow a slight operating leakage.

Hydraulic-packing pointers

First, clean stuffing box, examine ram or rod. Next, measure stuffing-box depth and packing set—find difference. Place ⅛-in. washers over gland studs as shown. Lubricate ram and packing set (if for water). If you can use them, endless rings give about 17% more wear than cut rings. Place male adapter in bottom, then carefully slide each packing turn home—don't harm lips. Stagger joints for cut rings. Measure from top of packing to top of washers, then compare with gland. Never tighten down new packing set until all air has chance to work out. As packing wears, remove one set of washers, after more wear, remove other washer.

Lubricate...

Adjust flow
from casing

Lantern ring

Lantern ring

...with these simple lubricators...

Spring loaded

Deadweight

Weighted lever

Gravity

For some types of jobs,

We've learned that most types of packings must leak slightly to lubricate the shaft. But sometimes you can't count on such leakage. And if fluid sealed is abrasive or corrosive, you want to keep it out of the packing. One answer is to use a *lantern ring*.

Typical lantern ring, first sketch, is a channeled ring, with radial holes, fitting loosely into the stuffing box. Connection from the casing feeds fluid under pressure to outer channel, radial holes carry it to inner channel. It then flows along the shaft in both directions, insuring lubrication.

When you need to keep abrasive or corrosive fluids out of the packing, you can't feed the lantern ring from the casing, must turn to an outside source. Water, grease, oil, other fluids, used for

this purpose, are usually called "sealing liquid." Under pressure, the sealing liquid bucks fluid trying to leak out of the casing. That reverses flow through packing on the casing side of the lantern ring.

Ring position is important: With lantern ring nearer casing, flow into casing is increased as sealing liquid meets less resistance, flow to atmosphere is decreased. For maximum flushing effect into casing, put lantern ring at bottom of packing set, with no packing between it and casing. This "flushing connection" is used on equipment handling highly abrasive fluids. If you change ring position, be sure ring lines up with sealing connection.

Feeding schemes for sealing liquid are numerous (sketches). For some

services, a spring-loaded grease cup is a simple solution. Dead-weight and weighted-lever lubricators are easy to maintain and work without trouble. Just set weight for pressure needed and keep lubricator filled. Large valves are often sealed against vacuum with a funnel filled with water. This also keeps packing from drying out.

Sealing fluid can be injected continuously or periodically, as by a pulsating lubricator or grease gun. Trouble with periodical injection is that fluid from casing gets into stuffing box as soon as injection pressure drops. Continuous sealing flow, at pressure only slightly higher than casing fluid, is best, prevents contaminating casing fluid with sealing liquid.

Oil-bottle hookup, sketch, uses casing

Chamber and fins

Lubricant

Cooling fins

Cooling
chamber

Cooling the packing set

Heat developed in a packing is proportional to product of friction and surface speed. Packing temperature rises until rate of heat dissipation equals rate of heat generation. You can hold down heat by dissipating it with a water-cooled shaft, or by water-jacketing the stuffing box.

Valves often have a so-called cooling chamber between packing and valve body. But there is little temperature drop at packing because most of heat flows through metal, first sketch.

Regulator valve, second sketch, is for extremely high pressures and temperatures. Here cooling fins around stuffing box of long valve stem dissipate heat. Because valve stem moves constantly, must not stick, cartridge of molybdenum-disulfide lubricant is fitted into stuffing box. It won't carbonize or oxidize, helps keep stem free.

High packing temperature leads to excessive thermal expansion that increases load, causing seizure. Cooling and dry lubricant may be only answer.

Oil bottle   Independent system   Differential regulator

packing needs to be lubricated from outside

pressure to inject sealing oil, needs no regulator or pump. Just watch gage glass, keep bottle filled with oil. This system uses little oil. When pump is shut down, drain water from bottle, fill with oil, and it's ready.

Circulating system has relief regulator set a few psi above casing pressure. To carry away heat, line from stuffing-box bottom is piped back to tank (dashed line). Fluid in tank can be cooled, if necessary. Differential regulator system keeps a set pressure between casing, gland.

Corrosion. Before injecting sealing liquid, you should know that fluids trapped in stuffing box often cause corrosion. This is known as crevice corrosion, often takes place with fluids that don't corrode metals in an open vessel,

but cause damage when trapped in clearance between packing and metal. Packing materials suspected of causing this action are copper and brass reinforcements, also graphite. Oily packings are less likely to give trouble.

One solution for packing that's used with water or aqueous solutions is to install it just before machine is used. Then remove it if unit is to be idle for long periods.

Applications that normally need a sealing liquid include:

1. All packing against gases—they leak excessively because of low viscosity. Also vacuum units, steam engines and turbines, h-p autoclaves where agitator enters gas space at top.

2. All packings for liquids of low viscosity or high volatility (such as gas-

oline) used at high speeds and high pressures. Examples: Lube oil injected into stuffing boxes of pumps handling propane; oil between double axial seals on water pumps, etc.

3. Packings for liquids carrying large amounts of dissolved or suspended matter that tends to drop out in passage through packing. Examples: Water-sealed boxes on centrifugal pumps handling caustic soda and on hydraulic turbines operating on muddy or contaminated water.

4. All packings for fluids of any kind that are poisonous, corrosive or too valuable to lose. Examples: Pure gasoline injected into boxes of pumps handling anhydrous hydrogen fluoride in the alkalization process; oil injected into ammonia-compressor rod packing, etc.

Should you swab rods, and oil new packings?

Metallic packing on reciprocating rod may need oil feed, needs no swabbing. Excess oil from swabbing many kinds of steam packings may do harm—isn't needed. You can swab V-type lip packings, but usually even this isn't necessary. Lubricant put into most asbestos packings is all that is needed on steam jobs. Here condensate has lubricating qualities. Laminated-type duck-rubber packings around rams of large hydraulic bloomer-mill shears in steel mills are lubricated by spraying ram with viscous, high-temperature oil every hour or so. But here radiant heat from hot steel is problem.

As for oiling new packing, there is up to 30% saturant in most jam-type packings—so why oil them? Main thing is to see that shafts are not rusty, not covered with abrasive grit and generally in good condition before applying new packing. Like any bearing, shaft should be oily and not dry. If in doubt, wiping the shaft with oily cloth may be good insurance when repacking a stuffing box.

Here's one way to convert

Follow these steps_____

1 Dismantle pump; remove burrs, sharp corners from shaft, also from the shaft threads. Shaft must be free of defects at O-ring location, see sketch

2 Machine the shaft sleeve according to drawing furnished by seal firm. If your pump's shaft has no sleeve, seal will be supplied to fit your present shaft

3 Here are mechanical seal parts, also gland backing-flange machined to fit the gland. Sleeve is machined and on the shaft, ready to receive the seal

7 Slip seal-ring with O ring over the shaft; don't harm O ring. Turn ring so it spreads the drive-spring open, is held firmly in place as shown, sketch

8 Shaft sleeve here has drive-spring with seal-ring, all rotating parts. Seal-ring face is lapped to within a few millionths of an inch; don't harm face

9 Place stationary unit (gland, gland backing-flange, seal-ring, gasket) on shaft. Take care not to harm lapped sealing faces; keep these parts clean

Shaft sleeve — Seal-ring — Gasket
Drive spring
Washer — O rings
— Sleeve nut
— Mating ring
O ring
— Gland
Spring sleeve — Gland backing flange

How seal works_____

▶ MECHANICAL SEAL has come to mean two flat surfaces with sealing faces at right angles to the rotating shaft. This makes it an axial seal, also. Because these two rings take all the wear, they save the shaft. One ring is fastened to the shaft and rotates, while the other is stationary. Usually one ring is metal with a hardened face, while the other is a composition carbon ceramic, or metal. Instead of a metal ring, some small pump shafts have a lapped shoulder that the carbon ring seals against.

One ring is mounted so it moves axially with the shaft. That allows its face to remain in contact with the stationary ring, regardless of shaft endplay, sealing-face wear or sealing face run-out. So this ring, usually called the seal ring, must be mounted flexibly.

The opposing ring, known as the mating ring, is mounted rigidly to the gland or pump housing. This rigid ring is backed up with a gasket or O ring to prevent leakage.

your pump to mechanical seals

4 Mating ring (carbon) is inside gland here and gland is on gland backing-flange. Drive spring is also assembled with seal ring to show how unit looks

5 Slide washer against sleeve shoulder, then slip spring-sleeve against the washer. Sleeve is split so drive spring will hold it tightly against the shaft

6 Place spring over split spring-sleeve with turning motion. Turn right-hand spring clockwise, left-hand spring on shaft's other end counterclockwise

10 Install pump rotor assembly in the pump casing. Center impellers in the pump casing recesses. Lock impellers, sleeves and nuts in place with screws

11 Cut pump-casing gasket so gasket extends beyond the stuffing-box face. Bolt down casing head, then cut off the gasket flush with face at stuffing box

12 With pump closed, gasket flush with casing to prevent leakage between upper and lower halves, assemble back-up flange and bolt the gland evenly

There are many types of mechanical seals—stationary, rotary, balanced, unbalanced, bellows, diaphragm. These and others on the market are covered in detail in POWER's 24-page report on Mechanical Seals.

While mechanical seals are relatively new in industry, most of us use them without knowing it. The water pump in your car and the one in our hot-water heating system at home are the best examples. You don't know you have them because they don't leak and so do not have to be repacked or adjusted.

One disadvantage of a mechanical seal for a large industrial unit is that it must be installed over the shaft's end. That means breaking the coupling and removing the motor. If a seal is installed for the first time, as in these photos, then the pump impeller must be removed as shown. This is best done during the usual annual overhaul, when pump is opened up anyway. Then this job can be done correctly.

An advantage is that these seals don't usually leak visibly. What leakage there is evaporates. But when they do start leaking, it's ordinarily the first sign of trouble. For pumping abrasive fluids, two seals are often used on a shaft, with a clean fluid circulated between them at a slightly higher pressure than the abrasive fluid. This keeps the harmful solids out of the sealing faces.

Seals are now built for shafts revolving at over 40,000 rpm. Fluids sealed include crude oil, gasoline, acids and all kinds of corrosive chemicals. Pump shown handles oil, kerosene and gasoline. Leakage here can mean a fire or asphyxiation in closed spaces. For pumps lifting liquids, leakage through packing on suction side often draws in air, reduces vacuum and increases pumping time unless a sealing fluid is used in lantern ring. With mechanical seals, stripping time is reduced because there is no air leakage.

Courtesy, Sealol Corp, Providence, R. I.

Sealing hot water with mechanical seals?

Answers to ten basic questions cover critical importance of operating temperature, effect of total dissolved solids present and the special role of water treatment chemicals

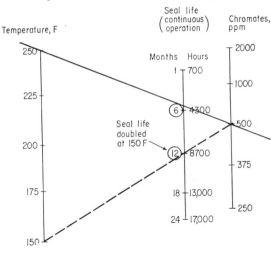

Data above based on following factors:
a–Up to 2-in. diameter shaft
b–Up to 100 psi
c–Up to 3600 rpm
Factors of assembly, mechanical malfunction, or foreign material in system are not included

Q What are the biggest problems when sealing hot boiler feedwater with mechanical seals?

A Hot boiler feedwater presents many unique sealing problems not encountered with other hot liquids. These problems have a definite relation to (1) water temperature, (2) chemical water treatment used, and (3) chemical properties of the water itself, which vary considerably depending upon the source of supply.

Q Which problem causes the greatest headache?

A Temperature is the biggest headache. Reason is that water at a temperature above its boiling point (212 F) is prevented from changing to steam at that temperature when under a higher pressure, as in a pump. In the area between the faces of any seal. a pressure drop (pressure gradient) takes place. Then as pressure drops, the water is allowed to form steam. And the steam opens the seal faces and thus allows leakage.

Q If the boiling point of water is 212 F, why must the seal faces be kept at 160 F, or less, to prevent leakage and ultimate failure of the seal?

A Experience dictates that a maximum temperature of 160 F at the seal faces be set because of the frictional heat generated at these contact faces. For example, if maximum water temperature is 190 F, and if 25 F of friction heat is generated at the faces, the resulting 215 F will generate steam. Of course frictional heat may be much higher than 25 F, depending on the installation.

1 Reducing seal face temperature from 250 to 150 F doubles seal life. Chromate content of water is also an important factor

Q Where does lubricity (viscosity) of water come into the picture?

A From the preceding question, we can see that the higher the water temperature at the seal faces, the poorer is the water's lubrication value. Mechanical seals must always have a pressure gradient film of liquid between their sealing faces. Water is not generally considered a good lubricant. The viscosity of water, at 70 F for example, is about one centipoise. And water gradually reduces in viscosity as the temperature rises. At 212 F, its viscosity is reduced to only 0.28 centipoise. At this low rating the water has lost almost all lubricating qualities and the seal faces will run dry in direct contact (loss of lubricating film gradient). Under these conditions the seal actually operates in a vapor, or steam phase. From this it can be seen that the lower the temperature of the water in the stuffing box, the longer the seal life.

Q If the temperature of the sealing faces must be maintained below the boiling point of water, how can hot water above the boiling point be sealed?

A Above 160 F the temperature deterioration is accelerated until the boiling point of water is reached at the contact mating faces. This takes place regardless of the pressure above vapor pressure in the system. The answer to longer seal life is to cool the sealing faces so they remain below 160 F. Four methods are shown right.

try these stuffing box cooling hookups, then carefully control chemical treatment

2 Hot water from pump discharge is cooled to the required temperature in heat exchanger before entering the stuffing box

3 Cooling water at higher than the pump's gland pressure is forced into the stuffing box from outside source, but must not fail

4 Pumping ring on pump shaft in stuffing box sends hot water from pump through heat exchanger before it cools the seal faces

5 Jackets in the stuffing box and in the gland keep sealing faces cool without cooling water mixing with water inside the pump

Q What is the second biggest sealing problem?

A The physical properties of the water itself cause many sealing problems. For one thing, formation of abrasive mineral is increased as temperature increases. Here is another good reason why water temperature must be cooled to a safe maximum. Aim of cooling is to prevent formation of minerals such as lime, calcium and black iron oxide particles at the seal faces. Black iron oxide is not only abrasive, but forms a hard black deposit over the seal parts and shaft, thus hindering seal performance.

Q What role does chemical treatment play?

A Chemical treatment is fed into the water to prevent formation of scale, pitting, etc, inside the system. While these chemicals are necessary for this service, they are often detrimental to seal life. Specifically, use of chromate treating compounds can be tough on seal faces. The nomograph, left, indicates expected seal life under conditions indicated. Obvious answer is to keep the chromate, or any chemical, at a level consistent with good treatment practice, yet not high enough to cause excessive wear of the seal faces.

Q Since most boiler water contains particles of iron oxide, how are these particles prevented from reaching the faces of the pump's mechanical seal?

A The stuffing box of a boiler feed pump is a natural trap for the collection of iron oxide, as well as for other abrasives. These particles are extremely harmful because some of them are so small that they migrate across the seal faces in the lubricating film. Others find their way into clearances of sliding parts and under the gaskets.

Installing a magnetic separator into the cooling water will remove most of the iron oxide particles. The separator contains Alnico (permanent) magnets. Water enters the separator through one of the taps (direction of flow is unimportant), impinges against a baffle, flows over a second baffle and leaves the separator through the outlet.

Q Since stellite is so extremely hard, won't a stellite sealing ring, running against carbon graphite, prevent excessive wear when sealing hot water?

A For lower pressure systems, stellite has been used. But if the pump should lose suction, even momentarily, stellite, when used for sealing water, will heat check and the seal will probably fail completely.

Q What are four common methods of keeping the wearing faces of mechanical seals below 160 F?

A There are many schemes; above are four popular ones. First method is to bleed discharge water from the pump through a small heat exchanger and run this cooled water into the flushing tap on the gland (Fig. 2).

Another way is to run cooling water from an outside source into the stuffing box (Fig. 3).

Third method (Fig. 4) circulates water with a pumping ring attached to the shaft in the stuffing box. Hot water circulates through a heat exchanger and returns into the packing gland to cool the sealing rings.

Some hot water pumps and boiler-feed pumps have water jackets cast into the stuffing box (Fig. 5). Here packing gland also has a cooling water jacket. Because stuffing box is a dead end for water forced past the neck bushing into the stuffing box, jackets do a good job.

Treat Your Compressor Seals Right

1 **SHAFT SEAL** fitted with coil spring to maintain pressure between rubbing, stationary faces. Friction ring, closely fitted to shaft, revolves with it

2 **BALANCED** packless shaft seal's two steel diaphragms run in oil

3 **BELLOWS** in shaft seal provides a flexible and gas-tight connection

▶ SHAFT SEALS on refrigeration compressors prevent leakage of (1) gas and oil out of the crankcase and (2) air into it. Seals on modern high-speed compressors are often the spring-loaded type, Fig. 1. Another popular design, Fig. 2, uses tempered-steel diaphragms. In both, friction surface is much less than in a stuffing-box-type seal. In Fig. 3, unit has a bellows element.

Coil Spring. In this type, Fig. 1, the entire seal assembly rotates with the compressor shaft. Contact between the seal-face surface and its mating stationary seal seat forms the running seal joint. Both surfaces are highly finished so they are smooth and flat.

The seal face loosely encircles the shaft and is held tight against the stationary seat by the coil spring. Any gas or liquid pressure in the crankcase also acts on seal face, pressing it to the seat. But the spring furnishes main force.

Seal-face back contacts front of the resilient friction ring. This ring is closely fitted to shaft and prevents leakage at this point. Expansion of the ring is restricted by the friction-ring band, keeping seal tight at all times.

Diaphragms, Fig. 2, give a positive balance between internal crankcase pressure and external atmospheric pressure. Case-hardened steel collars retain the diaphragms, which are immersed in oil. Total friction surface is said to be less than 5% of the conventional stuffing-box-type seal.

Bellows. Seal nose, Fig. 3, carries a narrow ring with an accurately finished surface that is made to bear against a hardened-steel collar on the rotating shaft. Collar sealing surface is also highly polished. Flange mounts the seal assembly and the bellows provides a flexible and gas-tight connection between flange and seal nose. Spring maintains a firm contact between seal nose and shaft collar.

Leaks. Shaft seals seldom develop leaks if the compressor is run right. In good operation one of the most important items is the right amount of lube oil in the compressor crankcase. Keep the oil clean at all times and be sure it is

the grade recommended by compressor manufacturer.

High-speed Freon compressors squeak or squeal when shaft seal is dry or scored. Stop the compressor as soon as you hear noises like this that do not come from drive belts or other parts of the compressor. Check crankcase oil level immediately. A damaged seal can lead to expensive repairs.

Once a week, during normal running, stop compressor and check seal for oil leaks. If much oil is found on seal exterior, test for refrigerant leaks, using a halide torch.

Seals like that in Fig. 1 may develop a slow leak when compressor is first started. Long storage or lack of oil between sealing surfaces is usually the cause. On most jobs, crankcase oil reaches the seal immediately after unit starts, establishes a tight seal again.

If leak doesn't stop within a short time after starting the compressor, shut the unit down and open seal for inspection.

Handle its parts carefully because seal faces must be perfectly smooth to be effective. Deep scratches or grooves usually mean that seal must be replaced or reconditioned.

Choosing Seal. Life of a compressor seal depends on the basic design, proper finish of sealing surfaces and right installation. Seal must be suitable for shaft speed because there may be many variations in a unit suitable for 400 rpm and one built for 1200 rpm. So be most careful when replacing a seal to see that it is right for the speed, pressure, temperature, fluid conditions on your job.

Maintenance. Your first rule in caring for seals is cleanliness, just as it is for ball and roller bearings. Don't unwrap a new seal until you have your work area spotless. Store seals only in a clean, dry place. Once the seal is installed, check it every week for liquid and gas leaks. Seals require little attention other than this; they'll run for years without trouble if properly lubricated.

15
Belting, V-belts

Handy Aspirin for

SLIPS AND SQUEALS?———————————— STRETCHES?————————

Slips and squeals?
Tighten loose belt
Use wider belt
Use thicker belt

Belt thins here
Pulley crown too high
$\frac{1}{8}$ in.
Reduce taper to $\frac{1}{8}$ in. per ft.

Belt stretches?
Dress dry shiny belt
Use wider belt
Use thicker belt

TIGHTEN loose belt, use wider or thicker belt. Keep crown taper to $\frac{1}{8}$ in. per ft or belt thins in center and slips

USE WIDER or thicker belt. Excessive stretch usually means that belt used is not strong enough for pulling the load

RUNS OFF PULLEY?————————————————————————

Runs off pulley?

Chalk line
Out of line

Crooked belt

Pulley crown too high

ALIGN PULLEYS with chalk line if they are out of line. Do this by holding line against rim of both pulleys as shown.

On short centers, use a straightedge. Check belt's side for straightness. If straight, the pulley crown is probably high

WHIPS AND FLAPS?————————————————————————

Whip and slap?

Tighten belt
Use thicker belt
Use wider belt
Straighten crooked belt

Shift bearing here for better support
Brace motor or driven machine
Lopsided pulley
Crooked shaft
Too much or too little slack

TIGHTEN BELT, use thicker or wider belt, or straighten belt. If none of these, support shaft bearing right or brace motor

or driven machine. Look for lopsided pulley, crooked shaft or too much or too little slack. This is the usual trouble

CRACKED OUTSIDE PLY?———————— CRACKED INSIDE PLY?————

Cracked outside ply?

Belt too tight
Pulley diameter too small

Cracked inside ply
Belt burned from slip

Tighten belt
Use thicker belt
Use wider belt
Decrease crown taper
Dress belt

LOOSEN tight belt by moving pulleys closer together or by splicing belt. Or use larger pulley to increase radius

TIGHTEN belt, use thicker or wider belt, decrease crown taper or dress belt. One of these will stop inside ply cracks

Leather Belt Headaches

RUNS CROOKED?

CROOKED RUNNING is hard on belts. Never force belts over pulley or over high pulley crown. That stretches it badly on one side. Take up evenly on both sides of clamp when joining the ends. Align the pulleys and then keep them so

RUNS TO ONE SIDE OF DRIVEN PULLEY?

TAKE SLACK out of belt. Or you may need a wider or a thicker belt. Quick way to check if belt is crooked is to turn it around. If it runs to opposite side of pulley, that's your trouble. And here again, check the pulleys for alignment

WEAVES BACK AND FORTH ACROSS PULLEY?

STOP WOBBLE by making pulley bearing on shaft snug fit. Or true up pulley in lathe, and remove high spot that causes belt to climb over it. There's no excuse for putting up with crooked belts, yet they cause weave that makes belt wear

PEELING GRAIN?

PREVENT belt slipping and squealing, clean belt with a commercial solvent, stop chemical fumes, scrape off loose grain, use right belt dressing and keep oil off belt.

You <u>can</u> stop

Watch belt tension for longer belt life

RELAXED BELT has straight side, as shown by ruler. Rubber tries to bulge in pulley groove but sheave sides keep it straight

BULGED BELT is bent as when going around sheave. Bulge gives belt the added friction it needs to grip sheave, keep from slipping

Here's how you can diagnose belt failures

Oil deterioration

Oil saturation ruined the rubber in this belt. For complete protection against oil install splash guards. Where oil cannot be avoided replace your belts with ones that are resistant to oil

Cover fabric ruptured

Cover fabric, left, was ruptured by prying belt over sheave during installation. Wood or metal falling into groove will do same thing. Avoid by moving motor up until belt fits into groove

Slip burn

Because this belt was too loose it didn't move; friction against sheave burned rubber. When belt finally did grab sheave, it snapped. So check tension. If OK, check drive for overloading

Snub break

Here's what happens when belt gets crack-the-whip action. Cover wear shows that belt was slipping badly, while clean break indicates a sudden snap. Keep belts taut — check tension

Abrasion wear

Sidewalls of this belt were worn away from foreign material and rust in the sheave until belt dropped to bottom of groove. Install dust guards. Belt must be kept taut in dusty places

Base cracking

This belt has been running against severe back-bend idlers, has been stored improperly or has been operating under high temperatures. Avoid ambients over 150 F and check storage

these V-belt headaches

Follow these points for trouble-free operation

First rule for long belt life is to maintain right belt tension. Loose belts slip, causing belt and sheave wear. Snapping action of loose belt adds sudden stress, often breaks belt

To test for tension press down firmly on each belt as shown in the photo at left. If there is proper tension you can depress the belt an amount equal to the thickness of the belt for each four feet of center-on-center distance. This is an important fact to remember

Keep sheave aligned or you'll have excessive belt and sheave wear. Unparallel shafts are common, cause belts to work harder on one side. In such a case belts wear faster and you have to replace entire set. Misalignment in sheave only is indicated by cover, sheave wear

Sheave grooves must be smooth. Dust, oil and other foreign matter cause pitting and rust. "Dished out" sidewalls ruin belts fast. A shiny pulley bottom shows that either sheave, belt or both are badly worn. Big headache with badly worn groove is that it causes belt to ride lower than others on the same sheave. Result: "differential driving," meaning that the belts riding high in the grooves are traveling faster than the low-riding belts. Here one belt on the sheave works against the other

Ply separation

Too small sheave caused the split along pitch line on this belt. Remember: every belt size has a minimum sheave diameter. Always check the drive design and install right one

Rupture

Many things can cause this condition. Rocks or tools falling into sheave grooves will rupture cords in the grommet or plies. Belts loose enough to twist in the groove can rupture cords

Worn belt sides

When the sides of your belt show wear, it may be from normal causes. But look for misalignment. Grit or dirt will also cause abnormal wear. Replace belt before the wear is excessive

Install belts right

MOVE DRIVER UNIT forward so V-belt can be slipped into the groove easily, as shown

DON'T FORCE BELT into sheave with screwdriver or wedge. Fabric, cords rupture

Figuring Centrifugal Force

▶ You KNOW ABOUT centrifugal force, but can you figure it? Let's try.

Car Rounding Curve. Fig. 1 shows three cases of a 2000-lb car rounding a curve. In case *A* the car goes 20 miles per hour around a 50-ft curve. Side thrust on the tires (another name for centrifugal force) is 1070 lb.

In case *B* the same car takes the same curve at 40 mph—twice as fast. But side thrust isn't just doubled, it's four times as great or 4280 lb.

Now what happens when this same car travels at same speed around a curve of twice the radius, Fig. 1*C*? Force drops to half or 2140 lb.

What's It Mean? In Fig. 1*B* we see that centrifugal force goes up as the square of the speed. It's four times for double speed, nine for triple.

Second, the force is inversely proportional to the radius of curvature. If you double the radius, you cut the force in half. This last rule compares two objects running at the *same linear speed* (not RPM).

Rule 1 is the formula. See if you can figure the side thrust on the cars in Fig. 1. Be sure to use the right *K*. My answers are rounded off.

In Terms of RPM. For most rotating machinery it's easier to figure centrifugal force than linear speed. Take a look at the 2-lb ball swinging around a shaft in Fig. 2. Compare these three cases with those in Fig. 1.

Going from Fig. 2*A* to *B*, we double the RPM and force goes up four times, just like doubling the carspeed in Fig. 1. Now compare Fig. 1*C* with Fig. 2*C*. In the first case doubling the radius halved the force. But in Fig. 2 doubling the radius doubled the force. Why? At constant RPM, increasing the radius directly increases linear rim speed.

Try the formula in Rule 2 for the case in Fig. 3. For even RPM the table is mighty handy. Use it to check my answers in Fig. 2.

If You're Talking Speed . . .

1 When a moving object rounds a curve, its centrifugal force goes up in proportion to the square of the speed and inversely as the radius of the curve the object is following

RULE 1. If **S** = speed of moving object, **R** = radius of curve it follows and **F** = centrifugal force in pounds per pound of object weight, then . . .

$$\ldots\ \mathbf{F = K \times S \times S \div R}$$

When **S** is in miles per hours and **R** is in feet, **K** is 0.0671. When **S** is in feet per sec and **R** is in inches, **K** is 0.375. Total force is **F** times pounds object weighs

If You're Talking RPM . . .

2 With rotating objects, centrifugal force goes up in proportion both to the radius and the square of the RPM. Increasing radius at constant RPM increases linear rim speed

Centrifugal force = 87.2 lb

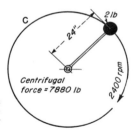

RPM	F, per lb, per inch radius
600	10.2 lb
900	23.0 "
1200	41.0 "
1800	92.1 "
2400	164 "
3000	256 "
3600	368 "

3 Here's what happens in an unbalanced rotating machine. Unbalance acts just like a weight set offcenter. Use the table (right) to get **F** for other speeds. For other weights, multiply by number of lb, also by inches of radius

RULE 2. If **F** is centrifugal force in lb per lb of weight rotating at radius **R** . . .

$$\ldots\ \mathbf{F = RPM \times RPM \times R \div K}$$

When **R** is in ft, **K** is 2920. When **R** is in inches, **K** is 35100. Again multiply **F** by the number of pounds the rotating object weighs to get total centrifugal force

16
Centrifuges, lube oil

Your Centrifuges: How

The Disk-Type Bowl --

▶ MIXTURE of oil, water and dirt is separated by centrifugal force into three layers in oil purifier bowl. Dirt, the heaviest, is thrown against bowl's rim. Water, the next layer, discharges continuously. Oil, being the lightest, collects and is discharged nearer center.

Disk-type centrifuge has larger diameter. Liquid is thrown about same distance as depth of liquid layer. In gravity settling, shallow vessels work faster and better than deep ones. Reason is that more time is needed if particles to be separated have greater distance to travel to bottom. Same is true in centrifugal bowl.

For best results under these conditions, and with lower centrifugal force because of larger diameter, this bowl has series of disks to separate liquid into thin layers. That creates shallow settling distances between disks.

Centrifugal purifier makes use of force thousands of times stronger than gravity. If separator is designed right, centrifuge does work that gravity separation alone can't—and much faster.

Disk-type bowl travels up to 7200 rpm, is designed to take advantage of well-known principle of thin-strata distribution. Oil is fed into the bowl through a strainer cup in top, and passes down through hollow shaft to bow's bottom. From there it rises into the bowl.

Oil is thrown out toward rim. Heavy impurities hug rim. Water is thrown out against dirt, travels up and out as shown. Clean oil floats up to the top. Liquid is divided into thin layers of 0.050-in. space between disks. Disks increase speed and quality of purification. Oil passes between disks, up through holes in disk, to outlet near center. Balance between columns of oil and water is attained by using discharge ring of right size. Light oil takes larger opening in the ring than heavy oil—usually.

1 Apply brake until bowl stops. Remove cover. Lock bowl, loosen bowl ring

MORE DISKS cause solids to be forced against them sooner, doing better job

The Tubular-Type Bowl --

This is the second type of liquid separator. A tubular bowl is hung from a flexible spindle, which is driven by belt and pulleys from an electric motor.

Dirty oil enters bowl at bottom, is separated, and oil and water discharge from the top into separate covers—the lighter oil into the upper cover and the heavier water into the lower. Sediment accumulates within the bowl and is periodically removed.

The bowl is designed to pass liquid from inlet to outlet with a minimum of turbulence that might interfere with separation.

Bowl turns at 15,000 rpm, developing high centrifugal force of 13,200 gravity. Bowl length, and so liquid travel, is many times settling distance. A simple one-piece, 3-wing device within the bowl brings the liquid up to speed smoothly, and keeps it from slipping within the bowl. Sketch shows straight-through path of liquid in bowl, and how particles of sediment are thrown against bowl wall.

Stress on a centrifuge bowl is proportional to peripheral speed. The smaller the bowl diameter, the greater the centrifugal force it can safely develop. The centrifugal force increases because there is greater change in the liquid's direction in the smaller bowl in a given time. There is a "sharper corner to turn."

Impurities commonly found in lubricating and fuel oil are water, sludge, dirt, carbon, bits of metal, and acid. All but acid and sludge, which are oil soluble, can be removed with either type of centrifuge.

LIQUID TRAVEL is up through long bowl. Cutaway section shows the working part of centrifuge and fins in tubular bowl

To Know and Clean Them

How to Clean---

2 Unscrew bowl cover. Remove shaft and disks as unit, put in pail of solvent

3 Scrape bowl shell with scoop to loosen and remove dirt collected against rim

4 Replace disks (wet with solvent) in bowl. Much dirt is already washed off

5 Reassemble bowl. Start machine without admitting oil. Run "dry" 5 min

6 Centrifugal force throws dirt off disks. Seal with water—turn on oil

How to Clean---

1 Remove the cover to get at the collected solids inside

2 Screw extension sleeve onto the bowl's bottom

3 Lift bowl with extension sleeve above block—drop

4 Remove sleeve, pull out caked sediment as here

5 Remove solids from fins, brush dirt from the bowl

6 Roll parchment-paper liner loosely into long bowl

7 Push in fins (3-wing assembly) as is shown here

8 Seat fins against bowl. Screw on—tighten bottom

307

STEEL DRUMS carry by far the major share of industrial lubricants entering the average plant. Careless handling or storage can completely ruin them and their contents

Here Are Storage and Handling...
POINTERS ON LUBES*

▶ MODERN OILS AND GREASES for lubricating services are refined and manufactured to meet rigid standards. Many oils carry special additives to equip them with precise properties for particular duties. Careless or even haphazard handling and storage at your own plant can destroy the very values you may have paid a premium to get.

Problems. For example, exposure to abnormal temperatures or contamination from moisture or dirt—two strong possibilities—can bring on leakage or produce deterioration that destroys a lubricant's suitability. We'll see why later. Of course, careless handling of the containers themselves may result in dents and split seams that take their toll in direct loss. Here are some simple rules for storage and handling.

Abnormal Temperature. Both oils and greases experience considerable change under temperature variations. Their effectiveness as lubricants is reduced by hot locations. These make heavy-bodied oils thinner, tend to evaporate the required moisture content of (1) soluble cutting oils, leaving behind a jelly-like mass (2) certain greases where oil separates out from the soap.

Cold areas, on the other hand, make lubricating oils and greases more difficult to handle. The oils flow more slowly, and the greases become stiffer. Oils containing fatty materials congeal them out of solution under low-temperature conditions. In addition, soluble oils, cold enough to freeze their moisture content, prove useless.

Somewhat along the lines of temperature effect is spoilage in long storage. Soluble cutting oils suffer most from this cause through a loss of moisture. But preventing this loss is more a problem of stock control than of storage or handling.

Contamination. An improperly protected lubricant supply is almost certain to experience dust and dirt contamination. No plant man worthy of the name needs to have pointed out to him the dangers of dirt in lubricating oils and greases.

Moisture contamination is a major problem, too. Oils are more likely to be contaminated than greases. Some oils are designed to function as a lubricating emulsion in the presence of moisture, or even small amounts of water, but water contamination during storage can render them useless for later service. This holds true for most steam-cylinder oils and similar ones meant for duties like lubricating the bearings of the wet ends of paper machines. Nonsoluble cutting oils and heat-treating oils suffer complete destruction if contaminated by appreciable quantites of water.

Still another form of contamination develops. This is the introduction or unwanted mixing of different oils when dispensing them within your plant. It stands to reason oils of different qualities when blended produce a third, entirely new oil. You must take certain precautions, and then exercise definite care to keep original oils effective.

(More on next page)

---

*Abstracted from *Technical Bulletin on Handling, Storing* and *Dispensing Industrial Lubricants*, by *Socony-Vacuum Oil Co, Inc.*

Draw on these easy-to-use tips based on in-plant...

STORAGE

Outdoor storage is never too good. Brand markings and labels wash off or become obliterated under weathering. Future dispensing is needlessly endangered and other lubricant contamination becomes more possible. Temperature changes from season to season can produce enough expansion and contraction stresses over a period of time to develop leaks in the container seams.

Moisture contamination is a potent threat with outdoor storage. Rain water collects readily inside the chime top, see first sketch. It is gradually sucked past the bung by the drum's breathing during alternate hot-and-cold periods.

If you must store outdoors, spread tarpaulins over the drums or erect a temporary shelter. Then if mechanical means are available, turn each drum over so the bung end is down. If you can't do this, lay the drums on their sides, and line them up so the bungs make up a horizontal line across the drums' middle.

Actually, there are four major recommendations that apply to outdoor storage: (1) Provide at least temporary shelter against the weather. (2) Place drums so they cannot breathe through the bungs. (3) Make sure bungs are made tight if you are going to move drums. (4) Before opening, dry and wipe bungs thoroughly as well as the surrounding surfaces.

An oil house is the ideal and logical place to store and dispense oils and greases of all sorts, as well as to carry out the cleaning of dispensing equipment. Size varies with the individual plant. Some plants require only a single oil house or even a part of a full-package warehouse. But no matter what its size, every plant needs an oil house of sorts.

Experience indicates the oil house should be closed to all but authorized personnel. Locate it as centrally as possible, yet in an area free from possible process contamination, as from cement dusts, chemical fumes, coke dust and similar byproduct wastes.

Storage takes many forms. Mechanical aids, in sketch below, are one. You can buy the storage racks commercially. Take steps to prevent any harm to personnel from toppling or rolling drums, barrels and containers. Further, it's smart to see that the drums are grounded electrically, second sketch, especially with flammable volatile solvents.

OUTDOOR STORAGE, never desirable, makes drum contents an easy prey to contamination by an outside moisture like rain

ELECTRICALLY GROUNDED connections on drums, especially if they hold volatile solvents, is a smart measure for safety

OIL HOUSE should be set up so all its facilities are permanently installed, and ample waste drains supplied as best storage bet

HANDLING

Handling the containers, when lubricants are both coming into storage and being dispensed, goes best with certain mechanical aids. They can be anything from the simple hand truck, sketch top of facing page, to chain falls and trolleys, sketch at right, or electrically operated elevators or lift trucks. Some companies, however, just roll tht drums from delivery point to storage.

By far the greatest volume of industrial lubricants arrives in drums shipped by railroad or motor truck. The load consists mostly of a single tier of drums, each standing on end. Unloading platforms, level with railway car or truck, make it easier to handle the drums and reduce possible damage risks to containers. Wooden or metal ramps serve well where there are no unloading platforms. Slide the drums endwise

MECHANICAL AIDS like chain falls and trolleys make handling of drums into and out of storage comparatively easy

to the ground. Never drop an oil drum from a truck or railway car onto a cushion of discarded automobile tires.

Major transporting activity of oil drums should be confined to trips from unloading area to the oil house. This method involves the handling of bulk quantities only. It entails stacking and storing, plus dispensing and disposal of waste oils and empty containers.

MANUAL LABOR, particularly if helped by hand trucks, can do effective job in the average small-to-medium-sized plant

DISPENSING

Hand oilcans, large dispensing safety cans, oil wagons, grease guns, lubricating cars and sump pumps comprise the more important dispensing and transporting equipment for grease and oil.

Clean this class of equipment regularly. But don't limit your cleaning to mere wiping with solvent-soaked rags. Instead, use generous quantities of solvent for both cleaning and rinsing. Baths of safety solvent (flash above 100 F) meet the usual safety regulations where ventilation is adequate and correct covers are provided. First sketch at right shows a good arrangement with two baths, one for cleaning and one for rinsing.

Actual transfer of oil from container to dispenser may entail nothing more than a hand-operated drum pump, shown in sketch, driven into the bung of the oil drum. Such a pump is a positive-delivery type capable of supplying measured oil quantities.

You can get pumps, however, with spring-action closable returns, and in air-operated or electrically driven designs. Drum faucets also prove efficient and economical transfer devices. They come in different sizes for fast, slow oils.

Because of its consistency, grease presents some special problems in transfer to dispensing equipment. A good metal paddle does a fine job particularly for heavier greases.

But the major problem comes from dirt and other contaminants that land on the surface of open drums and exposed grease. Take every care to keep drums closed when not in use. Extra precautions against stray dirt should apply while the grease is being removed.

Air-operated pumps do a fairly good job in pulling grease out of drums without exposure to the air. But you cannot always count on them for greases harder than No. 2 or 3 consistency.

Sump Pumps. Used oil has to be periodically removed from reservoirs and sumps. The sumps can be anything from the small reservoirs for textile-spindle and machine-tool-spindle housings to large systems like hydraulic machines.

Sump drainers can be homemade. One such example is shown in second sketch, right figure. It employs a heavy steel drum, a siphon, and compressed air to draw from sump.

Often you'll find it desirable to purify certain types of lubricating oils for reuse. Extent of the purifying equipment depends entirely on how much of a saving it gives you. A very simple arrangement is pictured in the second sketch, left figure. If your wants are greater, it may pay to set up a purifying section within your oil house.

All in all, though, lube oil-and-grease handling is a relatively simple operation that gives little trouble if it is planned and controlled properly.

HAND-OPERATED drum pump, *left*, is a good dispenser of lubes. keep it and other dispensers clean by solvent baths, *right*

SPENT OILS can be reclaimed in simple filter, *left*, or pulled from sump by compressed air siphon, *right*, for disposal

AS DISPENSING METHOD that is both quick and easy, use rocker-type racks and a set of drum faucets, as is shown here

17
Welding, hard surfacing

Joining Little Pipe to Big

▶ WHERE A SMALL SIDE OUTLET is welded into a much bigger pipe the difference in wall thickness may cause some trouble. Here's how to insure a good job.

MARK THE HOLE

1. Place the nipple in position and mark the hole on the header with a piece of sharpened soapstone. Center punch mark the soapstone circle with about ¼-in. spacing. Heat from the torch quickly destroys soapstone, but the center punch marks will remain visible.

CUT THE HOLE

2. Cut out the circle by holding cutting torch nozzle square with the pipe surface. Try to cut a good round hole just big enough to admit nipple after all oxide is knocked off the cut edge. Bevel the edges to a 45 deg angle, leaving a shoulder at the bottom end of cut about ⅓ the pipe wall thickness.

TRIM THE NIPPLE

3. Trim end of nipple to follow curve of large pipe. To mark off the nipple end, insert it in the hole resting against a bar clamped below inner surface of header. Edge of nipple should come just flush with lower end of bevel at side of hole, as pictured. Mark this curve permanently with center punch, and trim the end to fit with a cutting torch. Cut edges square; beveling nipple edges is not necessary.

WELD THE NIPPLE

4. Draw up bar until it rests snugly against inside surface of header and put nipple in position for welding. Tack weld it on one side. Straighten it into exact alignment by tapping with a hammer before making the second tack.

What's a little water leak, say a faucet dripping at the rate of one tumblerful every minute? Suppose this goes uncorrected for a whole year—525,000 minutes. Allowing 16 tumblers per gallon, that figures more than thirty thousand gallons down the drain.

NEEDLE VALVE for tough use finished with alloy of cobalt-chromium-tungsten

RIGHT

WRONG

OVERLAP hard-facing beads, then lay second-layer beads between first beads

SUBMERGED ARC is semi-automatic. Bare electrode wire is power-fed through blanket of alloy flux. Alloy pickup from flux gives up to 62 Rockwell C hardness

Hard Surfacing

1—Electric-Arc Welding Method

Two major hard-facing tools are the electric arc and the oxyacetylene torch. The electric arc deposits alloys fast so heat doesn't penetrate deep. But steel with more than 0.3% carbon may crack unless preheated. Because arc heat is intense, base metal melts near surface. This dilutes the first layer of hard-facing alloy with base metal—more than when a torch is used. It may take two or more layers to get undiluted hard facing.

Electric arc forms when two conductors of an electric circuit touch, then separate, and there's enough voltage to maintain current flow across the gap. Arc is struck between a carbon or metal electrode connected to one terminal of a generator, and the metal to be welded, which is connected to the other terminal. Arc size is varied by changing distance across gap, altering current output, or putting resistance in the circuit. Distribution of heat in the arc depends on current density in the electrode.

For *straight polarity,* connect electrode to generator's negative terminal; for *reverse polarity,* tie to positive.

Hard-facing electrodes are both bare and coated. The flux coating reduces spatter loss, assures good penetration,

prevents oxidation of deposited metal and helps stabilize the arc. Flux makes metal "wet" easily so weld metal flows and diffuses through melted flux. It also helps carry flame-produced oxide to the surface so it doesn't remain to contaminate the hard facing.

Current Settings. If you don't have electrode maker's data, use the table below as a handy guide:

Rod-size, in.	Current, amp
$\frac{1}{8}$	80 to 110
3/16	150 to 200
$\frac{1}{4}$	200 to 350

Low or medium current with a short arc of about 3/16 in. prevents loss of alloying metals. It also reduces dilution of alloy by giving minimum penetration. Increasing current increases penetration, softens deposit, and increases dilution with base metal.

Electrode size depends on nature of work and current that it is to carry. Because the electrode acts as an electrical conductor, be sure to use right current for each size electrode.

Good rule is to use largest convenient rod for the job. Large electrodes permit faster work because more metal is

deposited. They are cheaper and faster and their arc stability is often greater. But if material is crack-sensitive, use a small-dia electrode. That reduces heat input and also danger of distortion.

Use only dry electrodes. Best way is to make sure they are kept clean and are stored properly in a dry place. When using, be sure to avoid trouble by following the manufacturer's directions. Better hard-facing jobs will more than pay you for your troubles.

Submerged arc (photo) increases depositing speed 3 to 4 times because high current density with small wire gives high burnoff rate. For example, 600 amp on 5/64- or 3/32-in. wire is equal to 100 amp on 5/16-in. electrode. Alloy pickup is from granular flux, containing alloy needed, as flux pours from funnel and hides arc. Inexpensive wire can thus make hard-facing deposits of 36 to 62 Rockwell C hardness.

Twin carbon arc has two carbon electrodes in holder, gives softer arc and will even hard-face thin materials without burning through. Reason is that current flows from tip of one copper-coated and cored carbon to other, not through metal being surfaced.

WORN CAMSHAFT LOBES are built up with torch. Water keeps shaft from warping by carrying away heat. This repaired job will last much longer than a new shaft

DIESEL VALVE turns slowly while hard-facing rod is deposited around its seat

2—Oxyacetylene-Torch Method: It's Handy for Parts

Big advantage of surfacing by oxy-acetylene torch is that surfacing alloy can be applied by flowing alloy onto the base metal with a technique known as sweating or tinning.

These terms describe formation of a lower-melting-point element that enables alloy to flow onto base metal while hardly melting it.

Term "sweating" refers to action that takes place when applying high-melting-point ferrous, nickel and cobalt-base alloys because of using an excess acetylene flame. This flame carburizes the surface of base metal, which in turn lowers melting point of the film. Alloy flows easily over this thin liquid film.

Term "tinning" describes action when surfacing with bronze or copper alloys. Only here it's more like a cleaning action caused by flux or flame that prepares base metal ahead of the welding action. That allows capillary flow of low-melting-point alloys onto the higher-melting-point base metal.

Flame Types. To burn acetylene gas completely takes $2\frac{1}{2}$ times its own volume of oxygen. But about equal volumes of the two gases are fed through the torch. Other volume and half of oxygen is gotten from the air. The combustion of acetylene in oxygen produces carbon monoxide and hydrogen. Oxygen

for burning the CO and H_2 to carbon dioxide and water vapor comes from the surrounding air.

Hard-facing flames usually have an excess of acetylene. Sketch shows a neutral flame—equal volumes of acetylene and oxygen going into the torch. Increasing the acetylene builds up a white flare caused by unburned white-hot carbon particles. These free-carbon particles tend to carburize the surface of the base metal and also to reduce any oxides present to the metal form. So excess-acetylene flames are also called carburizing flames and reducing flames.

The usual hard-surfacing flame has a white flare about three times as long as the inner cone. This flame gives good sweating action because the free-carbon particles are absorbed into the base-metal surface, lowering its melting point. Melting of an extremely thin surface layer provides a liquid medium on which molten hard-facing alloy flows. Carburized base metal, to a depth of about 0.001 in., is absorbed by alloying with the molten hard-facing layer. So gas-welded deposits can be made with penetration less than 0.010 in. into the base metal.

How Much Acetylene? Excess-acetylene flame also prevents loss of carbon

content from welding rod and may even increase carbon content of weld deposit from 0.2 to 0.4%. Because some alloys absorb carbon from flame, hardness of deposit depends on excess acetylene used, and on cooling rate of finished work. But long excess-acetylene "feathers" are unsatisfactory because they are not as hot, put too much carbon into deposit and make poor welds.

Always keep inner cone from touching molten metal already deposited. Inner cone contacting puddle puts in too much carbon. That hardens deposit, which might cause cracks. Keep inner-cone tip 1/16 to $\frac{1}{8}$ in. away.

A slight reducing flame is used with rods having fairly high iron content, where dilution of facing alloy with base metal is not serious. When working on cast iron, watch corners and sharp edges, because cast iron has a lower melting point than many hard-facing rods.

When hard-surfacing with the gas torch, use only a slight excess of acetylene. Use flame to keep air from contact with molten metal. This prevents oxidation. Never remove flame completely until metal starts to solidify.

Which Way? Deposit hard-facing alloys with either forehand or backhand technique, sketch. Forehand is usually

EXCESS hard-facing metal is wiped away from this sprocket wheel while plastic

NO GRINDING needed; contour is formed by die while the alloy is still plastic

To Be Finished to Close Tolerances

recommended because penetration is controlled easier, rod is deposited faster with less scale in molten pool.

Backhand technique shown is also common with the sweating method. Direct excess-acetylene flame at a 30- to 60-deg angle to the surface. As you deposit alloy, move flame back to remelt rough spots, get a smoother finish.

Typical Job. Torch method is handy for doing parts that will be finished to close tolerances by machining or grinding. Cam shaft of internal-combustion engine, photo on facing page, forms a typical job. To repair worn lobes, partly submerge shaft in water so heat is carried away and shaft doesn't warp.

In photo, shaft is held in place with a plastic compound at end of pan. Get this at a supply house or substitute loose asbestos mixed with lube oil. Heat can also be removed by packing wet asbestos around shaft, building it up to area being surfaced.

Set torch to carburizing flame and hold against cam lobe until metal sweats. Extent of sweating area depends on size of welding tip, but it should extend a distance of ¼ in. around excess-acetylene flare.

Withdraw flame enough so end of rod is between flame's inner cone and the hot base metal. Apply the rod to

sweated area, with tip of inner cone just touching rod. As rod melts a pool forms, flowing over sweating area. Remember: Base metal is too cold if first few drops don't flow evenly.

Remove rod from flame and flow molten pool where wanted by playing flame on it. Return rod to flame and melt off more metal into pool. Keep flame on front edge of molten pool so flame keeps it molten and also sweats base metal in front of pool. Move rod in and out of flame as it moves steadily forward. Float off scale specks, don't cover them up.

Keep supplying the pool from the rod as the pool flows forward. Use rod so it deposits molten metal uniformly at a steady rate over area to be covered. On round work, this is easier if work is turning. Play the flame over the deposited alloy from time to time to remelt and smooth out any high or low spots, left behind.

Plastic Working. The sprocket wheel in the photo above is built up with an alloy that can be worked in the plastic state. Rod is used to wipe away excess alloy. Then, while alloy is still plastic, it's die formed. This establishes the contour without need for grinding. The alloy used is 75% nickel, 25% chromium and boron and wears very well.

Right Flame and Welding Methods for Job

Neutral

Reducing

Hard-facing

CARBURIZING. The flame welds but hardly dilutes facing alloy

SWEATING (top): Relationship of the tip, rod and base metal

BACKHAND (top) or forehand depends on welder's preference

3—Spraying Methods: They Use Alloys in Powder or Wire Form____

FIRST, adjust powder and flame with knobs on top of the gun. Then, open gun's trigger as shown. Powder is now spraying.

SECOND, build up corners of undercut by hand, while shaft revolves in lathe

THIRD, run gun back and forth by feed of lathe. Shaft builds up to size here

FOURTH, second stage is fusing alloy to shaft with large gas torch at right temp

MANY ALLOYS may be sprayed on by one of several methods. Spraying gives a thin, uniform coating—from a few thousandths thick up to 0.60 in. Contours and shapes are easily coated. Spraying also handles alloys not easily applied in rod form.

In one method, alloys are sprayed from a suitable gun, then fused with a torch. In another, powdered alloy and plastic binder form a wire that's fed through a typical metalizing gun. Resulting deposit is then fused. Still a third way uses a special torch to apply a coating and fuse it in one step. Let's take a look at each scheme:

Spraying Alloys Available in Powder Form

Powder spraying is done with a gun roughly like that for metalizing, but with a powder hopper and a carburetor for mixing powder with air. This applies alloy but does not fuse it. A separate torch is used to fuse sprayed alloy to base metal. It can also be done in a furnace or by induction heating.

Typical Job. Conveyor shaft, in photo, of cold-rolled steel, wears badly in center. First, undercut it so there is at least 0.030 in. for hard facing all around. Never make sharp corners in groove, or oxide film will develop porosity when fusing and the coating will peel back.

Blast with angular-steel or aluminum-oxide grit, not round shot. Caution: Don't touch base metal. If it must stay in lathe overnight, cover it.

In lathe, blow shaft with compressed air while it revolves, then wire-brush. Adjust to neutral flame of 13-psi oxygen and 11-psi acetylene. Have 75-psi air pressure to propel powder.

Spraying. Turn on powder, then open gun's trigger and build up corners of undercut, holding gun in hand. Next, set gun in holder attached to toolpost and adjust so spray is against shaft's center. Start thread-cutting feed and reverse at end of each stroke. This sprays even layers of alloy. Spray on enough alloy to allow for shrinkage. This depends on coating thickness. Check with set calipers.

Fusing. With shaft built up to right diameter, use large torch with neutral flame. Play back and forth over revolving shaft, heating evenly. Then hold flame at one end of shaft, photo, until shaft has a transparent appearance. Move flame slowly along shaft as alloy shines under flame. Now alloy is fused to shaft. With asbestos gloves, remove part from lathe and drop into a box of powdered insulation for slow cooling overnight. This step is important or deposit will crack.

Unless cooled gradually, hot alloy shrinks faster than inside of shaft, and cracks. If alloy is inside a sleeve, fast cooling pulls alloy away from wall.

Spraying Alloys in Wire Form

Besides the many wires made for various metalizing jobs, hard-surfacing wire is made of powdered alloys extruded with plastic binder. Plastic evaporates in metalizing-gun flame, so only the alloy deposits on base metal. Alloys may be same as for powder spraying—only difference is that wire is deposited with usual metalizing guns.

After spraying, deposit is similar to metalizing job, which means it's not fused. There are various ways of fusing: For small to medium work, use an acetylene torch. For larger work, use two or more torches, or put in a furnace.

Liner Job. Photo shows hard surfacing inside of a gas-engine cylinder liner. This isn't a common application and is probably one of the toughest to do right. But cylinder will have from three to four times more life than new.

First bake casting at from 450 to 500 F to burn oil out of cast-iron "pores." This is important on old oil-soaked castings. Baking takes about an hour, depending on heating method. On liner, Tempil-stick mark shows when right baking temperature is reached. This is a temperature-indicating stick that melts at given temperature.

Bore out to get rid of ring steps, fluted wear and scoring. Finish bore with a No. 18 or 20 thread final cut for anchoring alloys. Then grit blast. Preheat from 200 to 300 F. Keep gas torches on outside as shown to maintain this temperature while spraying.

Spraying. Metalizing tool shown is fed back and forth by lathe for smooth, even deposit. Spray cylinder up to original

WIRE BEING SPRAYED inside gas-engine liner is a powdered hard-facing alloy bound by a plastic that vaporizes in flame

diameter, plus about 0.030 in. for finishing. Because alloy used here is over 50 Rockwell C, and hard to machine, cylinder will have to be finished by grinding.

Fusing. After spraying, bring up to shiny red heat for fusing. For this job, a furnace is best, as it will give even heat at all parts of the liner. Then allow to cool gradually over a few days until room temperature is reached. If liner cools too fast, or unevenly, alloy coating pulls away from the liner. For most jobs, but particularly for one like this cylinder liner, slow, uniform cooling can't be overemphasized.

Spraying Powdered Alloys and Fusing Them in One Operation

This method feeds powdered hard-facing alloy onto the base metal through an oxyacetylene flame, as when spraying powder and wire. But here the process differs.

Powder, with flux if needed, is projected through the flame by an auxiliary gas, such as carbon dioxide, nitrogen or hydrogen, by means of a spray gun that is specially designed for the purpose.

Continuous Fusing. Temperature of oxyacetylene flame is adjusted so deposit is continuously fused or braze-welded to the base metal. This makes the process something like oxyacetylene hard surfacing, as the base metal is prepared and must be preheated and treated in the same way.

One advantage is that coatings as thin as 0.010 in. may be deposited. Up to 0.150 in. of coating has been applied. There is a full bond with base metal, as the deposit is completely melted without a second fusing operation.

Control. Accurate control of base-metal temperature is obtained by increasing or decreasing amount of flame, then proportioning the projecting gas (and alloy), and the processing gas. Both base-metal and alloy temperatures are controllable, all the way from melting temperatures down. This gives a high-quality metal-bonded hard surface.

Equipment. Photo shows torch, control box with air- and gas-pressure regulating valves, and hopper for powdered metals. Torch resembles usual welding torch but has three control knobs on the handle. These are part of a total of 16 controls on the apparatus for accurate regulation of all oper-

POWDERED ALLOY FUSES to base metal as deposited under regulation of controls that help operator make needed adjustments

ating conditions. Torch also applies plastics or synthetics.

Alloys containing beryllium, bismuth, carbon, cobalt, molybdenum, tungsten, etc, may be deposited with this equipment (not yet commercially available to the industry).

COKE CHUTE, hard faced with chromium-boride crystals applied in paste form, is expected to last about eight times longer

4—Spread Alloy in Paste Form as if Icing a Cake

Some hard-facing materials come in paste form. One of these is chromium-boride, which reads 68 to 72 on Rockwell C scale. It has good resistance to extreme abrasive wear found in materials-handling chutes, fan blades, pump impellers, scraper blades, crusher rings, screw conveyors, etc.

In addition to resisting abrasion, chromium boride is good against corrosion and impact, has hot-hardness qualities. It's easy to apply, by carbon arc or oxyacetylene torch.

Applying Paste. Clean base metal thoroughly. If there's scale or rust, remove by grinding, shot-blasting or sanding with abrasive cloth. Amount of preparation needed depends on previous service of material to be coated. Watch out for problems like oil-soaked castings. If not baked out, oil can cause hard-surfacing job to fail.

Ordinarily, preheat the work to about 100 F. Play oxyacetylene flame or blow torch back and forth across piece so it heats up slowly and evenly.

Paste comes in cans, mixed with water. If it's too thick, add water and stir to the right consistency. Then spread it on metal with a putty knife, a stiff brush or a piece of wood. Coating thickness runs about 1/16 in. Putting on the paste is much like icing a cake. Next, let paste dry. The preheat evaporates the water quickly.

Sweating. If metal to be coated is 14-gage or less, use oxyacetylene or carbon arc torch. Spread paste about 1/64 in. Heat base metal up to sweating temperature. Fusing of base metal and paste will show as a dark line ahead of the torch flame. Move torch across work, keeping the dark line ahead of the flame.

For metal of over 14-gage, use standard carbon electrode of 3/16, 1/4 or 3/8 in. dia, depending on thickness of base metal. Have straight polarity with just enough current to maintain arc. Hold carbon about 80 deg to the vertical.

Using Carbon Arc. Work with a quarter-moon weaving motion, not over 2½ times diameter of carbon. Point carbon in pencil sharpener or by holding against grinding wheel. If handled right, carbon melts so point stays as it should. Have about 3 in. of carbon sticking out of holder.

Keep black spot of carbon floating on the surface just in front of the point. After sweating-on first weave, pass across work and return with about half the width of the second weave overlapping the first. If you get porosity on the edge of the weave, pay no attention. It will disappear when you overlap the next weave.

Use paste made of chromium-boride crystals only on rough work that is not to be finished.

Sweating-On Powder. Chromium carbide comes as a powder, mixed with flux. Sprinkle it on the base metal and fuse with a carbon arc or carbon arc torch for a smooth, dense, abrasion-resisting surface, which can be as thin as 0.025 in.

Hardness depends on mixing with base metal, runs about 54-61 Rockwell C for one layer and 57-63 for second. Deposit develops full hardness in resisting scaling at high temperatures. Corrosion resistance is like that of stainless steel. Don't use this type of alloy where impact is great.

FOLLOW THESE EASY STEPS

1 Preheat base metal with torch to only about 100 F

Water —
Paste

2 If paste is too thick, mix with water to thin out

3 Spread on about 1/16 in. thick, as if icing cake

4 Let paste dry—the preheat does the job quickly

Sharpen like pencil
— Carbon

5 Use about 1/4-in.-dia carbon, pointed like pencil

About 3 in.
80°

6 Use straight dc polarity and hold carbon as shown

1/4 moon

7 Weave quarter moons, and use lowest amperage

8 Reverse, overlap the previous stringer, as shown

1 Grit-blast the tungsten-carbide inserts to remove the surface oxides and other films. Machine surface of the base metal, then do a thorough degreasing job

2 Flux surface of the base metal with brush. Before fluxing, clean in carbon tetrachloride. Degreasing, cleaning and grit-blasting must be done properly

3 Position silver-solder strips after fluxing. Place one strip on base metal, a copper screen on the solder strip and second solder strip on copper: flux

4 Put the tungsten-carbide inserts in place, on top of "sandwich" of silver-solder strips and copper screen. Six inserts are butted together on this job

5 Heating to 1350-1400 F by the propane-air burners against underside of base metal bonds the inserts. Test bond by tone of a hammer blow as shown

6 Attaching hard-faced peg to moving rotor of coal pulverizer. Coal gets pulverized by impact and attrition between moving and stationary pegs shown

5—Brazing Tungsten-Carbide Plates

Diamond-substitute alloys of Type III, A, used here, come as inserts of various shapes. Some contain 90 to 95% tungsten carbide. Remaining 5 to 10% is nickel, cobalt, iron or other elements to give the desired properties.

Braze these carbide plates to wearing surfaces because brazing heat (torch flame) will not melt the carbide materials. They do "wet" easily, so can be bonded to metal surfaces.

Because stationary and moving grinding pegs of the coal pulverizer shown take heavy punishment from impact and abrasive wear, they are hard faced with inserts of tungsten carbide. Here are steps in typical application:

Operations. (1) Machine surface of base metal to be hard-faced. (2) Degrease with trichlorethylene, clean with carbon tetrachloride. (3) Grit-blast tungsten-carbide inserts with silicon-carbide grit to remove surface oxides and other films.

(4) Flux machined surface of base metal with brush. (5) Cut strip silver (silver solder) to size and dip in flux. Then put in position on base metal. (6) Dip a copper screen or copper sheet in flux and place on strip silver. Because insert's coefficient of expansion is not the same as the base metal's, always put copper between. This also takes impact shock, as inserts are hard, with no "give." (7) Flux second strip of silver alloy and place on copper. (8) Put inserts in position on top layer of strip silver. (9) Braze by directing flame under base metal until both strips of silver melt. (10) Test bond by tapping inserts with a hammer. Tone indicates the quality of the bond.

Tungsten carbide can also be used to tip your maintenance tools. Technique is same as explained on this page.

More About the Electric-Arc Method: How to

Getting Ready

WEAR on this shaft comes from highly abrasive coal dust working into bearing

WHICH ELECTRODE?

Build up shaft with iron-base rods having less than 20% alloying elements. For a surface resistant to abrasion, finish with Type I, A, 4 iron-base rods having more than 20% alloying elements

Electrode dia, in.	Current, amp straight	Voltage, arc volts
1/16	20 to 40	17 to 20
5/64	25 to 50	17 to 20
3/32	30 to 80	17 to 21
1/8	70 to 120	18 to 22
5/32	120 to 170	18 to 22
3/16	140 to 240	20 to 24
7/32	170 to 300	21 to 25
1/4	200 to 350	22 to 26
5/16	250 to 450	23 to 27

Electrode dia, in.	Current, amp reverse	Voltage, arc volts
1/16	20 to 40	17 to 20
5/64	20 to 60	17 to 21
3/32	30 to 80	17 to 21
1/8	80 to 130	18 to 22
5/32	120 to 180	18 to 22
3/16	140 to 250	20 to 24
7/32	170 to 300	20 to 24
1/4	200 to 400	20 to 24
5/16	250 to 500	22 to 26

ELECTRODE size, amperage and polarity

Applying the Facing

4 Ventilate well if the welding must be done inside some confined space

5 Bead must spread as shown. Be sure it overlaps to avoid porosity, slag

6 Run two or three beads around circumference few inches at shaft's end

10 Run next bead on opposite side of shaft to distribute the heat evenly

11 Use bead sequence shown to prevent shaft from distorting permanently

12 Stagger the bead ends all around the shaft, as here, to strengthen pads

Put It to Work Hard-Facing a Worn Stoker Shaft

1 Clean shaft with safe solvent, such as trisodium phosphate; rinse and dry

2 With a gas torch, preheat part to be coated—just enough to remove chill

3 Set calipers for finished size plus 20% for finish machining, shrinkage

7 Turn shaft around and run some beads on other side to spread heat evenly

8 Run bead along shaft. Start from outer edge and work in from the end

9 Stop if you have magnetic blow (this happens with ac); start at other end

13 Laying second pad, with Type I, A, 4 alloy: put beads between first ones

14 First pad takes dc straight polarity, second pad, dc reverse polarity

15 Cool slowly by covering shaft with sheet or powdered asbestos, as shown

NOW Low-Temperature

EUTECTIC* PRINCIPLE

*EUTECTIC means easily melted. Pertaining to an alloy or solution having its components in such proportion that the melting point is lowest possible with these components

1 Every pure metal has its own melting point, 621 F for lead, 450 for tin

2 Melting point of solder varies, but may be much less than for tin or lead

Lead	Tin	Melts
75%	25%	514 F
50%	50%	415 F
37%	63%	360 F

6 New low-melting alloy forms and melts under rod at 1090 F; sinks in copper

7 Keep pressing down on rod. As new alloy forms, rod keeps sinking deeper

8 Rod breaks through. You did not have to heat copper to its melting point

WELDING ALUMINUM

1 Aluminum, welded usual way, needs fusion heat of 1218, can damage

2 To low-temperature weld, first apply flux to area that is to be welded

3 Play torch around area to be welded, not directly on it. That's important

Welding Saves You $$$

Eutectic principle

Pure copper (melting point 1981 F)

Aluminum flux

Pure copper

Aluminum welding rod (melting point 950F–1150F)

3 Eutectic point of both is 443 F, but lead melts 621 F—antimony 1169 F

4 Flame breaks through a sheet of pure copper at its melting temp of 1981 F

5 New flux fuses aluminum rod to hot copper. Press down on aluminum rod

WELDING CAST IRON

Melted area

Bead

Critical point is 1450 F at this distance from weld

-2300 F

Flux applied here

1450 F welding temp

At same distance from weld, temperature is only 800F

1 Critical point for cast iron cracking is 1450 F. It expands fast there

2 Fusion welding heats metal above its critical point of 2300 F, beyond weld

3 Heat to only 1450 F with low-temp method; keep below critical point

Flux is only temperature indicator for aluminum

Perfect fillet formed by rod all around, by capillary action without moving rod around area to be welded

Melting temp of aluminum 1218 F

At 950 F, apply welding rod

4 Melted flux (950 F) is temperature indicator. Apply the welding rod now

5 Perfect fillet is formed all around by holding welding rod in one spot

6 Low temp did not thin metal. Fillet is even and formed at all the joints

1 Start out by giving your plates double-vee butt, ⅛- to 3/16-in. clearance

2 Heat of first welding pass may buckle up plates. It shows stress is present

3 Turn plates over after the third pass; then clean and make ready for welding

4 First pass on the second side of plates pulls the plates back towards balance

5 Finished plate looks like this. Peen weld hard for any slight unbalance

Welding Plates Made Easy

You don't have to worry about heavy steel plates buckling under welding. Here's how to do it right

▶ Too often a plant maintenance man thinks that welding two flat steel plates together is a simple job. All you need do, he thinks, is slap the two together and weld away. Actually to weld a plate of any size, say 10 ft by 4 ft by ½ in. thick, to another of the same size and keep them straight is a matter of knowing stress and strain.

But there is a definite way of going about the job to both do it right and avoid undue stress. Plates must have a double-vee butt on the facing sides. Lay the plates flat, with a ⅛- to ⅛-in. space between them. This spacing makes sure you burn through to make a pure metal job.

Tack the plates together and proceed with the job. A straight edge will help you see which way the strain is pulling. Here's how you proceed:

1. If you have a dc welding machine, use straight polarity all through the job. An ac machine is simpler to handle, because you don't have to worry about any magnetic blow. Magnetic blow comes as a result of magnetizing the metal so much at one spot that the welding material is repelled there and won't stick. Welding from the other direction cancels this magnetic field.

2. Lay your first pass through the entire bevel with a ⅛-in. mild-steel elec-

trode, Fig. 2. It is safer to weld two feet and stop. Then move up about two feet and repeat. Now go back and weld the open spaces. This method, known as step welding, spreads heat evenly through entire plate.

3. Once you complete the first pass, clean the weld thoroughly with a wire brush and hammer. Now lay a continuous bead with a ¼-in. mild-steel electrode starting from left to right. Weld the whole length unless you come to a magnetic blow. Then start on the opposite end and weld until you complete the entire pass.

By this time, any stress in the two plates caused by the heat of the depositing metal will have buckled up the plates, Fig. 2. Clean the second pass and apply the third just as before.

4. After the third pass, turn the plates over, Fig. 3. With an air gun and a sharp gouging tool, such as a flat chisel, clean the weld until you see clear metal, and get rid of all the slag, excess metal and dirt deposited with the first pass.

5. Now you can make the first pass on this side. Use more heat than usual for the ¼-in. mild-steel electrode to make sure you penetrate well into the metal of the plates to form a solid piece. When this pass is finished, you'll find

the strain has pulled the plates back towards balance, Fig. 4. A straight edge will help you decide. If the plates haven't pulled too much, lay a second pass. (Clean all welds thoroughly before making additional passes). Second pass should bring plates in line.

If plates have stayed pretty well in line, you may apply a third pass. But if they are so close that additional heat will pull the plates past balance and buckle them the opposite way, peen the weld. To do it right, use the peening tool in an air gun. It evens the strain and helps hold plates in balance. You can now make a third pass.

6. After the third pass, check plate balance with straight edge. If they balance, you are ready for the next move. If they have buckled past balance, peen the weld hard to bring plates back in line.

You can then turn the plate back to its original side and make passes four and five to finish that side.

7. Turn the plate over again and finish the reverse side. Metal already deposited by this time should be enough to hold plates even and no unbalance will develop. When finished, Fig. 5, you have a straight plate 20 ft by 4 ft by ½ in. thick with a welded seam as pure as the plates' metal.

18
Rigging

ROPE for general industrial use is handy because it's flexible, light. But use a rope with safety factor of 5 to 1 for the load

"Pick It Up?"

Rigging jobs come up often around plants, but usually few landlubbers know too much about them unless they have sailed aboard ship. It takes know-how to work safely with manila rope, wire cable, chain hoists, while saving costly manhours

▶ "PICK IT UP, it ain't spit!" Sounds like something old Marmaduke Surface-blow might bellow at a well-intentioned but unhurried wiper boy on the SS *Suretosink*. But how many of us with plenty of plant experience know all we should about how to "pick it up"?

Safety committees and insurance companies put up enough posters on safe manual lifting so we ought to know the important things: Bend your knees, keep your back straight, push with your legs instead of lifting with your back—and if it's too heavy for one man, get some help. If it's still too heavy, then you've got to resort to some kind of rigging. *(More on next page)*

Handy Table of Rope Sizes and Strengths

3-strand rope, standard lay; use at least 5 to 1 safety factor

Nominal Size			Net wgt of 100 feet Lb	Minimum length in one lb (net wgt) Ft	Approxi- mate gross wgt full coils Lb	Minimum breaking strength of good rope	
Threads	Circum- ference In.	Diameter In.				Manila Lb	Sisal Lb
6-Fine	9/16	3/16	1.47	67.9	50	450	360
6	3/4	1/4	1.96	51.0	50	600	480
9	1	5/16	2.84	35.2	50	1,000	800
12	1 1/8	3/8	4.02	24.9	50	1,350	1,080
15	1 1/4	7/16	5.15	19.4	63	1,750	1,400
18	1 3/8	15/32	6.13	16.3	75	2,250	1,800
21	1 1/2	1/2	7.35	13.6	90	2,650	2,120
	1 3/4	9/16	10.2	9.80	125	3,450	2,760
	2	5/8	13.1	7.65	160	4,400	3,520
	2 1/4	3/4	16.3	6.12	200	5,400	4,320
	2 1/2	13/16	19.1	5.23	234	6,500	5,200
	2 3/4	7/8	22.0	4.54	270	7,700	6,160
	3	1	26.5	3.78	324	9,000	7,200
	3 1/4	1 1/16	30.7	3.26	375	10,500	8,400
	3 1/2	1 1/8	35.2	2.84	432	12,000	9,600
	3 3/4	1 1/4	40.8	2.45	502	13,500	10,800
	4	1 5/16	46.9	2.13	576	15,000	12,000
	4 1/2	1 1/2	58.8	1.70	720	18,500	14,800
	5	1 5/8	73.0	1.37	893	22,500	18,000
	5 1/2	1 3/4	87.7	1.14	1,073	26,500	21,200
	6	2	105.0	0.949	1,290	31,000	24,800
	6 1/2	2 1/8	123.0	0.816	1,503	36,000	28,800
	7	2 1/4	143.0	0.699	1,752	41,000	32,800
	7 1/2	2 1/2	163.0	0.612	2,004	46,500	37,200
	8	2 5/8	187.0	0.534	2,290	52,000	41,600
	8 1/2	2 7/8	211.0	0.474	2,580	58,000	46,400
	9	3	237.0	0.422	2,900	64,000	51,200
	9 1/2	3 1/8	264.0	0.379	3,225	71,000	56,800
	10	3 1/4	292.0	0.342	3,590	77,000	61,600
	11	3 1/2	360.0	0.278	4,400	91,000	72,800
	12	4	426.0	0.235	5,225	105,000	84,000

Wire Cable. How can you check a wire cable? You *don't* do it by running your hand over the wire—that's one of the easiest ways to pick up some nasty metal slivers. You *do* look the wire over completely. If more than 10% of the wires in one running foot are broken, the cable is unsafe.

Chain Hoist. A weak chain is not easily detected. It may look just a little old and rusty, yet may have reached its fatigue limit and be full of invisible cracks. Then it will suddenly break without warning. So, chain hoists and chains used for lifting should be inspected and tested for 150% capacity. Inspect them at least every 12 months. If you notice a kink in a chain when making a lift, lower the load at once and take out the kink. This is important.

Best bet for precision rigging work is a chain hoist on the hook of a crane lift. Then when you get close to landing the load, further lift or lower only with the chain hoist, never with the crane itself. We learned this the hard way.

We were landing the water-end head of a large condenser on a chain hoist on the big hook. It was let down nice and easy, and two bolts got caught. But to line up rest of holes, we had to lift it up just a hair and tap it over a bit. OK so far, but an eager beaver in the gang signaled the crane man to hoist. He did — with a vengeance. One bolt sheared off. The chain-hoist hook slipped out of the eyebolt. The chain whipped through the air while the big plate slid down and hung on the other good bolt. We were lucky. No one was hurt. But it brought home two more lessons:

1. Always mouse a hook if there is the slightest chance that the load might slip off.

2. Have one responsible man, and only *one*, give the necessary signals to the man handling the crane.

How to Take Care of Your Rope

Uncoil Rope Properly

Unwind right-laid rope counter- clockwise. Lay coil flat — pull end up through the center of coil

Dry Rope If It Gets Wet

Don't ever store wet rope. First dry thoroughly to avoid short life, loss of strength and even mildew

Overloading Is Risky

Safety factor is ratio of minimum breaking strength to load. Make 5 to 1 on new rope

Slack Off The Guys

Guy lines and other supports exposed to the weather tighten up from wetting; keep them slack

Sharp Bends Are Very Bad

Extra strain on outer fibers from sharp bends reduce rope's tensile strength. Pad corners as shown

Reverse Rope End-for-End

Localized wear in short section is bad. Reversing rope ends helps all sections get needed equal wear

Keep Rope Kink-Free

Twisting and kinks harm rope structure. Avoid these if you want most service from your good rope

HOW TO TIE THE RIGHT KNOT IN A JIFFY

CLOVE HITCH

FOR MAINTENANCE WORK around the plant, the clove hitch is one of the most useful knots. Make single loop with two hands. Then make a second loop with right hand. Now move second loop over first loop and slide over beam end or load

TIE CLOVE HITCH IN PLACE around pipe or beam like this. Throw loop over pipe, cross rope end underneath, then over pipe. Stick end of rope through cross-under loop. This knot won't slip and yet can be untied with no trouble at all

BOWLINE

HERE'S A MUST. To tie bowline quickly, tie an ordinary overhand knot. Flip end 1 over to form loop. Tuck end 1 behind end 2 and down through loop. One of many uses is to form eye on end of line as shown. Knot unties very easily

STOPPER HITCH

FOR TIGHTENING LINE already under strain, the stopper hitch is best. Coil rope around line to be tightened and allow for loop, with two half hitches as shown. This knot stays in place on cable or rope and won't slip on hard surface

SHEEPSHANK

TO SHORTEN a long line without untying or cutting rope, use the sheepshank. For example, if a long line is needed to tie on to a load and then, because of obstruction, line must be shortened, back up tractor and tie a sheepshank

SLING SHORTENER

SHORT SLING can be made from long sling easily when there isn't enough head room. Form into two loops and tie over-hand knot. Then place two loops over hook. This knot saves having too many slings of different sizes around the plant

HOW TO USE RIGHT KNOT FOR JOB

Blackwell hitch

Weavers knot

Pipe

Pole

Clove hitch

1 Knots are useful for operating engineers. Hoisting oil drums or barrels is safe with rope under barrel, two half hitches around barrel and a weavers knot. For hoisting small parts, use steel drum and sling on pipe shoved through drum

2 For hoisting heavy equipment in or out of plant, gin pole is easy to rig. Fasten guy ropes to any suitable anchor near base. At top, clove hitch around pole will hold. Some states require 4 guy lines on this job, others 5 guy lines

Timber hitch

Closed sling

End

Square knot

End

3 Hoisting piping or lumber saves time with this timber hitch. Load material over single line and wind end back around as shown. It won't slip after strain is taken and is easy to undo. Closed sling around load is another easy way

4 Another way to lift heavy loads is winding as many turns around load as necessary to be safe. Tie ends with square knot. This is only right way to tie square knot. It won't slip. No matter how tight knot gets, it's easy to untie

Blackwell hitch

Bowline

Bowline

A   B   C

Scaffold hitch

5 Blackwell hitch on hook keeps load from shifting and prevents rope falling off hook if strain is suddenly taken off rope. Bowline on hook doesn't slip and never gets tight. This is one of most useful all-around knots to use anywhere

6 Scaffold hitch prevents scaffolds from turning. Handy when painting plant walls and ceilings, or working on high suspended pipes. **A, B** and **C** show easy steps. Be sure to tie loose end above scaffold with bowline knot as shown here

1 Proper size is important. Hoist capacity must at least equal load weight

2 Rig the chain block and load carefully. Never use the load chain as a sling

3 A metal sling carries the load safely. Never use rope to tie load to the hook

4 Tail-chain anchor pin must be free of defects. Inspect it before using hoist

5 Load brake must be in good operating condition, never slipping or dragging

6 An open load hook is dangerous. It may drop load. Replace it immediately

7 Stand aside. Never get under the load. Why ask for a crushed or broken head?

8 Handle all chain blocks carefully for good service. Never drop or throw them

9 Report defective chain blocks promptly. Never use one; it may drop the load

9 steps to chain-block safety

Sketches, above, illustrate some of the essentials of safe chain-block operation. If your men follow them, they will obtain maximum safety and keep chain block maintenance to a minimum.

Additional points to consider include the following: (1) Always suspend chain block from the saddle, not the tip, of the hook. (2) After hoist is suspended, check load chain for twisting or kinking. Never put a strain on a kinked chain. Take up the slack slowly and see that every link seats properly. (3) Never use the load chain for a sling. This can damage the load chain. (4) Use hoists of larger capacity than the regular work requires. (5) Tag each sling to indicate the maximum load it can lift safely. Remember, increasing the angle between legs of a chain sling and the vertical increases the strain on the sling.

Follow manufacturer's instructions for maintenance and repair. Inspect all parts periodically. Replace worn, corroded or damaged parts. Inspect upper and lower load hooks. Replace if they are distorted, opened, elongated in shank, or damaged in any way.

Inspect load chain with a magnifying glass. Replace if worn, corroded or damaged. Look for bent links, cracks, scores or markings, corrosion pits, lifted fins. Lubricate unit as recommended by manufacturer.

19
Fire, accident prevention

Know How to Use the

TYPE OF FIRE

CLASS A: Fires of wood, paper, cloth, rubbish, etc. Use pump tank, soda acid or just water

CLASS B: Fires of oil, gasoline, grease, paint, etc. Use CO_2, dry-chemical or foam type

CLASS C: Electrical-equipment fires. Put out with CO_2, dry chemical or vaporizing liquid

Five Handy Fire Fighters

SODA ACID OR WATER. Use on class A fires only. Start at top of wall and work down, wetting surface as you go to knock down flame. Then go back and drench thoroughly to prevent reflash. Range of stream is from 30 to 50 feet

CARBON DIOXIDE (CO_2) is for class B and C fires. Direct the gas across the flaming area and sweep the fire before you. CO_2 is a nonconductor, leaves no residue, and does not affect equipment or even foodstuffs. Range is 3 to 6 feet

How to Operate

SODA ACID OR WATER. Turn upside down, bump on ground to create pressure. Main extinguishing agent is water, so keep away from the electrical equipment

CARBON DIOXIDE (CO_2). Unhook hose, pull out lock pin, squeeze trigger or turn handwheel. Discharge at fire base. Keep the stream range at about 3 to 6 feet

Right Fire Extinguisher

FOAM. Use n class B, alsoo on class A fires. Back the stream of foam off the opposite side of the wall (if possible) to build up a solid blanket of foam without splashing the burning liquid and without spreading it. Range is 30 to 40 feet

DRY CHEMICAL. For class B and C fires. Direct stream into nearest corner of fire (if possible) and sweep across flame. Chemical releases a smothering gas on fire; fog of dry chemical shields operator from the heat. Range is 8 to 12 feet

VAPORIZING LIQUID. For class B and C fires. Play stream against side of container or wall to break it into droplets so liquid vaporizes quickly. Remember, it's the vaporized gas and not the liquid that puts out fire so gas must contact fire

FOAM. Turn over, bump on the ground. Aim at base of wood fire, float foam on top. The foam is a flowing blanket of durable bubbles filled with an inert gas

DRY CHEMICAL. Remove ring pin, free hose, push lever down, squeeze nozzle-handle—or, if another make, unhook hose, pull lock pin and squeeze trigger

VAPORIZING LIQUID. Unlock by turning handle. Pump stream on burning material. Range of stream is 20 to 30 ft. V-L is nonconductor, won't damage equipment

How's your accident IQ?

There are ten basic causes of accidents. Use this checklist as a guide in evaluating the safety program in your own plant

I have found that when a plant encounters operating troubles, management always takes one course of action. "Find the cause of the problem." Will the same approach work in Safety? Yes! But let's look at what happens when an accident occurs and someone is injured. The injured employee's supervisor is required to investigate and fill out an accident report form. The usual report form has questions on location, equipment concerned, action taking place, and protective devices used.

Primary cause. Usually the supervisor must answer the question, *Primary cause of accident?;* and usually it is answered incorrectly.

Why is this the case? Since many supervisors do not investigate in depth, they come up with what they think is a "suitable answer." Another reason is that a supervisor may tend to "cover up" for the employee or for himself.

As a result of a program of "in depth" accident investigations, a list of ten basic causes of accidents was

drawn up, and is reprinted below. These were used to train supervisors in *accident investigations.* Please note the ten items concern "people." This is because our study indicated that most accidents are caused by people. And an equipment failure in most cases is the result of some person not doing his job. Thus the greatest benefit in a *safety program* is obtained by concentrating on the "people" aspect. Only then (in most cases) will equipment failures take care of themselves. Here is our list of the ten basic causes of plant accidents:

1.—**Disobedience of rules.** Most plants, large and small, today have *safety rules.* These may be plant-developed or supplied by the insurance carrier and vary from a few broad guide lines to a complete list of detailed rules. In either case, failure to comply will cause many accidents.

2.—**Ignorance of rules.** A new employee usually receives indoctrination which covers basic safety rules. But this teaching must be continuous, as employees will forget. And new equipment, new processes with new hazards create a pressing need for continuous updating of safety rules.

3.—**Inattention to work,** or failure to keep alert. Employees become preoccupied and their minds tend to wander. This may be due to personal problems at home, or perhaps thinking about a forthcoming hunting or fishing trip. Did you know that accidents actually increase when a company and union approach a strike deadline?

4.—**Lack of complete instructions.** Here the supervisor fails to realize the need for detailed instructions. He may assume the employee "knows how," and not realize that a few added words will help the man to perform his assignment safely.

5.—**Not following instructions.** Many times the supervisor will give detailed instructions; yet they are not followed. The man may feel he "knows how" and not realize the importance of following the instructions as they were outlined.

6.—**Interference by others.** This may vary from simple horseplay to the case where another employee closes a switch or opens the wrong valve. Signs and locks can help prevent this kind of accident.

7.—**Not using protective devices** or safety equipment. Goggles, masks, gloves, aprons, suits and other protective equipment are not comfortable to wear, but wearing them is certainly the lesser of two evils. The use of such protective equipment requires education, training and supervision.

8.—**Lack of skill or experience.** In any manpower pool there exist experience and skill in varying degrees. Thus the supervisor should take care to select personnel with the required skill and/or experience.

9.—**Poor judgment.** The supervisor and employee must visualize any potential hazards and take necessary precautions. This often is based on judgment. And we know that many accidents are caused by poor judgment.

10.—**Taking chances.** "I thought I could get by" is too often the answer of an injured employee. He truthfully tells that, being in a hurry, he "took a chance"—and lost.

When supervisors begin to dig a little deeper looking for the real cause, they will improve their accident IQ. Note that the word careless covers all 10 items. This is one good reason why it should never be used in a report. And remember that "be careful" is a meaningless statement in job instructions; be careful of *what?* ●

20
License examination questions, answers, and problems

WHEN MANAGEMENT and operating engineer work out answers to problems together, running steam plant isn't tough

WATER TREATMENT sampling and analysis today is key to good-quality steam and efficient operation of high-pressure boilers

STEAM Q and A

. . . so you don't need licenses in your locale? You still need these answers to run plants safely . . .

Q 1—What attention should be given water level in boiler when lighting off cold boilers? Also when cutting out boiler that has been steaming?

A—Before lighting off cold boiler, have water level about one inch from bottom in gage glass. As boiler heats, water expands and level reaches half glass or over by time boiler is ready to cut in.

When fires are extinguished, before cutting out boiler steaming on line with other boilers, bring water level to at least half a glass. Heat stored in hot brick walls generates steam for some time. Steam stop valve on boiler must remain open while steam's generated with fires out, or safety valves will lift. During this time steam flows to main header and water drops in glass, so feed the boiler from time to time.

When water level in glass stops dropping, boiler is ready to cut out by closing main stop. If steam stop valve is not closed then, water will rise in glass because boiler stops generating steam and condenses steam from other boilers connected to it.

Q 2—What data should boiler's nameplate have?

A—Boiler nameplates should have: (1) Manufacturer's name, (2) Boiler's working pressure, (3) Hydrostatic pressure boiler's tested for, (4) Tensile strength of boiler metal, (5) Steam-drum safety-valve and superheater safety-valve lifting pressures, (6) Year boiler was built.

Q 3—In converting from coal to oil, what attention should be given the furnace for burning fuel oil efficiently?

A—The burners must be set so oil does not impinge on the furnace walls or on boiler surfaces. When oil strikes these surfaces, carbon forms because oil does not have a chance to mix thoroughly with air. Then furnace smokes as more carbon forms. Oil impinging on brick walls thins bricks quickly, causing costly maintenance. Furnace walls must be thick enough to prevent radiation and keep furnace hot for efficient combustion.

Furnace must be free from air leaks. Besides cooling furnace, air leaking in distorts fuel-oil spray, often causing impingement.

Keep furnace free from oil deposits. These are caused by dirty burners not allowing tips to atomize oil properly.

Examine fires often. If flame is "dirty", pull burner and change tip.

Q 4—Explain what precautions to take when lighting off boiler after doing extensive brickwork.

A—New brickwall must be dried slowly. That's because walls are wet and sudden heating to high temperature causes early failures from moisture turning to steam, causing tiny ruptures. Start small wood fire in furnace. A slow wood fire dries a new wall in about 24 hours.

If this is not practical, use smallest tip in center burner to keep fire away from walls. Raise steam as slowly as possible. Operating engineers often overlook importance of thoroughly drying new brickwork, then wonder why walls fail.

Q 5—What should the operating engineer know about installing and operating steam reducing valves?

A—If possible, locate steam reducing valves at point where line has driest steam. That is usually highest point in line. Connect drain line from reducing valve to trap. Place gate valves on both sides of reducer with bypass and valve around them. Then reducing valve can be bypassed when servicing by opening bypass and closing gate valves.

Gate valves on either side of reducers do not obstruct steam for normal operation, as would globe valves. Make sure reducing valve is installed with steam flow according to arrow on valve.

Before turning steam on new reducing valve the first time, drain line thoroughly. Crack steam valve to line on high pressure side to build up pressure gradually against reducer. Then adjust spring tension to desired pressure by watching pressure on reducer's low side.

Q 6—What are some causes of water bobbing up and down in boiler gage glass? How you would correct fault?

A—This condition is usually from foaming, especially in smaller boilers where contaminated condensate returns to boiler. Foaming can also be from excessive water hardness, high water density in boiler or from impurities forming scum on boiler water surface. In severe cases, water boils violently and carries over into steam line.

First step is to open surfaceblow valve to blow scum from waters surface. CAUTION! *Keep eyes on gage glass while blowing and hand on valve.* This habit keeps you from walking away from boiler and forgetting to shut off valve. You may have to raise water in glass and give additional blows to stop foaming.

If test shows boiler-water concentration is high, raise water to near top of glass and open bottomblow valve. Repeat feeding and blowing until water tests right and foaming stops. CAUTION! *Cut out fires while giving bottom blow to watertube boiler.* Make thorough check to find source of contamination. Grease extractors may be faulty or loaded with oil.

Q 7—What do you check when vacuum in steam condensing system drops?

A—When vacuum drops, first check the condenser-cooling water circulator. It may have slowed down. Then check water level in condenser's hotwell.

High water submerges condenser tubes, reducing their cooling surface.

Condensate pump's running too fast and keeping condenser's hotwell dry will also lower vacuum in systems where condensate pump discharges directly to open collecting tank.

Losing loop seal in intercondenser drain loop lowers vacuum. If this happens, close throttle valve in loop until seal returns to normal. If no sight glass in loop, liquid height can be located by feeling temperature of piping.

Air-ejector strainer may be clogged with dirt. Or there may be an air leak in condenser shell. Also low steam pressure on vacuum side of turbine sealing glands allows air leak into condenser.

Q 8—Horizontal steam engines are said to "run over" or to "run under". What does that mean?

A—These terms are used to indicate the circular motion of the crank. When an engine "runs over", crank is high when traveling away from cylinder. When engine "runs under", crank is low when traveling away from cylinder.

Q 9.—Name some causes that might make a duplex steam feed pump stop running.

A—One of the most common reasons for duplex pumps stopping is steam leaking into cylinder at both ends of the piston while piston is in middle of stroke. This is caused by ends of valve wearing round against a shoulder worn on steam valve chest. Valves must be free of shoulders and slightly longer than distance between outside of valve ports for this reason.

Another common reason for stopping is valve-stem threads wear off at valve

nuts. Then stem slides through nut without moving valve, and pump stops. Soft packing in liquid end is often installed without enough end and side clearances. Then hot water swells packing so piston can't move. At times steam leakage past piston and rings is so bad that pump stops. Then shoulders must be ground from pump and new piston rings and piston installed.

Q 10—Why is vacuum for reciprocating steam engines carried lower than for steam turbines?

A—Expanding steam to a very low pressure (high vacuum) in reciprocating engines is uneconomical for several reasons. High vacuum brings low-pressure end of engine down to very low temperature. Also steam expanded to very low pressure (high vacuum) occupies a lot more space; low pressure cylinder would have to be built extremely large. The cylinder condensation loss from steam washing the cooler cylinder surfaces would also be higher. The lower condensate temperature would need more preheating before entering boiler. Also more power would be used for turning main circulating pump faster.

A good practical vacuum for reciprocating engine is about 26 inches, which is a back pressure of about two pounds absolute.

Steam turbines can make good use of a higher vacuum because a turbine expands the steam down to the exhaust pressure, whereas pressure in steam engine cylinder is always well above back pressure when exhaust valve opens. A practical and economical turbine vacuum is about 28.5 inches of mercury.

Q 1—What is a resonance tube?

A—Photo shows resonance tube used to diagnose turbine "ailments". Turbines in normal operation have characteristic hum. One tip-off of trouble inside is unusual noise. Lengthening or shortening the tube "tunes it in" to noise frequency. Trouble sounds are charted and faults found by dividing the sound velocity by the frequency.

Q 2—Explain need for turbines strainers and their maintenance?

A—Strainer in photo is in main line ahead of turbine steam chest. Strainer catches particles carried over with steam such as valve parts, cotter pins or disks that could wreck turbine blades. This one is 50 per cent plugged with boiler compound that carried over. Inspect strainer at least once a year.

TURBINE Q and A

Q 3—Give four important steps in warming up condensing steam turbine?

A—First, start the lubricating system and check oil flow to bearings. Also test low-oil alarm in system. Then engage and start the turbine's jacking gear.

Second, start condensate pump, cut in gland steam, and cut in second-stage air ejector to bring vacuum to about 20 in. Start main circulating pump.

Third, open all turbine and main steam-line drains. Next, bypass the main stops and warm up main lines slowly. Close drains when water is blown from lines.

Fourth, warm up the turbine slowly by cracking the warm-up valve. With jacking motor turning rotor and steam blowing through, turbines using moderate temp should be warmed in several hrs. Disengage jacking motor and open

throttle slowly. Check lube-oil system again.

Q 4—What are two main types of turbines and how do they differ?

A—Two main turbine types are impulse and reaction. In the impulse type, steam expands in nozzles leading from steam chest and is directed at the moving blades or buckets. Diaphragms separate the stage wheels. In the reaction turbine, steam expands in both the moving and stationary blades. In other words steam pressure drops in passing through both stationary and moving blades.

Q 5—How are turbine shafts sealed against loss of vacuum?

A—Two major types of sealing glands are used. One has soft-metal labyrinth rings, the other carbon rings.

On the high-pressure end steam leaks between gland rings and shaft, drop-

ping in pressure as each ring throttles it. Because steam must leak past a number of rings it drops in pressure each time, until reduced to a few pounds. On high-pressure units steam is bled off between rings and piped to low-pressure section of turbine to do work.

On the vacuum end, steam enters packing gland at about 8 psi and flows in two directions, toward the turbine, and out to the atmosphere. Steam tending to blow out of gland prevents air flowing in.

In labyrinth-type glands, the labyrinth rings clear the shaft by several thousandths of an inch. Carbon rings ride on shaft when turbine is stopped. Then when steam is turned on glands, the rings are raised off the shaft by steam blowing between shaft and ring. Clearance between carbon rings and shaft is then also a few thousandths of an inch.

Q 6—What precautions should be taken with a shut-down turbine?

A—The interior surfaces of shut-down turbines are subject to corrosion, especially if moisture is present or turbine is in a damp place.

After turbine is stopped, run air ejector for about an hour, keeping a moderate vacuum (about 12 inches). It's good practice to turn air ejector on for about fifteen minutes every day when turbine isn't running. This draws moisture out of casing.

While running air ejector, open the lines between turbine drains and ejector to remove moisture. Each day jack the turbine over a few revolutions and bring it to rest at a different position. Be sure to pump oil through bearings while jacking. Circulating oil prevents it guming and keeps water from collecting in bearings; a combination that corrodes shaft.

Q 7—What is a double-flow turbine?

A—The steam enters double-flow turbines at the center and divides, flowing in opposite directions. Each side is, in effect, a separate turbine, both mounted on a single shaft.

End thrust is divided about equally between the two halves, exerting little pressure on the thrust bearing. This naturally cuts down friction and reduces the size of thrust bearing needed.

Exhausts are connected together. This turbine is adapted for low-pressure work, because two sets of shorter blades can be used for passing larger steam volume.

Double-flow turbines are also used as the low-pressure unit of compound machines. The high-pressure unit is usually single-flow impulse-reaction.

Q 8—Explain how turbine's speed is controlled?

A—Closing or opening valves controls amount of steam flowing through turbine. A governor regulates the turbine's speed in normal operation by opening or closing these valves ahead of nozzle groups.

An overspeed trip protects the turbine against run-away speeds by closing the main throttle or a clapper valve to shut off all steam to turbine. This is a small weight in shaft that's thrown against trip by centrifugal force.

Q 9—What is velocity compounding in an impulse turbine?

A—In velocity compounded stage there are two or more rows of blades secured to one rotor disk. Stationary blades are attached to the casing between the moving blades. These stationary blades reverse the steam flow and direct it against the next moving row. Velocity compounding is usually found in the first stage.

Advantage of velocity compounding is that shorter turbine is used because this stage absorbs much kinetic energy.

Q 10—What is a pressure stage and how does it differ from a velocity stage?

A—A pressure stage consists of one or more nozzles which speed up the steam before it passes through the single row of moving blades. All pressure drop takes place in the nozzles. Steam pressure remains constant while passing through moving blades.

In a velocity-compounded stage the nozzles speed up the steam faster than in a pressure stage. Steam pressure is constant after leaving nozzles.

Q 11—What is an extraction turbine?

A—Steam passes through turbine's stages, expanding and doing work. When steam expands it decreases in pressure through each successive stage. Extraction ports bleed steam from one or more stages, depending on pressure or pressures needed.

Piping carries this bled steam for feedwater heating or for process. Regulating valves, either automatic or manual, adjust steam flow so turbine has enough steam for its load. Because bled steam has done work in turning turbine, its cost for process use is less than live steam. Also called a bleeder turbine.

Q 12—Describe how turbine disks are constructed for impulse turbines of the type shown here.

A—They are steel forgings, machined from a solid forging or built up from several small forgings.

Disks (wheels) are rough machined, annealed, heated and measured for trueness at 1000 F. Maximum distortion should be less than 0.001 in. from true center at any point. Then they are finish machined and tested for cracks.

The disks are shrunk or pressed on shaft of impulse turbines, with a spacer ring separating each wheel. Rotor is then bladed and dynamically balanced.

Q 13—Explain most serious energy losses in steam turbines.

A—In impulse turbines, steam leakage between diaphragms and shaft is important. In reaction turbines, leakage over blade ends through radial clearance passages averages about 5 per cent.

There is also frictional resistance between steam and nozzle sides, impact loss as steam enters moving blades, and eddying and fluid friction in blades. Friction caused by steam against wheels and blades is sometimes great, depending on steam density and wheel speed. Bearing friction is another loss.

Boiler Questions and Answers

These questions all refer to horizontal return tubular (HRT) boilers. They are similar to those asked in some of the better-arranged state examinations for licenses to operate steam boilers. Use them as an aid in interviewing and selecting power-house employees, and as a review of important features of boiler-room operating practice

Q—Why are diagonal stays not used instead of through-to-head stays below the tubes to brace the tube sheet in the same manner as above the tubes?

A—Because there is usually insufficient room for the proper number without placing them too close together, or using sizes larger than practical. Also, diagonal stays would tend to hold loose scale and sludge and prevent its free movement to blowdown.

Q—How much space would you want in front of an hrt boiler in planning its installation?

A—Enough room for tube replacement.

. Q—If you looked into the furnace and saw the bottom of the shell bulged, what would you do?

A—Shut the boiler down immediately and then have it inspected by an authorized inspector. Follow his recommendations before returning the boiler to service.

Q—Why are boilers over 72-in. diameter required to be supported by the outside suspension type of setting?

A—Because weight of the larger boilers may exceed safe load on brick-work. Crushing or buckling of walls might result from load of a large boiler full of water.

Q—What are two dangerous conditions offered by a weakened rear arch?

A—(a) If the arch collapses wholly or in part, upper part of the rear tube sheet may become overheated and damaged. (b) Anyone walking on top of the arch may cause it to collapse.

Q—What precaution is required with a flush-front-set hrt boiler?

A—The front arch protects the dry-sheet and front head seam from damage due to overheating. The arch should be kept in good condition.

Thousands of HRT boilers such as these are serving America today in industrial establishments, buildings and institutions everywhere

Q—Would loose brick lying in the firebox be of any interest to you? Why?

A—They might be from the closing-in line, or from some point where they are supposed to protect part of the boiler from overheating. Also, they might be from a point where their loss would weaken the walls.

Q—Name three points where brickwork should protect an hrt boiler from overheating.

A—Rear arch, closing-in line along sides of the shell, and front arch in the flush-front-set boiler.

Q—How and under what conditions may the segments of the heads above the tube dispense with diagonal or through stays?

A—In boilers not exceeding 36 in. diameter or 100 psi maximum allowable pressure, segments may be braced by stiffening with channel irons or angle irons (riveted back to back) riveted to the tube sheet. Specifications for structural form should conform to ASME Code rules.

Q—What dangers are there in settling of boiler foundation or supporting walls?

A—Severe stresses may be set up in piping, connections to the boiler, or in the boiler itself. Brickwork protecting parts from overheating may be dislodged. Sludge may settle to the front instead of to the rear and cause overheating. If serious, walls may become weakened and collapse.

Q—What would you look for at the holes in the brick wall where the blowdown pipe comes through?

A—To see that the pipe is not resting on the bottom of the opening (an indication that the boiler is probably settling) and to see that the pipe has sufficient clearance to allow free expansion and contraction.

Q—What is the lowest safe water level?

A—1 in. above the top row of tubes —at the lowest end.

Q—What is a strong-back?

A—It is a bar bolted to the tube sheet that prevents the sheet from buckling when stays are being installed before tubes are installed.

Q—If a boiler is pitched forward, slightly, what should be done?

A—Unless the installation was designed this way with the blowdown in front, the boiler should be reset with proper pitch towards the rear as soon as possible. In the meantime, the boiler should be shut down frequently and the manhole or handhole in the front head removed. All sediment should be washed out. Frequency of these washouts depends upon rapidity with which sediment accumulates.

Q—What is the difference between a through brace and a through-to-head brace? Where is each used and why?

A—The through brace has washers and nuts on each end. The through-to-head brace has nuts and washers on one end; the other end is forged into an eye that is held by a pin or other construction to hold it clear of the rear head. Through stays may be used above the tubes in the rear; outside nuts are protected from burning off by the rear arch.

Q—If a fusible plug is used, where should it be located?

A—Near the centerline of the rear tube sheet, not less than 1 in. above the top row of tubes.

Q—In answer to last question, why near the centerline?

A—Because this is the highest part of the rear arch and this extra space is needed often to use a wrench.

Q—Where would you look to find the maximum pressure for which an hrt boiler was built?

A—On the standard stamping on the front tube sheet above the tubes.

Q—Why do flat segments of the tube sheets or heads require bracing?

A—Because internal pressure tends to bulge these areas outwards into spherical shape.

HOW SERIOUS IS AN ELECTRIC SHOCK? WELL, THAT DEPENDS ON . . .

. . .CURRENT FLOWING
- Moisture present
- Pressure between contact surfaces
- How much of body touches "hot" line

. . .PHYSICAL CONDITION OF THE VICTIM

. . .CURRENT PATH THROUGH BODY

Body resistance in ohms:

Dry skin	100,000 to 600,000
Wet skin	11,000
Hand to foot (internally)	400 to 600
Ear to ear	100

. . .TOTAL TIME CURRENT FLOWS THROUGH BODY

HOW ELECTRIC SHOCK AFFECTS YOU

SAFE

1 ma (milli-ampere) or less	Can't be felt
1 to 8 ma	Shock not painful; can let go. Muscular control OK.

DANGEROUS

8 to 15 ma	Painful shock; can let go. Muscular control not lost.
15 to 20 ma	Painful shock; control of muscles lost. Cannot let go.
20 to 50 ma	Painful, severe muscular contractions, hard to breathe.
50 to 100 ma (possible) 100 to 200 ma (certain)	Ventricular fibrillation (A heart condition resulting in instant death—no known remedy).
200 ma and over	Severe burns, severe muscular contractions. Chest muscles clamp heart and stop it for duration of shock. (This prevents ventricular fibrillation).

THIS DIESEL will give trouble-free efficient service if the operating engineer knows equipment and keeps on his toes

DIESEL Q and A

FUEL-INJECTION system is the heart of any diesel. Operator must know it thoroughly before making the needed adjustments

Save diesels "timing" like this with your "time"

Diesel makes food do work in "one gulp" in cylinder

Q 1—If your four-cycle engine's camshaft is removed and the timing completely lost, how would you time the engine in a hurry?

A—A four-cycle engine's intake valve opens from 10 to 20 degrees before piston reaches top dead center (TDC) and exhaust valve closes from 10 to 20 degrees past top dead center.

For quick timing, jack piston of any cylinder to top dead center, with the camshaft gear out of mesh. Then turn camshaft until the inlet valve just starts to open and the exhaust valve just starts to close. Mesh gears while camshaft is in this position. Timing will usually be correct to one gear tooth.

Q 2—Give a short definition of an internal-combustion engine?

A—An internal combustion engine is any engine that changes the heat en-

ergy of fuel into mechanical energy within a cylinder. This includes gas turbines.

Q 3—If one cylinder of a diesel engine receives more fuel than the others, how is engine's operation affected and how would you prevent more fuel in one cylinder?

A—The cylinder receiving more fuel is affected in a number of ways, depending on conditions. That in turn affects the engine in various ways.

Excess fuel in one cylinder causes incomplete combustion in that cylinder because too much oil is injected for the air supplied. Then incomplete combustion causes a smoky exhaust. If excess fuel isn't stopped soon, carbon builds up around piston rings, sticking them in grooves. If rings stick, piston won't compress enough air (because

Pyrometer saves engine "blowing it's top" suddenly

of air leaking past rings) to keep temperature right in cylinder for proper combustion. That causes more smoke. Then stuck rings also cause combustion gas to blow past piston. Hot gas blows lube oil off cylinders, scoring them and even causing piston to seize.

Excess fuel may also cause combustion knock. Because of power loss in one cylinder from incomplete combustion, fuel is wasted. With remaining cylinders doing more than their share, engine starts overheating.

Best way to avoid excess fuel is to check cylinder exhaust temperatures with a pyrometer every hour. Then adjust fuel oil so all cylinder exhaust temperatures are within 10 F.

Q 4—What are the average compression pressures and temperatures in a full diesel engine?

A—Air is compressed from 400 to 500 psi in the cylinder, depending on engine design. At these pressures, it ranges from 800 to 1000 F. These temperatures are high enough to ignite fuel entering cylinder for combustion.

Q 5—How is diesel engine's power output increased without increasing the engines revolution or size?

A—The most popular method is by supercharging. This means delivering air above atmospheric pressure to the cylinder. Then same engine burns more fuel and increases mean effective pressure, without increasing cylinder size, or turning faster.

With supercharging, instead of engine drawing in atmospheric air, air is forced into cylinders by blower up to about 10 psi above atmospheric pressure. Then engines output is increased about 40 percent.

Supercharging is widely used at high altitudes. Because air is less dense, engine needs more air to burn same amount of fuel than at sea level.

Q 6—What is meant by injection lag and how is it caused?

A—Fuel is injected into the cylinder about ten degrees before piston reaches top dead center. But the fuel-pump cam is timed to force oil through the pump about 30 degrees before top dead center. This difference of 20 degrees is the injection lag.

Injection lag is caused by: (1) Expansion of fuel-oil discharge lines under high pressure, resulting in need for additional fuel volume to be displaced. (2) Compressibility of fuel. When fuel is near 3000 psi, it is compressed about 1 percent of its volume. (3) Leakage past the fuel-oil plunger.

So fuel-oil lag is actually a kind of lost motion and is compensated for by

pumping oil into the fuel-distributing system earlier than needed in the cylinder.

Q 7—Why are some diesel pistons oil cooled instead of water cooled?

A—Pistons are water cooled in the large older crosshead type marine engines. Modern engines use oil to prevent the crankcase lube oil being contaminated with water in case of leaks or cracks. When lube oil is contaminated with water, oil emulsifies and causes burned-out bearings.

Q 8—Name some causes of smoky exhaust.

A—A smoky exhaust is caused by one or more faults. Perhaps the most common is from fuel distributed unequally to cylinders. Or the engine's load might be excessive. If the cylinder or exhaust valve receives too much lube oil, engine will smoke because of burning lube oil. Also if fuel injection is late. Water in fuel or fuel not suitable for engine causes smoke. Fuel injectors not adjusted right are a sure cause. Weak or broken fuel-valve spring is another. Obstructed exhaust piping and low compression in cylinders are also common causes.

Q 9—Where should thermometers be installed on diesel plants?

A—There should be a thermometer at the following: Engine cooling-water inlet and outlet. Engine lube-oil inlet and outlet. Lube-oil sump tank. Air-compressor cooling water inlet and outlet. Cooling water outlet from each cylinder. Lube oil cooler's cooling-water inlet and outlet. In some installations thermometers are also placed on scavenging air.

Q 10—Why must fuel injectors be kept in perfect working condition?

A—A faulty injector is like a boiler's faulty oil-burner tip. Unless the oil sprayed into the cylinder is finely atomized, the oil cloud does not burn properly and some of the fuel is wasted in smoke.

Some diesels do have a coarse spray and still burn oil efficiently. But they depend on precombustion or turbulence chambers to mix the oil with air before entering cylinder.

The cloud of tiny oil drops sprayed directly into the cylinder ignites when heated to the right temperature. The smallest drops naturally heat up faster and therefore burn first. Because the larger particles heat slower, they ignite later and need more time to burn. If the particles are too large, they are exhausted before they burn completely. That wastes fuel and increases consumption. That's why oil injector maintenance is so important to efficiency.

Fuel "burns up" from punch-drunk air in cylinder

Supercharger puts extra "wallop" in diesel engines

Injection lag is fuel "lost motion" in the fuel system

Oil keeps hot piston from "freezing" in cylinder

Smoke can mean the operator is in a "fog" or groggy

Buying thermometers save "buying parts" for engine

Breaking up oil saves "breaking owner" in the end

WhyAmmonia Soaks Up Heat—

These questions and answers give a clear picture of what happens to liquid refrigerant and how it does its job while being pushed from one pressure to another by a compressor

1 Under constant pressure heat boils ammonia without raising its temperature above that corresponding to pressure

▶ The what and how of refrigeration in a few simple explanations:

Q 1—Explain the term, one ton of refrigeration.

A—The American Society of Refrigerating engineers defines the standard commercial "ton" of refrigeration as the transfer of 200 Btu per min, or 12,000 Btu per hour. It is the basis of all refrigeration calculations, whether for cold storage, air conditioning, ice making or ice-cream manufacture. When a person says he has a 70-ton ice plant—that is something else. He means that he can make 70 tons of ice in 24 hours, and the rated refrigerating tonnage of his plant, as defined above, might be 150.

Q 2—What does superheat mean, and what causes ammonia to become superheated in the evaporator coils?

A—Pressure and temperature of saturated ammonia have a definite relationship. The temperature corresponding to a pressure of 20 psi is 5.5 F. This should be the gas temperature as it leaves the evaporator coils. However, a thermometer in the discharge line may show a temperature of 8 or 10 F. The difference is superheat.

If we could remove the gas as soon as it forms, giving it no chance to remain in the coils and pick up heat,

we would not have superheat at this point.

It is impossible to eliminate superheat entirely. A poorly-insulated suction line also helps superheat the gas.

Q 3—What happens to ammonia as it passes through an expansion valve? Is its boiling point changed?

A—When ammonia passes through an expansion valve, it expands because it passes from a higher to a lower pressure. As pressure drops, boiling temperature falls and some or all of the ammonia, depending on conditions, changes from liquid to gas.

Q 4—Why is it that a change in a substance's heat content does not always show up as a change in temperature?

A—Every substance has a definite boiling point, and when we say "boiling point" we mean the temperature at which the substance changes from liquid to gas. Boiling point changes with pressure.

A drum of ammonia at 70 F shows a pressure of 114 psi. If it has previously been stored in a cooler place, boiling takes place until it reaches 70 F and a pressure of 114 psi.

Suppose we put a relief valve, set at 114 psi, on this drum and spray hot water over it, Fig. 1. The ammonia

starts to boil and the gas thus formed escapes through the relief valve which holds a steady pressure.

As long as any liquid remains, its temperature holds at 70 F, all the supplied heat being absorbed in changing the liquid to a gas.

Q 5—What is the specific heat of ammonia? Also, explain how such heat affects the use of this liquid as a refrigerant.

A—Specific heat is actually a ratio, but we use the term to indicate the quantity of heat needed to raise one pound of a substance one F, or to lower one pound the same amount. Specific heat varies with temperature.

Generally speaking, when the ratio between latent heat and specific heat of the liquid is large, we have an ideal refrigerant.

Liquid ammonia is in this class, its specific heat is 1.129 at 70 F, and that of its gas at constant pressure corresponding to 86 F, is 0.4011.

Ammonia refrigerant serves to convey heat away from some substance that we wish to cool or freeze.

When liquid ammonia expands from a higher to a lower pressure, in an evaporator coil, we have to remove some of the heat content.

For example, if liquid reaches the

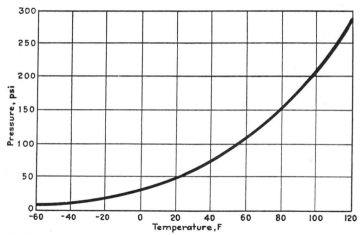

2 Ammonia vapor pressure-temperature characteristics. Its pressure increases slowly at low temperatures but more rapidly as the temperatures get higher

3 Latent heat of vaporization and heat content of saturated ammonia vapor and liquid. Heat contents of vapor and liquid are the same at critical temperature

when used in speaking of ammonia.

A—Sensible heat, applied to ammonia (or any other substance) shows up as temperature registered on the ordinary thermometer. The term, used without considering other factors, simply tells whether ammonia is hot or cold—as we think of these terms in everyday life. It is not a measure of total heat or Btu in ammonia.

Q 8—Give a simple explanation of the pressure-temperature relationship of ammonia in a closed cylinder?

A—Pressure of ammonia in a closed cylinder increases with its temperature —slowly at low temperatures, but rapidly at high temperature, Fig. 2.

Q 9—What is the critical temperature of ammonia and how does it affect the operation of a refrigerating plant?

A—Critical temperature is 280 F. At this temperature, latent heat becomes zero, and heat content of the liquid and the vapor is the same, Fig. 3. As the pressure at this temperature is well over 1600 psi, it is far above the normal operating range.

Q 10—What happens to ammonia in the evaporator coils?

A—Ammonia in the evaporator coil boils and does work of refrigeration. Actually what happens is this: Assume that condensing pressure—gage pressure of liquid going to the coils—is 185 psi, and that evaporator-coil pressure is 20 psi. Referring to ammonia tables we find: (1) heat content of the liquid at 185 psi (199.7 absolute) is 150.5 Btu per lb, (2) heat content of the liquid at 20 psi (34.7 absolute) is 48.9 Btu per lb.

Then heat removed to cool liquid to 5.5 F (20 psi) before doing work, is 150.5 — 48.9 = 101.6 Btu per lb. Since latent heat at 20 psi is 564.6 Btu per lb, and heat removed is 101.6 Btu per lb, the heat absorption-ability left to do work is 463.0 Btu per lb.

Q 11—How do you determine the amount of ammonia that must circulate per ton of refrigeration?

A—Find how much work or refrigerating effect we can expect from each pound of ammonia. This is the difference between the heat content of saturated liquid at condenser pressure and saturated vapor at suction pressure. It can also be expressed as the difference between the quantity of heat removed by the condenser water and the work done by the compressor.

Since one ton of refrigeration equals 200 Btu per minute, divide this number by the refrigerating effect of the ammonia in Btu per pound to get the ammonia need in lb per min per ton.

expansion valve at 96 F and the evaporator coils are at 5 F, the liquid must be cooled to 5 F before it performs any useful work.

Temperature difference, 91 F here, multiplied by the specific heat, gives the quantity of heat to be removed. This can also be found in ammonia tables.

Q 6—Explain the term latent heat, when applied to ammonia.

A—Latent heat is the number of Btu that must be applied to and absorbed by ammonia to change it from liquid to gas. If a quantity of liquid ammonia is just at its boiling point (−28 F at atmospheric pressure) each pound must absorb 589.3 Btu before it will change to a gas.

This is not a fixed quantity, but varies with pressure and temperature. For example, liquid ammonia at 70 F requires 508.6 Btu per lb to change it to a gas. The U. S. Bureau of Standards published Tables of Thermodynamic Properties of Ammonia; text books and manufacturer's catalogs contain tables of these characteristics.

Q 7—Explain the term, sensible heat,

How to Charge an Ammonia System

▶ IF LARGE AMMONIA SYSTEMS were like sealed units kitchen refrigerators, operating engineers wouldn't have to worry about charging them. But their life isn't that simple. Leaks waste ammonia, or oil and water find their way in, so the operator must constantly check quantity and quality. When ammonia gets low, he must add it. When charge gets too impure, he must draw it off and replace with fresh ammonia.

Like all other jobs around the plant, there are right ways and wrong ways to charge ammonia systems. Here are pointers on the right way.

How Shipped. Manufacturers ship dry (anhydrous) ammonia in steel cylinders holding 150, 100, 50 and 25 lb net. Regardless of size, liquid ammonia fills only 85% of the cylinder at 70 F. Extra space allows for expansion of ammonia at higher temperatures, say, up to 125 F.

Cylinders will stand normal expansion of ammonia, but not the powerful forces built up when exposed to high temperature. Keep cylinders away from steam lines, hot exhaust pipes and heating places—and never apply a torch.

Take Care of Drums. Ammonia manufacturers test their cylinders regularly to be sure they can stand high pressure, and that rough handling and rust haven't made them too weak for safety. Plant operators can help by emptying drums as soon as received and returning them promptly. If necessary to store drums, keep them in doors, never outside in direct sunlight or on ground where moisture starts rusting.

When a cylinder reaches the plant, weigh it first, then test. Makers of ammonia guarantee both quality and quantity; to be sure, check both—it's easy and worth while. Weigh cylinders with protecting cap before and after emptying. Check gross and empty weights against shipping charge.

Quality Test. It's lucky that the test for dryness and purity of ammonia is also the simplest. With reasonable care, nothing more than a small flask (4–8 oz) one old stopper and sampling pipe, you can duplicate the accuracy of the manufacturer's test.

The sketches show the details. Sampling pipe can be tubing, as shown, or ordinary black pipe, but must be clean. Fitting a shield (of metal, wood or cardboard) near the end will keep im-

These clear and detailed practical instructions are based on information from National Ammonia Division, E I du Pont de Nemours Co, and Henry Bower Chemical Mfg Co

STEPS IN SAMPLING

1 Collect 3- to 4-oz clean sample
2 Cap sample in flask with this stopper and bent tube
3 For slow evaporation keep sample cool. After evaporation, inspect flask for traces of oil or water

purities out of the flask when filling.

Place cylinder in horizontal position with dipper pipe pointing down (valve outlet or valve stem will point upward, depending on type of valve). Open cylinder valve slowly to allow first gas, then liquid, to flow, sweeping impurities out of sampling tube.

Stand flask under tube with mouth against shield as shown, and let in about 3- to 4-oz of ammonia. Then remove flask quickly and cap with a one-hole stopper, fitted with glass tube bent as shown in Fig. 2.

Lipson Sampling. The chief trouble with tests may come from ammonia's

CHARGING

4 Before and after weights of tank are only sure measure of amount of ammonia delivered to the system

habit of picking up moisture. Leaving the flask open, even for a short time, picks up a surprising amount, so be sure to close the flask quickly.

The tube in the stopper carries away gas during evaporation and points down, so that froth collecting at the mouth won't fall back into the flask. Better still, avoid testing on humid days; in such weather the amount of water in the flask at the start may be enough to make the test worthless.

Ammonia will evaporate with the flask standing in the open air, but frost collecting on the glass will slow down evaporation. For a better job, hang flask in can or bucket of brine by some simple clamp, Fig. 3. Or hold it under a stream of tap water. A low-temperature bath gives more accurate results—shouldn't be much above 32F.

After the ammonia has evaporated, dry the outside of the flask and look at what's left inside, if anything. Water will show as a single drop at the bottom, or as a lot of tiny drops. Oil may show as a film or a small puddle. Reject the cylinder if you find the slightest trace of oil or water.

After checking for oil and water, wave the flask gently through the air to sweep out ammonia gas; then cautiously smell the flask. A nauseating odor points to pyridine, an organic impurity.

How to Charge. Ninety-nine times out of a hundred or better the ammonia will be found to be pure, and you're ready to charge. Most large plants take the ammonia charge to a connection between "king" valve and expansion valve. Fig, 4 shows how to connect the cylinders at this point. The purge valve helps, but isn't absolutely needed. Raise the bottom end of the cylinder about three-quarters inch and be sure the dipper pipe points down.

If charging line and connections are

tight, close king valve and open cylinder valve and charging valve. After king valve is closed, pressure between it and expansion valve will drop to suction pressure, allowing liquid ammonia to flow into system. To be sure cylinder is discharging into system, put it on scale and watch change in weight. As cylinder empties it gets cold near valve from evaporation of last few ounces of liquid.

If no scales are available, this cooling effect will show emptying. But don't depend on it; always weigh cylinder after charging to be sure it is empty.

Sometimes a mistake in hooking up equipment, or some other trouble, causes reversed flow—from system to cylinder, overfilling it. Exposing such an over-filled cylinder to even slight overheating might bring on a dangerous explosion —better be safe than sorry.

When cylinder is empty, close cylinder and charging valve and open king valve. Lead pressure from charging line to purge valve. If no purge valve is installed, crack the union carefully.

Charging With Gas. To charge a system with gas, instead of liquid ammonia, turn cylinder over so inside dipper points up. Ammonia may be taken from the cylinder at the rate of 1½ to 2 lb per hr; heating cylinder with warm water will speed up the flow, but is not recommended.

Always check to make sure that gas flows out of cylinder and does not condense into it. Again, weighing the cylinder before and after will show for sure.

How much ammonia should a system contain? For the first charge or a complete recharge, best answer comes from the engineer who designed and installed the system. If the figure isn't handy, you can calculate amount needed from data given in refrigerating handbooks.

In a running plant, the gage glass on

receiver tells the story; receiver should be about one-half full. Without gage glass, you have to fall back on watching operation. If only lower coils frost when expansion valve is wide open and plant is not overloaded, the system needs ammonia. If a clicking sound is heard in expansion valves, it checks the lack of ammonia.

HOW TO PREVENT ICE ON

Induced Draft Towers

1—Run two-speed motors at half speed to retard ice formation. Fans use 15% full-speed power.

2—Shut down some fans (if multi-cell unit).

3—Cover louvered area's upper portion to limit air flow.

4—Reduce tower's water flow and stop one or more fans (if multi-cell unit).

5—Bypass water to part of cooling tower and stop one or more fans (if multi-cell unit).

Louvers and Filling

1—Reverse fan's motor (for limited time only) to blow warm air out louvers. Caution: Some speed reducers don't supply oil to upper bearing when reversed.

2—Stop some fans temporarily, but don't stop water. When cells thaw out, start fans and repeat process in other cells.

3—Cover upper portion of louvered area to limit air flow.

CLOSED SHELL AND TUBE ammonia condensers placed on roof. Oil separator traps oil in the gas; drains to the sump

WHAT'S WRONG HERE? Never place the stop valves under the safety valves in any system. This hookup violates every code

REFRIGERATION Q & A

27 questions often asked for unlimited refrigeration engineer's license

Q 1—What is a refrigerating system?

A—An apparatus or combination of mechanical equipment in which a refrigerant is circulated for cooling, or extracting heat from spaces or bodies.

Q 2—Name two types of refrigerating systems in general use?

A—Compression, absorption systems.

Q 3—In what type of systems are generators, analyzers, and weak-liquor coolers used?

A—In the absorption system.

Q 4—In what type of system is aqua ammonia used?

A—In the absorption system.

Q 5—What is a refrigerant?

A—A chemical compound or agent to produce refrigeration. Brine is not a refrigerant.

Q 6—What is a compressor as used in refrigerating work?

A—Compressor is a mechanical apparatus to draw evaporated gas from low-pressure side of system and to discharge it to high-pressure side, in a condensing

system. Unit may be reciprocating, rotary or centrifugal.

Q 7—A certain refrigerating system has a normal operating suction pressure of 150 psi gage and condensing pressure of about 1000 psi. What refrigerant is used? Ammonia, carbon dioxide or Freon?

A—Carbon dioxide (CO_2).

Q 8—Name several types of condensers in general use?

A—Shell and tube, double pipe, atmospheric and evaporator.

Q 9—Describe briefly an evaporating condenser.

A—It has a series of tubes or coils, usually the fin type, enclosed in a sheet-metal housing. Cooling surface is often divided into two sections. Hot discharge gases enter the top section, where part of the sensible heat from the hot gases is removed. Water is sprayed over lower section in sufficient quantities to keep the surfaces wet. A fan or blower drives the blast of air over the cooling surface, causing rapid evaporation. This condenser has advantage of large capacity in small area, eliminates cooling towers.

Q 10—What is meant by the term "direct refrigeration"?

A—A system where the refrigerant is circulated direct to the spaces or substances to be cooled. It is usually called a direct-expansion system.

Q 11—Is a direct-expansion system, with evaporating coils in the air-circulating system, permitted in public buildings?

A—No. But certain types of self-contained portable units with solid soldered connections are approved. See your local code.

Q 12—How are refrigerating systems usually classified as to capacity?

A—This varies in the different municipal codes. In New York City, *Class A* is a common system containing 1000 lb of refrigerant or more and having a capacity of 30 tons or over. *Class B* contains less than 1000 lb and more than 20 lb and has a capacity of less than 30 tons refrigeration. *Class C* contains not more than 20 lb refrigerant.

Q 13—What are hydrocarbon refrigerants?

AMMONIA MIXER is important safety device for dumping large quantities of ammonia in emergency. Water absorbs the ammonia, washes it down into the sewer

discharge line. Consult your local code.

Q 23—Can you pack an ammonia stop-valve stem while under pressure?

A—Ammonia valves have a machine surface on top of valve disk and bottom of valve bonnet. When valves are wide open, these surfaces seat, cutting off pressure from the valve stem, permitting packing under pressure.

Q 24—In placing stop valve in compressor discharge lines, what precaution must be taken?

A—Valve must be so placed that discharge gases enter the valve under the valve disk. Reason: Should the valve disk become loose from the stem, it cannot drop down and close off the line.

Q 25—What is a pressure-limiting device?

A—Many codes require a high-pressure cutout, so arranged that when a predetermined pressure is reached, controllers or oil switches in the power lines to the motors will trip and cut off the power. Some codes require a remote pushbutton station from which the entire plant can be closed down in a fire or other emergency.

Q 26—What is an ammonia mixer and what is its use?

A—There are several types. One is manually operated, another automatic. Each consists of a system of valves so arranged with connections to the ammonia, water-supply and sewer systems, that the entire ammonia charge may be mixed with water and discharged into the sewer system in a fire. Some codes require that these manually operated mixers be mounted outside the plant, in suitable heavy-metal cabinet. Keys to cabinets are supplied to local fire stations. See local code for special requirements.

Q 27—State briefly how you would withdraw the ammonia charge from the plant, for reuse or transportation.

A—In pumping out a plant, refrigerant must be discharged into drums, approved by Interstate Commerce Commission. Place empty drums on an accurate scale, with flexible metallic hose, between liquid ammonia connection and drum for accurate weighing. Never fill drums to more than 90% of capacity by weight, which would allow for a slight inaccuracy in weighing and prevent overfilling, as gas space must be left for expansion. Drums should never be placed outside in the sun. Only such quantities of refrigerant are kept in storage in drums, as the local code permits. Most metropolitan areas restrict this quantity to less than 500 lb, some permitting only 300 lb,

A—They contain hydrogen and carbon, like ethane, propane and butane. These refrigerants, together with menthol chloride are flammable when mixed with air.

Q 14—Are hydrocarbon refrigerants permitted in Class A systems?

A—Many codes prohibit flammable refrigerants in large systems. See your local code.

Q 15—Describe method of detecting leaks in a system using Freon.

A—A halide torch is commonly used. This torch assembly has a copper plate enclosed in a metal shell, heated by an acetylene gas flame. The flame from this reaction plate changes to a greenish color when brought into contact with Freon gas. Leaks as small as .01% by volume may be detected.

Q 16—How are small ammonia leaks detected?

A—Gas leaks in open air are usually checked with sulfur sticks. These sticks consist of a heavy cotton cord coated with sulfur. When lighted and brought into contact with ammonia gas, the smoke given off turns white. Leaks in brine systems can be detected by a number of reagents.

Q 17—State a general rule for locating safety valves.

A—All pressure vessels, fired or unfired, that can be valved off as a sep-

arate unit, like brine coolers, receivers, accumulators and air tanks, must be protected by proper safety valves.

Q 18—What is maximum pressure to set safety valves on ammonia accumulators and shell-and-tube brine coolers?

A—150 psi gage.

Q 19—What pressure is allowed on ammonia system's high-pressure side?

A—250 to 300 psi.

Q 20—Where should safety valves discharge?

A—Discharge may be connected to low-pressure or suction side of the system or to atmosphere, depending on type and size of plant. Safety valves on brine coolers are usually discharged into the main suction line, just above the stop valve.

Q 21—State two other locations that require safety valves.

A—Ammonia compressors and generators must have pressure-relief valves connected below the main discharge valve, arranged to discharge into the main line above the stop valves. Many modern units have relief valves built into the header assembly.

Q 22—What other valves are required in the discharge line above the main stop valve?

A—A suitable check valve should be installed above the compressor, in the

REFRIGERATION Q & A

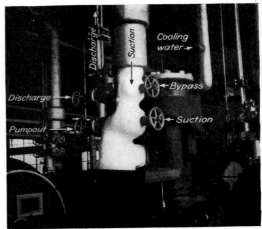

AMMONIA COMPRESSOR has header with a discharge, suction, pumpout, bypass, safety valve, all cast in one compact unit

18 questions often asked for unlimited refrigeration engineer's license

Q 1—What are the general requirements for ventilation in refrigerating machinery rooms?

A—This varies with amount of refrigerant in system, type of refrigerant, and whether ventilation is natural circulation or forced-draft. Thus, class-A system with over 1000 lb refrigerant needs (a) exhaust fan with capacity of at least 2000 cu ft per min or (b) window areas to outside air of 25 sq ft if on opposite walls or (c) 60 sq ft if windows open on one side of machinery room. See your local code.

Q 2—Can the ammonia condensers or discharge line be pumped out by the standard type of ammonia compressor in case of leaks?

A—Most compressors are fitted with bypass and pump-out valves, on machine header, besides the main suction and discharge valves. For normal operation, suction and discharge valves on header are open. Reversing their position permits pumping gas from discharge line, discharging it into the suction side of the system.

Q 3—Are open-flame lights permitted in the machinery rooms?

A—Arc lights or other open-flame lights are not permitted.

Q 4—How many helmets and gas masks are required in a refrigerating machinery room? Where should they be located?

A—Class-A systems, with over 1000 lb of refrigerant, should have two masks in the machinery room, near main exit, and one or two masks at convenient points, throughout the plant. All codes approve masks approved by U.S. Bureau of Mines for refrigerant in use.

Q 5—If you are testing a new system, using anhydrous ammonia as a refrigerant, what test pressures are required?

A—Most codes require air pressure of 300-psi gage on high-pressure side, 150-psi on low side. (IMPORTANT: Take great care in pumping air pressures on systems for testing.)

It is not advisable to use large ammonia compressors for pressures in excess of 150 psi. Above this point, use a small air compressor, designed for high-pressure work, and lubricated with air-compressor oil. Best procedure is: Open all valves on low-pressure side of system. This permits placing maximum allowable pressure on brine coolers and accumulators to equalize with that of high-pressure side. Entire system may then be tested at 150 psi.

Next step, close off low-pressure system and blow out air pressure, leaving a valve open at some point to atmosphere. This prevents possibility of placing a higher pressure than would be safe on low side of system, while testing high-pressure side. High pressure is then pumped up to the required pressure and tested for air leaks. Soapy water applied with a soft paint brush at all joints is a simple way to locate small leaks.

After checking over entire system by this method, blow out air pressure, close off atmospheric connections to low side and see that expansion valves are open.

Then pump a vacuum on the system, exhausting all air possible. Large compressors may be used for this purpose. System is now ready for charging. It is good policy to charge in only a small amount of ammonia, enough to raise gas pressure to 75 to 100 psi, and then again fairly test the system for small gas leaks. This may seem like a drawn-out process but, carefully carried out, it saves a lot of time and trouble.

Q 6—Are stop valves permitted between safety valves and receivers, accumulators and other pressure vessels to permit safety-valve repairs while system is under pressure?

A—No valves of any type are permitted whereby safety valves may be closed off, rendering them useless.

Q 7—If you were building a liquid ammonia receiver, say 16 in. in diameter welding in minus or plus heads, would you place the pressure on the concave or convex side?

A—On the concave side; 40% greater pressure is allowed on the concave side.

Q 8—Are brass and copper pipe fittings suitable for ammonia systems?

A—No. Ammonia fittings are made of cast steel or forgings. All piping should be steel or wrought iron.

Q 9—State size of safety valves required on 100-ton-capacity system using carbon dioxide.

A—One-half to ¾ inch.

Q 10—What type valves are required for liquid level gage glasses?

A—Gage glass valves must be fitted

Labels on diagram:
- Discharge line from compressor
- Safety valve vent to atmosphere or to discharge header
- Equalizing line from receiver
- Cooling tower
- Circulating water out
- Safety valve
- Suction line from tower
- Pump out line from compressor
- Purge connection
- Liquid line
- Drain valve
- Receiver
- Liquid drain
- Circulating water in

SHELL-AND-TUBE condensers have large capacity on small floor area. This is simple piping layout for easy operation and maintenance. Flanged heads remove easily

with a ball check, which closes automatically in case of glass failure. Metal screen or slotted metal casing must protect glasses.

Q 11—Are ammonia gas and air mixtures considered explosive?

A—Most authorities place explosive limits between 17 and 25% ammonia by volume. Most serious explosions from flame coming in contact with bad ammonia leaks have been caused by mixture of ammonia gas, air, and oil vapor. There is no danger from explosion in cutting into ammonia lines, receivers or accumulators with acetylene torches or making welded connections if system is open to atmosphere, ventilated.

Q 12—What is latent heat?

A—Latent heat is required to change the state of a body without change of temperature, as in changing melting ice at 32 degrees to water of the same temperature.

Q 13—Is it possible in an atmospheric cooling tower to reduce temperature of the circulating water below that of surrounding atmosphere?

A—Yes. In a well-designed cooling tower, temperature of water off the tower often approaches wet-bulb temperature; 10 to 15 degrees below dry-bulb temperature is not uncommon, depending on relative humidity.

Q 14—Why are holes cast into the rims of flywheels and motor rotors on large units?

A—To help turn machine by hand when making repairs and adjustments. (NOTE: 60% of applicants state that these holes are for balancing purposes.)

Q 15—What would cause the discharge-gas temperature, leaving a com-

pressor, to suddenly become abnormally high without any noticeable change in suction or discharge pressures?

A—There are several causes: (a) broken suction or discharge valve (b) broken piston rings, causing heavy blowby or leaky bypass valves (c) passing hot discharge gas back into suction side of compressor.

Q 16—What would cause discharge temperatures to drop below normal, discharge line to become cold and compressor to frost over heavily?

A—Excessive amount of liquid refrigerant passing expansion valve, causing liquid to slop over from evaporating side to the compressor.

Q 17—If you were asked to take charge of a large refrigerating plant that had been closed down for some time, state briefly how you would proceed, what precautions you would take.

A—Assuming the plant had enough refrigerant in the system and was tight, first make and inspect all auxiliaries, such as water-supply system, pumps, brine coolers or expansion side of system and condensers. Check position of all valves on high- and low-pressure sides, see that they open and close easily.

If synchronous motors are used on high voltages, take Megger tests to determine condition of insulation on motors and feeders. Check contacts on oil switches and breakers, remove covers from vertical-enclosed machines, check condition of connecting-rod bearing bolts.

Check entire oiling system. Bar over by hand large reciprocating machines, one complete revolution, and check piston clearances. Check all rod adjustments and packing boxes. In starting compressors, open suction valves slowly, as liquid refrigerant may be lying in the suction lines. After starting, check all bearings to see that oil rings are turning, as they often fail to start when units are shut down for prolonged periods. See that your plant is supplied with adequate safety equipment—gas masks, proper fire extinguishers, first-aid kits.

Q 18—What is the purpose of a safety head in a vertical, single-acting ammonia compressor?

A—Safety head is a false head, resting on a shoulder on top of the compressor cylinder. Discharge valves are fitted into this head. Heavy springs, placed between the main cylinder head and the safety head hold the latter in position. In case of broken suction valves or heavy slug of liquid refrigerant, this head rises, preventing serious accident to the compressor.

How important is lube life?

Many plant engineers fail to recognize the growing importance of lubricant life—and not simply as a means to trim their lube bills. Here's the reaction we got to questions asked of field experts...

Q. *Why is lubricant life more important today?*
A. Mainly because of labor saving says the user, and the oil supplier agrees. Even though automatic lubrication has come a long way, there are plenty of bearings and gear housings that need manual lubrication. Stretching the time between lube applications here can save the user real money.

Q. *What about lubricant cost—is it important?*
A. Yes, but not as important as the labor saving. The oilers' payroll in at least one plant is two-thirds higher than the lubricant bill, and that doesn't include storage and handling costs.

Q. *How do you go about boosting lube life?*
A. First, get some basic data from the oil suppliers and equipment manufacturers on lubricant life under good operating conditions. Dig back into your plant records and check lube schedules for your actual working conditions. Cross check lubrication frequency in the different areas of your plant to see if bearings and gears working under similar circumstances are getting the same treatment. Talk the problem over with engineers from other plants with the same problem.

Tie all this information together and judge for yourself what your frequency rates should be. Stick in a safety factor and don't double the rates right off the bat; take it in steps.

The oil supplier goes along with this approach, stressing the caution needed. A program designed to extend application periods on a large scale in a plant needs close supervision since the trouble caused by oversights will cost more than the saving in lubes and manpower.

Q. *What are some of pitfalls in extending lube life?*
A. Parts failure. Oil and grease breakdown caused by overwork leads to serious repair bills.

For example, in a recirculation system, oxidation brings on deposits which can clog lines or filters, resulting in reduced or cut off oil supply. Recirculation oil systems have a twofold purpose—to lubricate and cool. Cutting down lube flow might not hurt lube action, but reduced cooling effect can have fast, serious results.

Leaving oil in too long increases viscosity and boosts heating of the parts. Also, chemical reaction rates double every 18 F rise, and experimenting with a high-temperature application is risky.

Q. *How can you check on lube performance?*
A. If it's an oil, you can have it analyzed. This is costly, but about the only sure way to tell if oil is breaking down. Visual inspection of many oils won't tell you much until it's too late.

With greases, just looking at them will usually let you know whether they need to be replaced. They turn dark, get stiff and may show a separation of the oil and soap.

Q. *What do you check before changing frequency?*
A. If high temperatures, moisture or leaks are causing frequent lube applications, there isn't much point in doing anything until these troubles are cleared up. Often the choice of a different grease may give you longer life automatically, since it will resist the effects of moisture or heat.

Q. *Where has extending lube life paid off?*
A. The user we talked with got the best lube frequency change with motor bearings, pillow blocks and gear housings. User greases his motors and certain pillow blocks once a year and that's it.

He uses the expendable theory on some pillow blocks figuring that it's cheaper to replace them now and then rather than to lose product because of over greasing.

Oil supplier believes rolling-contact bearings and gears are easier places to extend lube life than sleeve bearings. He says having an all-loss or recirculating varieties to do the job.

Q. *Are there any new greases or oils on the market that promise longer life?*
A. Both user and oil supplier say "yes." User has switched over to EP grease entirely for all but the very special applications, allowing them to use one grease now where it previously took six different varieties to do the job before.

These greases cost more than ordinary grease but they cut oiler mistakes since there's only one grease to use. They have a tougher film, giving an added operating safety factor. As far as giving longer life, the oil supplier says the EP's can do two jobs: (1) Tackle the tough assignment and extend frequency rates from, say, once a day to once every three days. (2) Replace many types and do a good job over a wide range of applications. Either way the user can cut both his lubrication bills and the time needed for the oilers to do their job.

ONE RATED BOILER HORSEPOWER (A) is 10 sq ft of water-heating surface. One developed horsepower **(B)** is the evaporation of 34½ lb of 212F feedwater into steam at 212F and atmospheric pressure. Around 1870 a 1000-sq ft boiler (100 rated boiler hp) **(C)**) could develop 3000 lb of steam per hr (100 developed boiler hp) to supply engine of 100 brake hp

Let's Talk "Boiler Horsepower"

▶ Many engineers, including me, have publicly denounced "boiler horsepower" as an old-fashioned and cumbersome measure that should be abolished. To my mind there is no doubt about the logic of this position, but I must admit that the old habit will be hard to break. Anyway, I'm going to tell what boiler horsepower is and how it got that way.

History. Back about 1870 an ASME Committee decided that a good non-condensing engine of that day would take about 30 lb of steam per hour to generate one mechanical horsepower— also that the average boiler of that period could generate about 3 lb of steam per square foot of water-heating surface.

Then 10 sq ft would generate 30 lb, so they concluded that the 10 sq ft should be rated as one boiler horsepower and that 30 lb of steam per hr, approximately, should be considered one developed boiler horsepower.

This last was more precisely stated as 34½ lb of steam per hr "from and at 212", meaning the evaporation of 34½ lb of water at 212 F into steam at the same temperature at atmospheric pressure.

Boilers don't usually run under these conditions, so the founding fathers of boiler horsepower invented a "factor of evaporation" to convert "actual evaporation" into "equivalent evaporation from and at 212".

Factor of Evaporation. For example, if the feedwater temperature is 184 F, and the steam is at 100 psi gage and 500 F, this factor is 1.16. This means that it takes 1.16 times as much heat to generate one pound of steam under these conditions as "from and at 212 F."

Now suppose a boiler operating under these conditions has 2000 sq ft of heating surface and generates 65,000 lb per hr. The rated boiler hp will be 2000 ÷ 10 = 200. The 65,000 lb per hr will be equivalent to 65,000 × 1.16 = 75,400 lb "from and at 212".

Dividing 75,400 by 34.5, we get 218 developed horsepower. Then 100 × 218 ÷ 200 = 109 "per cent rating".

Now I admit that there is nothing hard about this arithmetic except figuring the factor of evaporation. But I won't try to explain that here, for why should readers bother to learn an old fashioned method?

No Sense in This. Now let's apply a little common sense to what has been going on here. Ten square feet of water heating surface in a modern boiler will generate anywhere from 50 to 500 lb of steam per hour (the 500 lb in some big central-station boilers). Also, some large modern turbines use as little as 5 lb of steam per hp-hr instead of 30.

So we have a peculiar situation in which one boiler horsepower can serve anywhere up to 100 mechanical horsepower. Which doesn't make sense.

Under today's conditions rated boiler horsepower has become just a cumbersome way of stating the water-heating surface.

Developed boiler horsepower is a complicated way of reporting how much steam the boiler is generating.

The Right Way. Because of this ridiculous situation the makers and users of large boilers said goodbye to boiler horsepower long ago. Both the capacity and actual loads of such units is clearly stated as so many pounds of steam per hour—simple and sensible.

In the smaller plants boiler horsepower still hangs on from force of habit, and probably won't die out completely for years.

If you wonder why people keep on doing things the hard way just remember that English is easier to write than Chinese, but it's hard to sell the idea to the Chinese.

① Work = AB x mean pressure x piston area

② Steam cylinder diagram

③ Shaded area is work delivered to piston

④ Shaded area is net work

⑤ Get MEP from *average* panel height

Indicated Horsepower Explained

Most of us know that *work* is merely *force* times the distance moved. If you pull 50 lb to drag a box across the floor, and if you drag it 100 ft, the work is 50 × 100=5000 ft lb.

The *power* of a steam or internal combustion engine may refer to either the *brake horsepower* delivered at the shaft or the *indicated horsepower* as determined from the diagrams drawn by an engine indicator that shows the pressure at all points of the stroke.

The *indicated horsepower* of an engine is merely the net power delivered to the piston faces by the expanding steam or gas in the cylinder.

The amount of work so performed in one minute, divided by 33,000, will be the indicated horsepower.

The first and most important step is finding the net useful push on the piston. Remember that the work delivered to the piston during the working stroke is partly offset by the work the piston returns to the steam or gas during the exhaust stroke.

In Fig. 1 pressure falls as shown by the curve when the piston face moves from *A* to *B*. Find the average pressure from the average height of the curve between *A* and *B*. This average pressure, in pounds per square inch,

multiplied by the area of the piston face in square inches, gives the average force or push on the piston. This force, multiplied by length *AB*, gives work done during the stroke.

Fig. 2 shows a steam-engine indicator diagram. The steam delivers work to the piston during the power stroke and takes back part of this work during the exhaust stroke.

In Fig. 3 the shaded area, all the way down to the atmospheric-pressure line, represents the work done on the piston during the working stroke.

In Fig. 4 the black area is the work returned during the exhaust stroke; so the shaded area in Fig. 4 is the net or useful work per working stroke.

Perhaps a simpler way to look at it is to consider that the average height of the diagram from the top line to the bottom line represents the net useful pressure. This is merely the average pressure on the working stroke minus the average pressure on the exhaust stroke and is called the mean effective pressure (MEP).

To get the MEP use a planimeter. If your plant does not have this very useful instrument, just figure the average height of a lot of narrow strips as shown in Fig. 5. Then the

indicated horsepower for each piston face in the engine will be $P \times L \times A \times N \div 33{,}000$, where P is the *MEP*, L is the length of the stroke in *feet*, A is the net area of the piston face in square inches (subtract area of rod in case of crank end) and N is the number of working strokes per minute.

Problem: What is the indicated horsepower of a simple double-acting steam engine with the following data?

Piston diameter	10 in.
Rod diameter	2 in.
Stroke	12 in.
Speed	350 rpm
Head-end MEP	42 psi
Crank-end MEP	43 psi

Solution:

Head-end area	=	87.5 sq in.
Rod area	=	3.1 sq in.
Crank-end area	=	84.4 sq in.
Stroke	=	1 ft

Head-end indicated horsepower = $P \times L \times A \times N \div 33{,}000 = 42 \times 1 \times 87.5 \times 350 \div 33{,}000 = 38.9$ hp.

Crank-end indicated horsepower = $P \times L \times A \times N \div 33{,}000 = 43 \times 1 \times 84.4 \times 350 \div 33{,}000 = 38.5$ hp.

So the total indicated horsepower is $38.9 + 38.5 = 77.4$ hp.

--- Volume of water at 212°F ...

··· produces 1,600 volumes of steam of 212°F and 14.7 psi absolute

2 At atmospheric pressure and 212 F one volume of water boils to 1600 volumes of steam. Expansion in evaporation is less at higher pressures

1 At standard sea-level air pressure (30-in. barometer, 14.7 psi abs) water boils at 212 F. If pressure is cut to 0.5 psi abs, water boils at 79 F

3 Heating one pound of water at atmospheric pressures takes one Btu per degree-F rise. So heating from 80 F to the boiling point at 212 F takes 212 — 80 = 132 Btu. To convert water to steam takes 970 Btu per lb

Heating and Boiler Water

▶ THE STANDARD PRESSURE of the atmosphere at sea level is taken as 14.7 psi abs (14.7 pounds per square inch absolute). At this pressure the barometer reads 30 in. of mercury.

The temperature at which water boils rises with the pressure. At the standard atmosphere of 14.7 psi abs, water boils at 212 F, Fig. 1, top.

In airplanes, and on high mountains, the pressure is less, so water boils at a lower temperature. For example, Fig. 1, center, at 11.5 psi abs (23.5-in. barometer) water boils at 200 F.

If vacuum equipment brings the pressure down to 0.5 psi abs (that is 1-in. barometer, or 29-in. vacuum) water boils at 79 F. You can hold your hand in it comfortably, Fig. 1, bottom.

EXPANSION

When heated from 32 F to 212 F, liquid water expands about four per cent. Here are weights per cubic foot at various temperatures:

Degree F	Pounds per cubic foot
32	62.4
70	62.3
100	62.0
150	61.2
200	60.1
212	59.8

At 212 F one pound of water occupies only 0.0167 cu ft. When evaporated, this expands 1600 times to form 26.7 cu ft of steam at 212 F and 14.7 psi abs, Fig. 2.

Steam is compressible, about like air, so one pound weight of steam occupies less and less space as the pressure rises. At 100-psi gage pressure (114.7 psi abs) one pound of steam occupies 3.9 cubic feet. At 445 psi gage saturated steam weighs one pound per cubic foot.

For practical purposes, at temperatures under 400 F, one Btu (British thermal unit) is the heat required to warm one pound of liquid water one degree F. So, to heat 2000 lb of water from 80 F to 180 F (100 degrees rise) takes 100 × 2000 = 200,000 Btu. If this same water cools from 180 F to 80 F, it will give off 200,000 Btu.

In Fig. 3, to heat one pound of water from 80 F to 212 F takes 212 — 80 = 132 Btu. If further heat is supplied, with the fluid under atmospheric pressure through the floating piston, the water will start to turn into steam, raising the piston.

LATENT HEAT

When 970 Btu has been supplied, all the liquid will have been converted into saturated steam.

In this case the total heat required to convert one pound of water at 80 F into saturated steam at 212 F and 14.7 psi abs is 132 + 970 = 1102 Btu. The 970 Btu is the "latent heat of vaporization" for this pressure, temperature.

The ABC of Combustion

ELEMENTS

	NAME	SYMBOL	VALENCE	WGT
	Hydrogen	H	I	I
	Carbon	C	4	12
	Oxygen	O	2	16

PRODUCTS

	NAME	FORMULA	WGT
	Water	H_2O	18
	Carbon dioxide	CO_2	44
	Carbon monoxide	CO	28

REACTIONS

2 H	+	O	=	H_2O
2 Parts	+	16 Parts	=	18 Parts
I Lb	+	8 Lb	=	9 Lb + 61,000 Btu

C	+	2 O	=	CO_2
12 Parts	+	32 Parts	=	44 Parts
I Lb	+	$2\frac{2}{3}$ Lb	=	$3\frac{2}{3}$ Lb 14,500 Btu

▶ THE BURNING of coal, oil or gas is a chemical reaction. These fuels contain many different atoms in many different combinations, but the only important heat-producing atoms in commercial fuels are carbon (C), hydrogen (H).

The oxygen for combustion comes from the air supplied to the furnace. Air is 23% oxygen by weight and 21% by volume. The remainder of the air is mostly nitrogen which takes no actual chemical part in combustion.

The "Elements" table (top left) shows the atoms of hydrogen, carbon and oxygen. These pictures are highly idealized, but are all right for the present purpose. In this table "Wgt" means the atomic weight (weight of one atom), using hydrogen as the standard. Thus carbon weighs 12 and oxygen 16.

Valence explained. "Valence" takes some explaining: Try picturing the valence number as the number of hooks or bonds available to tie this atom to other atoms. The plus bonds pictured will pair with the minus bonds on other atoms.

Under "Products" the water molecule contains two atoms of hydrogen and one of oxygen, so its formula is H_2O. Note that all the bonds are paired up (plus to minus)—also that the weight of the molecule is 18, the sum of the weights of its atoms.

In the formula for carbon monoxide, two of the carbon hooks are unused, indicating that carbon monoxide is the product of imperfect combustion.

Burning H and C. Now turn to "Reactions" to see how hydrogen and carbon burn. The first line shows how two atoms of hydrogen combine with one of oxygen to make one of water (H_2O). The formula balances, meaning that the weight of products equals the weight of combining elements. Also every atom on the left shows up on the right, and every valence bond is paired.

The same is true of the combining of carbon and oxygen to form CO_2.

To fix these important relations in mind why not turn this book face down and see if you can sketch (from memory and common sense) these atoms, compounds and combustion reactions?

ANTHRACITE
COAL

BITUMINOUS
COAL

FUEL
OIL

CO₂ vs Excess Air

▶ MANY OPERATING ENGINEERS are directly responsible for the operation of boilers fired by coal, oil or natural gas.

It is not always realized that it takes more than ten tons of air to burn one ton of 13,500-Btu coal. All this air eventually goes up the stack, although in a different chemical form.

To estimate the theoretical amount of air required to burn one pound of coal, divide the heating value of the coal by 1350. Thus, for the 13,500-Btu coal just mentioned, the theoretical air requirement is $13,500 \div 1350 = 10$ lb. Now furnaces and firing can never be perfect, so no actual furnace could burn this coal with only ten pounds of air. There must be some excess.

Twenty or thirty per cent excess might be reasonable in a small boiler furnace. With 30-percent excess air the furnace would use up 13 lb of air for every pound of coal fired. Since no matter is destroyed, and since very little material goes through the grate,

nearly 14 lb of hot gas would go up the stack for every pound of coal fired.

To heat one pound of flue gas one degree it takes about one quarter Btu. So, consider a case where the boiler-room temperature is 80 F and the stack temperature is 530 F. Then the temperature rise is $530 - 80 = 450$ degrees, and the stack loss in the hot dry gas is $450 \times \frac{1}{4} \times 14 = 1575$ Btu per pound of coal fired.

If, through poor equipment, poorly operated, the stack temperature were 580 F and the excess air were 80 per cent, the air supplied per pound of coal would be 18 lb and the stack gases would weigh 19 lb per pound of coal fired. The temperature rise would be $580 - 80 = 500$ deg. Then the loss in the hot dry flue gas would be $19 \times \frac{1}{4} \times 500 = 2375$ Btu per lb coal.

The standard way to determine excess air is to analyze the flue gas for CO_2, using an "Orsat" or a CO_2 recorder The sample is taken at the point where

gas leaves the boiler setting. Flue temperature should be measured at the same point. Knowing the CO_2 percentage, you can closely estimate the excess air from one of the charts above.

Then you can use a short cut to approximate the percentage of the coal's heat lost up the stack in the hot dry gases.

Here's the Rule: For the theoretical amount of air allow 1.8 per cent loss for every hundred degrees rise above boiler-room temperature.

For example, if the flue temperature is 520 F when the boiler room is 90 F, the rise will be $520 - 90 = 430$ deg. With the theoretical air supply the stack loss would be $4.3 \times 1.8 = 7.7$ per cent.

Increase this figure proportionately to allow for the excess air. For example, if the fuel is anthracite coal and the CO_2 runs 12 per cent, excess air is 60 percent from chart. Then the actual flue loss is $1.6 \times 7.7 = 12.3$ per cent.

How to Figure Water Heating

These quick and simple rules tell dollar-wise operators how much steam it takes to heat any given amount of water from any given temperature to any other in any kind of heating unit

Fig. 1—Heating by mixing. The condensed steam cools to 180 F, so the heat supplied by 1 lb of steam is the total heat of the steam minus the heat in condensate at 180 F.

Fig. 2—Heating by coil in tank. In this case the heat given up per pound of steam is the heat in the steam minus heat in condensate at 210 F.

Fig. 3—Heating in counterflow exchanger. Here condensate is cooled to temperature of cold entering feedwater, so the heat given up per pound of steam is substantially higher than in Fig. 2, a gain of about 140 Btu per pound of steam.

▶ FROM ONE ANGLE OR ANOTHER, every operating engineer is interested in water heating. This article will show how much steam is needed to heat any given weight of water from any temperature to any other temperature. It's really simple if you don't have to split hairs.

How Much Tubing? A more difficult question—one that's often asked—is how many square feet of steel, copper or brass tubing are required to heat such and such a quantity of water per hour to such and such a temperature using steam at a specified temperature. There is generally no way to get a reliable answer to this question, except by cut-and-try with actual tubing.

The reason is that conditions of water flow and arrangement of the coils make a whale of a lot of difference in the so-called "coefficient of heat transfer." This coefficient is merely the amount of heat that will pass through 1 sq ft of tube surface in one hour for each degree F of temperature difference between the steam inside and the water outside.

Professionals who are designing a certain kind of heat-transfer equipment day after day gradually learn what coefficients will give them fairly reliable results. But the operator generally lacks this specialized design experience and the heat-transfer books won't help him much.

As I suggested in the beginning, the operator rigging steam coils to heat water may be forced to guess the amount of piping on the low side and then add piping if the original installation can't handle the load.

How Much Steam? Now let's get back to the original subject of this article. How much steam does it take to heat water? In round numbers, 1 Btu (British thermal unit) will raise the temperature of 1 lb of water one degree F. This applies closely regardless of the starting and final temperature.

For steam a rough-and-ready figure, plenty good enough for many rough estimates, is 1000 Btu given up for every pound condensed. This 1000 Btu is not too far off when the steam is supplied at pressures under 30 lb, as is usually the case when heating water.

Problem: How much steam will it take to heat 5000 gal of water per hour from 60 F to 120 F?

Allowing 8.3 lb per gal, the weight of water to be heated is 41,500 lb. The temperature rise from 60 F to 120 F required 60 Btu per pound, so the total heat to be supplied is 41,500 × 60 = 2,490,000 Btu, say 2,500,000 Btu.

Since the condensing of 1 lb of steam gives up 1000 Btu, you divide 2,500,000 Btu by 1,000 to get the steam required — 2500 lb. per hour. If this steam costs 50 cents per thousand, the total steam cost for heating the water will be 50 cents × 2.5 = $1.25 per hour.

Actually, the condensation of 1 lb of steam rarely gives up exactly 1000 Btu. The amount given up varies with the steam pressure and with the final temperature to which the condensate is cooled.

Simple Steam Table. The abbreviated steam table here shown will handle almost any everyday problems of this sort. The table shows, for example, that saturated steam at 40 lb gage pressure has a temperature of 287 F. This is the temperature at which the

STEAM TABLE FOR WATER HEATING

Pressure psi, gage	Saturation temp, deg F	Heat (enthalpy), Btu per pound		
		Saturated water	Evaporation	Steam
0	212	180	970	1150
5	228	196	960	1156
10	240	208	952	1160
20	259	228	939	1167
30	274	243	929	1172
40	287	256	920	1176
50	298	267	912	1179
60	308	277	905	1182
70	316	286	898	1184
80	324	294	892	1186
90	331	302	886	1188
100	338	309	881	1190

water will boil into steam and at which it will condense from steam into water. The condensate from steam at 40 lb gage pressure can't be cooled below 287 F unless the water is first removed from immediate contact with the steam.

The table further shows that to heat water from 32 F up to this saturation temperature of 287 F requires 256 Btu. At this point, note that if you subtract 32 from 287 you get almost exactly this same figure of 256. For all practical purposes, with low and moderate steam pressures, the heat required to warm water up to the boiling pressure equals the boiling temperature minus 32.

After the water has been heated to the boiling temperature corresponding to 40-psi it takes an additional 920 Btu to evaporate it into steam. The sum of the 256 and 920 lb is 1176 Btu, the total heat (entirely) of the 40-psi steam.

This table works equally well for boiling and for condensing. When condensing, you start with a pound of saturated steam at 40-psi gage pressure, 287 F, with a total heat of 1176 Btu. If this steam condenses without cooling, it gives up 920 Btu. You then have a pound of water at 287 F containing 256 Btu. From then on, the steam gives up 1 Btu for every degree it drops in temperature.

There are two common ways to figure the heat given up by a pound of condensing steam. One is to take the heat in the initial steam and subtract from it the heat in the final condensate. The other is to set down the heat of evaporation (or condensation) and add to this the number of degrees the condensate is lowered below the saturation temperature.

Direct Steam Mixing. For a starter, consider the problem pictured in Fig. 1. The tank contains 5780 lb of water at 60 F. Steam is to be bubbled into this mass at 5 lb gage until the water is heated to 180 F. Temperature rise is 120 deg. Multiplying this by 5780 gives a total of about 694,000 Btu.

The final temperature of the condensed steam is 180 F, so its final heat is 180 − 32 = 148 Btu. According to the table, 5-lb steam contains 1156 Btu. The difference given up per pound is

then 1156 − 148 = 1008 Btu. Dividing this into 694,000 Btu gives 689 lb of steam.

Coil Heater. Fig. 2 shows a slightly different case. The same weight of water (5780 lb) enters at 60 F, leaves at 200 F, after being heated by saturated steam, entering the coil at 10 lb gage. The condensate is discharged at 210 F.

In this case, the temperature rise of the water is 200 F − 60 F = 140, and the total heat required is 140 × 5,780 = 809,000 Btu. From the table, the initial heat in the steam is 1160 Btu per lb. The final heat in the condensate at 210 F is 210 − 32 = 178 Btu. Then the heat given up by each pound of steam before it leaves the coil is 1160 − 178 = 982 Mtu. Dividing 809,000 Btu by 982, we find that 824 lb of steam will be required.

Counterflow Heater. If the tube is well insulated against heat loss, the arrangement shown in Fig. 3, or some other in which the water to be heated flows countercurrent to the cooling condensate, gives the greatest efficiency because it cools the condensate down almost to the temperature of the entering cold water, let's say within 10 deg. In the case pictured, the condensate leaves at 70 F with only 70 − 32 = 38 Btu. The heat given off by the steam per pound is 1160 − 38 = 1122 Btu. Then the steam needed to heat 5,780 lb will be only 809,000 ÷ 1122 = 721 lb.

Standard Electronics Questions and Answers, STEPHEN M. ELONKA AND JULIAN L. BERNSTEIN

    Volume I, Basic Electronics: Direct Current; Magnetism, Inductance, and Capacitance; Alternating Current; Vacuum Tubes; Semiconductors and Transistors; Amplifiers; Power Supplies

    Volume II, Industrial Applications: Oscillators; Special Circuits; Transducers and Sensors; Control Systems; Closed-circuit and Color TV; Industrial Processes and Devices; Test Equipment

Standard Instrumentation Questions and Answers, STEPHEN M. ELONKA AND ALONZO R. PARSONS

    Volume I, Measuring Systems: Instrumentation and Process Characteristics; Temperature Measuring Systems; Flow Measuring Systems; Pressure and Vacuum Measuring Systems; Level Measuring Systems; Analysis Measuring Systems; Physical and Electrical Measuring Systems; Readout and Display Systems

    Volume II, Control Systems: Information Handling Systems; Automatic Control System Characteristics; Pneumatic Operated Automatic Control Systems; Electric and Electronic Operated Automatic Control Systems; Program, Ratio, and Cascade Control Systems; Final Control Actuators and Elements; Process Computers; Educational Opportunities

Figuring Efficiency

1 Man pulls two feet to lift load one foot. Here the efficiency is 90%

2 Efficiency of this motor-generator set is 76.7%

3 Mechanical efficiency of steam engine is 92.9%

4 Overall efficiency of this diesel generating set is 33.4%

5 This small steam plant, exhausting to waste, figures only 4.98% efficiency

▶ ONCE YOU GET the basic idea, efficiency is easy to figure. I mean almost any kind of machine efficiency, with a few exceptions that I won't try to cover. If a machine could give out as much as went into it the efficiency would be 100 per cent. That never happens although big generators and hydro turbines get close to this perfect mark.

A machine that delivers half its input has an efficiency of 50%. One that delivers three quarters has an efficiency of 75%, and so on.

The input and output being considered may be work, heat, mechanical power or electrical power. The important thing is to figure both the input and output in the same kind of units, whether they be horsepower, kilowatt hours, Btu or foot pounds.

EXAMPLES

Input and output are in like units in Fig. 1 and 2. In Fig. 1 both input and output are measured in foot pounds. The man has to pull 50 lb to lift 90 lb and he pulls 2 ft for every 1 ft the load rises. Then:

Input = 50 × 2 = 100 ft lb
Output = 90 × 1 = 90 ft lb
Efficiency = 90 ÷ 100 = 0.90 = 90%

In the mg set, Fig. 2, both input and output are kilowatts. By feeding 103 kw to the motor you get 79 kw out of the generator. Then:

Efficiency = 79 ÷ 103 = 0.767 = 76.7%

In Fig. 3 the two units might appear to be different since input is *indicated* horsepower and output is *brake* horsepower. Yet both are mechanical horsepower. The indicated horsepower is the mechanical power delivered by the steam to the piston and the brake horsepower is the mechanical power delivered by the engine to the load. So the efficiency of the engine in converting cylinder power into useful power (called the mechanical efficiency) is:

39 ÷ 42 = .929 = 92.9%

The case of the diesel generating set, Fig. 4, is somewhat different. Here input is heat and output is electrical. For every kilowatt hour delivered this engine burns 0.53 lb of 19,300-Btu oil. Then the input per kilowatt hour is 0.53 × 19,300 = 10230 Btu.

Table shows that one kilowatt hour equals 3413 Btu, so:

Input = 10,230 Btu
Output = 3413 Btu
Efficiency = 3413 ÷ 10230 = 0.334 = 33.4%

For a final problem, assume that the simple steam plant shown in Fig. 5 blows all exhaust to waste and that, when delivering 82 hp, the fuel burned is 310 lb per hr of 13,500 Btu coal.

This figures 310 ÷ 82 = 3.78 lb coal per bhp hr, so:

Input = 3.78 × 13,500 = 51,030 Btu
Output = 2544 Btu (see Table)
Effi. = 2544 ÷ 51,030 = 0.498 = 4.98%

Conversion Factors

1 kilowatt = 1.3415 horsepower
= 738 ft lb per sec
= 44.268 ft lb per min
= 2.656.100 ft lb per hr
= 56.9 Btu per min
= 3413 Btu per hr

1 horsepower = 0.7455 kw
= 550 ft lb per sec
= 33.000 ft lb per min
= 1.980,000 ft lb per hr

1 horsepower = 424 Btu per min
= 2544 Btu per hr

1 Btu = 778.3 ft lb

1 kilowatt hr = 3413 Btu
= 1.342 hp hr

1 horsepower hr = 2544 Btu
= 0.7455 kw hr

Torque, Speed

1 One horsepower does useful work
2 One horsepower delivered as heat

F = force

Radius or lever arm is R feet or r inches

N = rev. per min.

3 Rim speed is 6.283 × R × N

1 ft. lever arm

50 lb. pull

4 Here torque is 50 ft lb or 600 in. lb

Motor — *Generator*

5 At 1750 rpm, 36 in.-lb is 1 hp

24"

50 lb.

190 rpm

6 24 × 50 × 190 ÷ 63,000 = 3.6 hp

▶ MORE OFTEN THAN NOT, power is delivered through rotating parts rather than in a straight-line push. In either case horsepower is the force in pounds times the distance through which it acts in one minute, divided by 33,000, because one horsepower is 33,000 ft lb. per min.

Fig. 1 shows a hoisting drum turning with a rim speed of 330 ft per min, winding up a weight of 100 lb at the same speed. Work done is 330 × 100 = 33,000 ft lb per min. or 1 hp.

Now look at Fig. 2. The two spring balances pull on the leather-strap brake.

The net drag on the drum surface is 140 − 40 = 100 lb. The surface moves 330 ft in one minute, so the work done is 330 × 100 = 33,000 ft lb per min. Again this is one horsepower, but the power is delivered as heat rather than as "useful work."

TORQUE

Now let's consider torque and the relation of torque and rotative speed to power delivered. Torque is nothing but twisting effort measured as the product of the pull by the length of the lever arm of the pull, Fig. 3.

The lever arm may be measured in feet, in which case the torque is in foot pounds. Often it is more convenient to measure the lever arm in inches, in which case the product of lever arm in inches by the force in pounds gives the torque in inch pounds.

For example, in Fig. 4, if man pulls 50 lb at a point 12 in. from the center of the fitting the torque is 1 × 50 = 50 ft lb, also 12 × 50 = 600 in. lb.

TORQUE AND POWER

Suppose (Fig. 3) that we have the radius of a wheel and the rotative speed in RPM. In one revolution every point on the rim moves 2 × 3.1416 × R feet, where the R is the radius in feet. This is 6.283 × R feet per revolution.

If N is the rotative speed in *RPM*, the distance traveled by the rim in one minute will be 6.28 × R × N feet. If the force at the rim is called F the horsepower will be 6.28 × R × N × F ÷ 33,000.

Note that this boils down to R × N × F ÷ 5250. Now R × N is nothing but the torque in foot pounds. Call this T. Then horsepower is T × N ÷ 5250.

If the radius is measured in inches the product of radius and force is the torque in inch pounds (call it t) and horsepower will be t × N ÷ 63,000. For convenient reference the torque-speed-power formulas are repeated in the box. Let's apply these to the one-horsepower motor generator set pictured in Fig. 5. The set runs at 1750 rpm. Find the torque. Then t = 63,000 ÷ 1750 = 36 inch pounds.

It may be worth remembering that when any machine's shaft, delivering one horsepower, is running at the common electrical speed of 1750 rpm, the torque or twist in the shaft equals a one pound pull at the end of a wrench a yard long.

Fig. 6 shows a "prony-brake" measuring the power of a small engine. Here torque t = 24 × 50 = 1200 in. lb. At 190 RPM the brake power is 190 × 1200 ÷ 63,000 = 3.6 hp.

Formulas for Rotary Power

R = radius of torque arm, feet
r = radius of torque arm, inches
F = force at end of torque arm, pounds
T = R × F = torque, foot pounds
t = r × F = torque, inch pounds
N is rotative speed, rpm
P is horsepower

Then:

P = T × N ÷ 5250
P = t × N ÷ 63,000

Table 1—Saturation—Temperatures

| Temp | Abs press | Specific vol | | Enthalpy ("Heat") | | |
| | | Sat liquid | Sat vapor | Sat liquid | Evap | Sat vapor |
F	Psi					
32	0.08854	0.01602	3306	0.00	1075.8	1075.8
40	0.12170	0.01602	2444	8.05	1071.3	1079.3
50	0.17811	0.01603	1703.2	18.07	1065.6	1083.7
60	0.2563	0.01604	1206.7	28.06	1059.9	1088.0
70	0.3631	0.01606	867.9	38.04	1054.3	1092.3
80	0.5069	0.01608	633.1	48.02	1048.6	1096.6
90	0.6982	0.01610	468.0	57.99	1042.9	1100.9
100	0.9492	0.01613	350.4	67.97	1037.2	1105.2
110	1.2748	0.01617	265.4	77.94	1031.6	1109.5
120	1.6924	0.01620	203.27	87.92	1025.8	1113.7
130	2.2225	0.01625	157.34	97.90	1020.0	1117.9
140	2.8886	0.01629	123.01	107.9	1014.1	1122.0
150	3.718	0.01634	97.07	117.9	1008.2	1126.1
160	4.741	0.01639	77.29	127.9	1002.3	1130.2
170	5.992	0.01645	62.06	137.9	996.3	1134.2
180	7.510	0.01651	50.23	147.9	990.2	1138.1
190	9.339	0.01657	40.96	157.9	984.1	1142.0
200	11.526	0.01663	33.64	168.0	977.9	1145.9
212	14.696	0.01672	26.80	180.0	970.4	1150.4
220	17.186	0.01677	23.15	188.1	965.2	1153.4
240	24.969	0.01692	16.323	208.3	952.2	1160.5
280	49.203	0.01726	8.645	249.1	924.7	1173.8
300	67.013	0.01745	6.466	269.6	910.1	1179.7
340	118.01	0.01787	3.788	311.1	879.0	1190.1
380	195.77	0.01836	2.335	353.5	844.6	1198.1
400	247.31	0.01864	1.8633	375.0	826.0	1201.0

Table 2—Saturation—Pressures

| Abs press | Temp | Specific vol | | Enthalpy ("Heat") | | |
| | | Sat liquid | Sat vapor | Sat liquid | Evap | Sat vapor |
psi	F					
0.50	79.58	0.01608	641.4	47.6	1048.8	1096.4
1.0	101.74	0.01614	333.6	69.7	1036.3	1106.0
5.0	162.24	0.01640	73.52	130.1	1001.0	1131.1
10	193.21	0.01659	38.42	161.2	982.1	1143.3
14.7	212.00	0.01672	26.80	180.0	970.4	1150.4
15	213.03	0.01672	26.29	181.1	969.7	1150.8
20	227.96	0.01683	20.089	196.2	960.1	1156.3
25	240.07	0.01692	16.303	208.5	952.1	1160.6
30	250.33	0.01701	13.746	218.8	945.3	1164.1
40	267.25	0.01715	10.498	236.0	933.7	1169.7
50	281.01	0.01727	8.515	250.1	924.0	1174.1
60	292.71	0.01738	7.175	262.1	915.5	1177.6
70	302.92	0.01748	6.206	272.6	907.9	1180.6
80	312.03	0.01757	5.472	282.0	901.1	1183.1
90	320.27	0.01766	4.896	290.6	894.7	1185.3
100	327.81	0.01774	4.432	298.4	888.8	1187.2
110	334.77	0.01782	4.049	305.7	883.2	1188.9
120	341.25	0.01789	3.728	312.4	877.9	1190.4
130	347.32	0.01796	3.455	318.8	872.9	1191.7
140	353.02	0.01802	3.220	324.8	868.2	1193.0
150	358.42	0.01809	3.015	330.5	863.6	1194.1
200	381.79	0.01839	2.288	355.4	843.0	1198.4
250	400.95	0.01865	1.8438	376.0	825.1	1201.1
300	417.33	0.01890	1.5433	393.8	809.0	1202.8
350	431.72	0.01913	1.3260	409.7	794.2	1203.9
400	444.59	0.0193	1.1613	424.0	780.5	1204.5

How to Use the Steam Tables—Part I

THIS IS THE FIRST of three simple, practical pages showing how an operating engineer can use steam tables to figure out problems involving heating water, making steam and superheating.

Values of the various properties of water and steam have been determined on the basis of experimental data. There are several sets of tables, but this one is from "Thermodynamic Properties of Steam," by J H Keenan and F G Keyes, published by John Wiley & Sons, Inc. (now out of print). However, the ASME and others have prepared similar tables, which may be found in standard engineering reference books.

Both Tables 1 and 2 give the properties of water and of saturated steam. The only difference is that you enter 1 with the boiling *temperature* and 2 with the boiler *pressure* (in psi absolute).

For example, Table 1 shows that, for water to boil at 100 F, the absolute pressure must be 0.95 psi abs.

Table 2 shows that at 40 psi abs water boils at 267 F. Note that we don't usually bother to put down all the digits given in the table. Most practical work doesn't require such hair-splitting accuracy. An operating engineer will rarely need to figure water temperature closer than the nearest degree, or heats or enthalpies closer than the nearest Btu.

After the first two, the columns are the same in both tables. "Sat liquid" means liquid water at the saturation or boiling temperature. "Sat vapor" means steam at the boiling temperature.

When water is boiling in a closed container, both the water and the steam over it are in a saturated condition.

The steam generated by a boiler without a superheater is saturated. For steam, saturated means steam that contains no liquid water, yet is not superheated (still at boiling temperature).

Note that the absolute pressure is gage pressure plus about 15 lb.

Now, in Table 2, try reading across the line for 50 psi abs (35 psi gage).

Boiling temperature is 281 F. At this temperature one pound of water fills 0.0173 cu ft and one pound of saturated steam 8.51 cu ft. Specific vol is in cubic foot per pound of water or steam.

It takes 250 Btu to heat the pound of water from 32 F to the boiling point and another 924 Btu to evaporate it, making a total of 1174 Btu.

"Enthalpy" is what used to be called "heat" in the old steam tables. It's given in Btu per lb. They would have labeled the last three columns: "Heat of the liquid", "Heat of Vaporization" and "Total Heat".

Table 1—Saturation—Temperatures

Temp	Abs press	Specific vol		Enthalpy ("Heat")		
F	Psi	Sat liquid	Sat vapor	Sat liquid	Evap	Sat vapor
32	0.08854	0.01602	3306	0.00	1075.8	1075.8
40	0.12170	0.01602	2444	8.05	1071.3	1079.3
50	0.17811	0.01603	1903.2	18.07	1065.6	1083.7
60	0.2563	0.01604	1206.7	28.06	1059.9	1088.0
70	0.3631	0.01606	867.9	38.04	1054.3	1092.3
80	0.5069	0.01608	633.1	43.02	1048.6	1096.6
90	0.6982	0.01610	468.0	57.99	1042.9	1100.9
100	0.9492	0.01613	350.4	67.97	1037.2	1105.2
110	1.2748	0.01617	265.4	77.94	1031.6	1109.5
120	1.6924	0.01620	203.27	87.92	1025.8	1113.7
130	2.2225	0.01625	157.34	97.90	1020.0	1117.9
140	2.8886	0.01629	123.01	107.9	1014.1	1122.0
150	3.718	0.01634	97.07	117.9	1008.2	1126.1
160	4.741	0.01639	77.29	127.9	1002.3	1130.2
170	5.992	0.01645	62.06	137.9	996.3	1134.2
180	7.510	0.01651	50.23	147.9	990.2	1138.1
190	9.339	0.01657	40.96	157.9	984.1	1142.0
200	11.526	0.01663	33.64	168.0	977.9	1145.9
212	14.696	0.01672	26.80	180.0	970.4	1150.4
220	17.186	0.01677	23.15	188.1	965.2	1153.4
240	24.969	0.01692	16.323	208.3	952.2	1160.5
280	49.203	0.01726	8.645	249.1	924.7	1173.8
300	67.013	0.01745	6.466	269.6	910.1	1179.7
340	118.01	0.01787	3.788	311.1	879.0	1190.1
380	195.77	0.01836	2.335	353.5	844.6	1198.1
400	247.31	0.01864	1.8633	375.0	826.0	1201.0

Table 2—Saturation—Pressures

Abs press	Temp	Specific vol		Enthalpy ("Heat")		
psi	F	Sat liquid	Sat vapor	Sat liquid	Evap	Sat vapor
0.50	79.58	0.01608	641.4	47.6	1048.8	1096.4
1.0	101.74	0.01614	333.6	69.7	1036.3	1106.0
5.0	162.24	0.01640	73.52	130.1	1001.0	1131.1
10	193.21	0.01659	38.42	161.2	982.1	1143.3
14.7	212.00	0.01672	26.80	180.0	970.4	1150.4
15	213.03	0.01672	26.29	181.1	969.7	1150.8
20	227.96	0.01683	20.089	196.2	960.1	1156.3
25	240.07	0.01692	16.303	208.5	952.1	1160.6
30	250.33	0.01701	13.746	218.8	945.3	1164.1
40	267.25	0.01715	10.498	236.0	933.7	1169.7
50	281.01	0.01727	8.515	250.1	924.0	1174.1
60	292.71	0.01738	7.175	262.1	915.5	1177.6
70	302.92	0.01748	6.206	272.6	907.9	1180.6
80	312.03	0.01757	5.472	282.0	901.1	1183.1
90	320.27	0.01766	4.896	290.6	894.7	1185.3
100	327.81	0.01774	4.432	298.4	888.8	1187.2
110	334.77	0.01782	4.049	305.7	883.2	1188.9
120	341.25	0.01789	3.728	312.4	877.9	1190.4
130	347.32	0.01796	3.455	318.8	872.9	1191.7
140	353.02	0.01802	3.220	324.8	868.2	1193.0
150	358.42	0.01809	3.015	330.5	863.6	1194.1
200	381.79	0.01839	2.288	355.4	843.0	1198.4
250	400.95	0.01865	1.8438	376.0	825.1	1201.1
300	417.33	0.01890	1.5433	393.8	809.0	1202.8
350	431.72	0.01913	1.3260	409.7	794.2	1203.9
400	444.59	0.0193	1.1613	424.0	780.5	1204.5

How to Use the Steam Tables—Part 2

As on the preceding page, we reproduce above a few lines from Table 1 and Table 2.

The most common type of practical problem involving the steam tables is figuring how much heat it takes to convert feedwater at any given temperature into saturated steam at any given pressure or temperature.

In all our sample problems we shall juggle the data to fit the few lines in the sample tables. The complete tables will give you any pressure or temperature you may need.

When heating water and making steam, the heat *supplied* is the "enthalpy" at the finish minus the enthalpy at the start. Or call it "heat," instead of enthalpy, if you find it simpler.

Take the case of a boiler generating saturated steam at 135-psi gage (150-psi abs). The enthalpy, or heat, of the final steam is 1194 Btu per pound.

The amount of heat required to produce this steam in an actual boiler will depend on the temperature of the feedwater.

Suppose the feedwater temperature is 180 F. Table 1 shows that the heat in the water is 147.9 Btu, say 148 Btu. Then the heat supplied to turn this water into steam is merely the difference, or 1194 − 148 = 1046 Btu.

It's an easy step from this to figuring the boiler efficiency. Let's say the boiler generates 10 lb steam per pound of coal burned and the coal contains 13,000 Btu per pound. Then, for every 13,000 Btu put in as fuel, there is delivered in steam, 10 × 1046 = 10,460 Btu.

The efficiency of any power unit is its output divided by its input, or (here)

10,460 ÷ 13,000 = 0.805 or 80.5%.

For most purposes you won't need Table 1 to get a close value of the heat of the liquid. Just subtract 32 from the water temperature.

For example, the enthalpy of water at 180 F is the heat required to raise it from 32 F to 180 F, or a difference of 148 deg. This takes about 148 Btu.

It won't work out quite so closely for very high temperatures. Take the case of water at 300 F. Table 1 gives 269.6 Btu, while our simple method gives 300 − 32 = 268 Btu, but still close enough for most purposes.

Below 212 F the rule is extremely accurate, never out by more than one-tenth of a Btu.

Condensing of steam in a condenser, heater, radiator or process is nothing but heating and evaporation worked backward.

TABLE 3—SUPERHEATED STEAM

Abs Pressure, psi (Sat Temp)		Sat Liquid	Sat Vapor	Temperature, degrees Fahrenheit							
				300	400	500	600	700	800	900	1000
15	v	0.016	26.29	29.91	33.97	37.99	41.99	45.98	49.97	53.95	57.93
(213.03)	h	181.1	1150.8	1192.8	1239.9	1287.1	1334.8	1383.1	1432.3	1482.3	1533.1
20	v	0.016	20.09	22.36	25.43	28.46	31.47	34.47	37.46	40.45	43.44
(227.96)	h	196.2	1156.3	1191.6	1239.2	1286.6	1334.4	1382.9	1432.1	1482.1	1533.0
40	v	0.017	10.498	11.040	12.628	14.168	15.688	17.198	18.702	20.20	21.70
(267.25)	h	236.0	1169.7	1186.8	1236.5	1284.8	1333.1	1381.9	1431.3	1481.4	1532.4
60	v	0.017	7.175	7.259	8.357	9.403	10.427	11.441	12.449	13.452	14.454
(292.71)	h	262.1	1177.6	1181.6	1233.6	1283.0	1331.8	1380.9	1430.5	1480.8	1531.9
80	v	0.018	5.472		6.220	7.020	7.797	8.562	9.322	10.077	10.830
(312.03)	h	282.0	1183.1		1230.7	1281.1	1330.5	1379.9	1429.7	1480.1	1531.3
100	v	0.018	4.432		4.937	5.589	6.218	6.835	7.446	8.052	8.656
(327.81)	h	298.4	1187.2		1227.6	1279.1	1329.1	1378.9	1428.9	1479.5	1530.8
150	v	0.018	3.015		3.223	3.681	4.113	4.532	4.944	5.352	5.758
(358.42)	h	330.5	1194.1		1219.4	1274.1	1325.7	1376.3	1426.9	1477.8	1529.4
200	v	0.018	2.288		2.361	2.726	3.060	3.380	3.693	4.002	4.309
(381.79)	h	355.4	1198.4		1210.3	1268.9	1322.1	1373.6	1424.8	1476.2	1528.0
300	v	0.0189	1.5433			1.7675	2.005	2.227	2.442	2.652	2.859
(417.33)	h	393.8	1202.8			1257.6	1314.7	1368.3	1420.6	1472.8	1525.2
400	v	0.0193	1.1613			1.2851	1.4770	1.6508	1.8161	1.9767	2.134
(444.59)	h	424.0	1204.5			1245.1	1306.9	1362.7	1416.4	1469.4	1522.4
500	v	0.0197	0.9278			0.9927	1.1591	1.3044	1.4405	1.5715	1.6996
(467.01)	h	449.4	1204.4			1231.3	1298.6	1357.0	1412.1	1466.0	1519.6
600	v	0.0201	0.7698			0.7947	0.9463	1.0732	1.1899	1.3013	1.4096
(486.21)	h	471.6	1203.2			1215.7	1289.9	1351.1	1407.7	1462.5	1516.7
800	v	0.0209	0.5687				0.6779	0.7833	0.8763	0.9633	1.0470
(518.23)	h	509.7	1198.6				1270.7	1338.6	1398.6	1455.4	1511.0
1000	v	0.0216	0.4456				0.5140	0.6084	0.6878	0.7604	0.8294
(544.61)	h	542.4	1191.8				1248.8	1325.3	1389.2	1448.2	1505.1
1200	v	0.0223	0.3619				0.4016	0.4909	0.5617	0.6250	0.6843
(567.22)	h	571.7	1183.4				1223.5	1311.0	1379.3	1440.7	1499.2
1400	v	0.0231	0.3012				0.3174	0.4062	0.4714	0.5281	0.5805
(587.10)	h	598.7	1173.4				1193.0	1295.5	1369.1	1433.1	1493.2

How to Use the Steam Table—Part 3

THIS THIRD EXAMPLE will finish up our lesson on the practical use of steam tables to help you generate steam efficiently. Table 3 also is reproduced by permission of the publisher of the preceding two tables; you'll find equivalent data in many engineering reference books.

First column gives the absolute pressure and (directly below it in parenthesis) the corresponding saturation temperature or boiling temperature.

In the next column v and h stand for volume of one pound and its heat content. For example, at 150 psi absolute the volume of one pound is 0.018 cu ft for liquid water and 3.015 cu ft for the saturated steam. Corresponding heat contents of per pound are 330.5 Btu and 1194.1 Btu.

The temperature columns give the volume and heat content per pound for superheated steam at the indicated temperature. Take the case of steam at 150 psi and superheated to a total temperature of 600 F. Look in the 600 F column opposite 150 psi. The volume is 4.113, as against 3.015 for saturated steam at the same pressure. This is natural because steam expands like a gas when you superheat it.

Also the heat content is naturally higher—1325.7 Btu instead of 1194.1 Btu.

Note that this table gives the actual temperature of the superheated steam rather than the degrees of superheat, which is a different thing. If the steam has been superheated from a saturation temperature of 358 F to 600 F, the superheat is 600 − 358 = 242 degrees.

These superheat tables are just as easy to use as the saturation tables described in the two previous installments.

Problem: How much heat does it take to convert one pound of feedwater at 205 F into superheated steam at 150 psi abs and 600 F? Heat in steam is 1325.7, Call it 1326. Heat in the water is 205 − 32 = 173 Btu. Then heat required to convert one pound to steam is 1326 − 173 = 1153 Btu.

How to Read a Line Chart

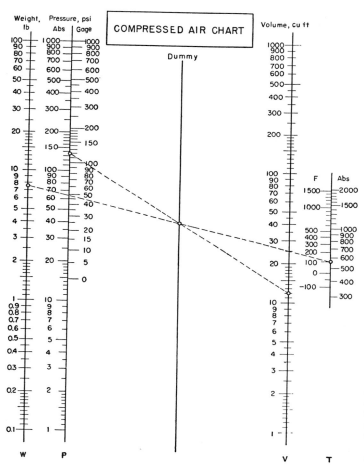

COMPRESSED AIR CHART

HERE we will kill two birds with one stone—get a handy chart for our files and learn how to use all charts of this very useful type.

For air, as for any other gas, there is a definite relation between weight, pressure, volume and temperature. If you know any three of these you can figure the fourth from a formula given in the thermodynamics books. If you don't want to sweat over an algebraic formula you can get the same answer from the chart on this page.

There isn't space here to explain how I plotted this chart, but here's how you use it:

Draw one dotted line from pressure on P to volume on V. Draw another from weight on W to temperature on T. In all cases the two lines must intersect at the dummy line.

If you know P and V connect them first.

If you know W and T connect them first.

This will locate the intersecting point on the dummy. Then draw a line from the remaining known quantity through the dummy intersection to the scale for the unknown. Where your line cuts that scale read the value of the unknown. If you proceed as shown by the dotted lines you can't miss.

Various engineering reference books contain hundreds of charts of this general type for solving all kinds of practical mathematicial problems. Usually dotted lines are drawn to show how to use the charts.

Charts of this kind, in which you draw a straight line across several separate scales, are known as *collinear diagrams*.

These are great time savers for busy engineers. You can take any scale as the unknown. If you complete the indicated construction, you always end up with a line crossing the unknown scale where you can then read the answer.

TO USE THIS CHART: Connect **P** to **V** and **W** to **T**. First connect two of the known quantities (**P** and **V** if you know both or **W** and **T** if you know them). Draw a second line from third known item through intersection of first line and dummy. Continue second line until it crosses column of unknown; then read your answer. Let's try a problem: **Problem:** A 12-cu ft tank holds air at 120 psi gage and 100 F. What is weight? **Solution** (shown by dotted lines): Connect 120 psi gage on **P** to 12 cu ft on **V**. From 100 F on **T** draw line through dummy intersection and extend until it cuts line **W**. Answer: 7.6 lb air in tank

Tension, Compression and Shear

Tensile Stress. When a tensile test specimen, Fig. 1, is stretched to the breaking point the material tends to *neck down. Tensile strength* of the material in *pounds per square inch* (psi) is the total breaking tension divided by the *original* cross-section.

At any tension short of breaking, the *unit stress* is the actual total tension divided by this same original cross-section.

Up to a certain unit stress, called the *elastic limit,* the metal stretches in exact proportion to the applied load and the unit stress. Beyond that the metal starts to *give* a little and stretches more and more for a given increase in unit stress.

Compression. The *compressive strength* of a material is tested by crushing a short block. Total crushing load, divided by original cross-section, is the compressive strength in pounds per square inch.

Shear. In the same way, if the section is sheared across (as in a power shear) the *shearing strength* is the total force required divided by the cross-sectional area sheared.

Factor of Safety is ultimate strength divided by actual stress.

Bolts and Pins. Fig. 2 shows two applications. What is unit tensile stress at thread root of bolt at left? Cross-sectional area at the thread root of a 1-in. (U. S. Standard) bolt is 0.551 sq in. Then unit stress is 3,000÷0.551 =5440 psi. For Class *B* bolt steel of 60,000 psi tensile strength, factor of safety is 60,000÷5440=11.0.

The ½-in. pin at right, Fig. 2, has a cross-sectional area 0.5×0.5×0.785= 0.196 sq in. Since pin is in *double shear*, total sheared area is 2×0.196 =0.392, so shearing unit stress is 2500 ÷0.392=6,380 psi.

Stress in Shell. In a seamless or welded tube or drum under internal pressure, the tendency to split lengthwise is double the tendency to split around. Thus, unless the circular seams are very weak, such shells or drums *always* fail by splitting lengthwise (longitudinally).

Figure the shell strength without seams, then correct for the efficiency of the longitudinal joint. Consider a half slice of shell one inch long, Fig. 3. Fluid pressure tends to split this into two halves, both of which are shown. If internal diameter (inches) is *D* and fluid pressure (psi) is *P*, the force on the shaded *piston area* is *P*×*D*. This force is resisted and balanced by the pull of the two metal sections *A*. This resisting force is 2×*t*×*s*, where *t* is shell thickness and *s* is unit stress. From this relation (See notes in Fig. 3.) we derive the shell formula *P*= 2×*t*×*s*÷*D*.

If diameter is 30 in., thickness 0.5 in. and safe unit stress 15,000 psi, safe pressure for a seamless shell is 2×0.5 ×15,000÷30=500 psi. If the shell has a longitudinal riveted seam of 80% efficiency, the allowable pressure for 15,000 psi stress is 0.80×500=400 psi.

Heat Stresses. Almost any form of iron or steel (except cast iron) has a *modulus of elasticity* around 30,000,000. That means that a stress of 1 psi will stretch it 1/30,000,000 part of its own length. This relation holds right up to the elastic limit.

To put it another way, stretching iron or steel 1 in. per 100 ft. (1200 in.) produces a tensile stress of $\frac{1}{1200}$ ×30,000,000=25,000 psi. The same increase in length can be caused by a temperature rise of 120 F (say, from 80 to 200 F).

It follows that if a piece of steel rod or pipe of any length is heated to 200 F, then rigidly held at both ends, and cooled to 80 F the resulting tensile stress will be 25,000 psi, Fig. 4. This figures about 200 psi for each degree of change.

1 Tensile test piece, before and after pulling

2 Practical problems in tension and shear

Steam force on shaded area equals total tension in two metal sections, so P×D=2×t×s, then P=2×t×s÷D (See text)

3 How to derive formula for shell stress in pressure vessels

To expand 1 in. per 100 ft requires 120 deg F temp rise or tension of 25,000 psi

4 Thermal tension in rigidly held iron or steel rod is about 200 psi per degree change

Short Cut to Square Roots

I offer to show how to get square roots without tables or slide rule—an easier method than the one taught in schools.

The following rule applies only to numbers between 1 and 100. Later I'll show how to bring greater or smaller numbers into this range.

THE RULE

1—By inspection set down the square root *to the nearest whole number only.* This will fall between 1 and 10.

2—Divide this into the original number and carry the answer (quotient) to *one decimal place only.*

3—Average the original whole-number square root and this quotient to *one decimal place only,* and call this average the *approximate square root.*

4—Divide the original number by the approx square root and carry the quotient to *three decimal places.*

5—Average this latest quotient and the approx square root to three decimals and call this the "exact" square root.

Except for an occasional number between 1 and 3, the "exact" root thus obtained will never be in error by more than one in the third decimal place.

Problem: Find square root of 11. Nearest whole-number root is 3. Then 11 ÷ 3 = 3.6 plus

Average of 3 and 3.6 is 3.3 = approx. root.
Then 11 ÷ 3.3 = 3.333
Average of 3.3 and 3.333 is 3.317 = "exact" square root of 11.

Problem: Find square root of 42.62. Nearest whole-number root is 6.
Then 42.62 ÷ 6 = 7.1 plus
Average of 6 and 7.1 is 6.5 = approx root
Then 42.62 ÷ 6.5 = 6.557
Average of 6.5 and 6.557 is 6.528 = "exact" square root.

NUMBERS 1 TO 3

For any numbers between 3 and 100 this rule will never give an error of more than one in the third decimal place. Between 1 and 3 error may sometimes be slightly greater. For example, the square root of 2 by this rule would be 1.416 instead of 1.414.

Where you want high accuracy in the square root of numbers between 1 and 3 insert the following step between step 3 and step 4 in the rule already given:

3A Divide the number by the approximate square root obtained in 3, and carry the quotient to *two decimal places only.* Average this quotient and the approx-square root of 3 *to two decimal places* and use the average as a *corrected approx square root* in step 4.

Problem: What is the square **root** of 2.304?
Nearest whole number is 1, say (or you can use 2 if you prefer).
Then 2.304 ÷ 1 = 2.3 plus
Average of 1 and 2.3 is 1.6
Then 2.304 ÷ 1.6 = 1.44
Average of 1.6 and 1.44 is 1.52
Then 2.304 ÷ 1.52 = 1.516
Average of 1.52 and 1.516 = 1.518 = "exact" square root.

NUMBERS BEYOND RANGE

For numbers greater than 100, or less than 1, move decimal point left or right two digits at a step until the number is in the range. **Then take square root by rule already given.**

Finally move the decimal point in the opposite direction a single digit for each two digits moved in the original number.

Problem: Find the square root of 23,000. Moving point *two double steps* to *left* gives 2.3. Square root of 2.3 is 1.517. Move point back *two single steps* to *right* to get 151.7 as the square root of 23,000.

Problem: Find the square root of .000032. Move point three double steps to right to get 32. Square root of 32 is 5.657. Then move point left three *single* steps to get 0.005657 as the square root of 0.000032.

SQUARE ROOTS

No.	1.0	1.1	1.2	1.3	1.4	1.5	1.6	1.7	1.8	1.9
Rt.	1.000	1.049	1.095	1.140	1.183	1.225	1.265	1.304	1.342	1.378
No.	2.0	2.1	2.2	2.3	2.4	2.5	2.6	2.7	2.8	2.9
Rt.	1.414	1.449	1.483	1.517	1.549	1.581	1.612	1.643	1.673	1.703
No.	3.0	3.1	3.2	3.3	3.4	3.5	3.6	3.7	3.8	3.9
Rt.	1.732	1.761	1.789	1.817	1.844	1.871	1.897	1.924	1.949	1.975
No.	4.0	4.1	4.2	4.3	4.4	4.5	4.6	4.7	4.8	4.9
Rt.	2.000	2.025	2.049	2.074	2.098	2.121	2.145	2.168	2.191	2.214
No.	5.0	5.1	5.2	5.3	5.4	5.5	5.6	5.7	5.8	5.9
Rt.	2.236	2.258	2.280	2.302	2.324	2.345	2.366	2.387	2.408	2.429
No.	6.0	6.1	6.2	6.3	6.4	6.5	6.6	6.7	6.8	6.9
Rt.	2.449	2.470	2.490	2.510	2.530	2.550	2.569	2.588	2.608	2.627
No.	7.0	7.1	7.2	7.3	7.4	7.5	7.6	7.7	7.8	7.9
Rt.	2.646	2.665	2.683	2.702	2.720	2.739	2.757	2.775	2.793	2.811
No.	8.0	8.1	8.2	8.3	8.4	8.5	8.6	8.7	8.8	8.9
Rt.	2.828	2.846	2.864	2.881	2.898	2.915	2.933	2.950	2.966	2.983
No.	9.0	9.1	9.2	9.3	9.4	9.5	9.6	9.7	9.8	9.9
Rt.	3.000	3.017	3.033	3.050	3.066	3.082	3.098	3.114	3.130	3.146

No.	10	11	12	13	14	15	16	17	18	19
Rt.	3.162	3.317	3.464	3.606	3.742	3.873	4.000	4.123	4.243	4.359
No.	20	21	22	23	24	25	26	27	28	29
Rt.	4.472	4.583	4.690	4.796	4.899	5.000	5.099	5.196	5.292	5.385
No.	30	31	32	33	34	35	36	37	38	39
Rt.	5.477	5.568	5.657	5.745	5.831	5.916	6.000	6.083	6.164	6.245
No.	40	41	42	43	44	45	46	47	48	49
Rt.	6.325	6.403	6.481	6.557	6.633	6.708	6.782	6.856	6.928	7.000
No.	50	51	52	53	54	55	56	57	58	59
Rt.	7.071	7.141	7.211	7.280	7.348	7.416	7.483	7.550	7.616	7.681
No.	60	61	62	63	64	65	66	67	68	69
Rt.	7.746	7.810	7.874	7.937	8.000	8.062	8.124	8.185	8.246	8.307
No.	70	71	72	73	74	75	76	77	78	79
Rt.	8.367	8.426	8.485	8.544	8.602	8.660	8.718	8.775	8.832	8.888
No.	80	81	82	83	84	85	86	87	88	89
Rt.	8.944	9.000	9.055	9.110	9.165	9.220	9.274	9.327	9.381	9.434
No.	90	91	92	93	94	95	96	97	98	99
Rt.	9.487	9.539	9.592	9.644	9.695	9.747	9.798	9 849	9.899	9.950

Useful Facts From Geometry: 1

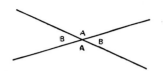

When two straight lines cross, the angles of the same letter are equal

When a straight line crosses two parallel lines, the angles of the same letter are equal

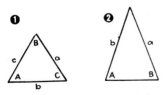

1 The interior angles, A, B, C of any triangle add up to 180 degrees. 2 The interior angles A, B, C, D of any quadrilateral (figure with four straight sides) add up to 360 degrees.

In any right triangle the two angles not right always add up to 90 degrees

1 Any triangle with three equal angles, A, B, C, has three equal sides, a, b, c, and vice versa. Such triangles are called "equilateral." 2 Any triangle with two equal angles A, B, has two equal sides, a, b, opposite the equal angles, and vice versa. Such triangles are called isosceles"

Any angle inscribed in a semicircle, as indicated here, is a right angle

In any right triangle the square C on the hypotenuse equals A + B, sum of the squares on the legs. This holds either for the indicated areas or for the numerical squares of the sides, a, b, c. Thus, if a is 5 ft and b is 7 ft, c is $\sqrt{5{\times}5 + 7{\times}7} = \sqrt{25 + 49} = \sqrt{74} = 8.60$ feet

A triangle with sides proportional to 3, 4, 5 (say 3, 4, 5, or 6, 8, 10, or 9, 12, 15, or 30, 40, 50) will be a right triangle with the right angle between the 3 side and the 4 side. Proof: $3{\times}3 + 4{\times}4 = 5{\times}5$

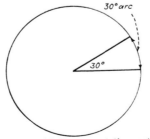

An angle with its apex at the center of a circle is measured by the inter-

cepted arc. Thus, a 30-deg angle will intercept a 30-deg arc, and vice versa

An angle with its apex anywhere on the circumference of a circle will be half the arc intercepted on the circumference, and vice versa. Here, each 30-degree angle intercepts a 60-degree arc

Equally spaced parallel lines will divide a line crossing them at any angle into equal parts. Thus if the a's are equal, the b's will be equal, and the c's will be equal.

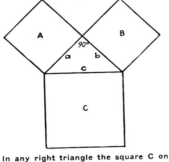

Unequally spaced parallel lines will divide a line crossing at any angle into segments proportional to the spaces between the parallel lines. Thus 1 is to 3 is to 6 is to 2 as a is to b is to c is to d or e is to f is to g is to h

Useful Facts From Geometry: 2

If a plane figure is enlarged without change of shape (as by photography, photostating or pantograph) the area increases in proportion to the square of any selected dimension. Try this for squares of 1, 2, 3 and 4 sides. Areas are 1, 4, 9 and 16, as in Fig. 1.

FIG. I

Areas

Side-->1   2   3   4

For a circle, the area is always less than the square of the diameter. Yet the area is still proportional to the square of the diameter. The 4-in. circle has exactly 16 times the area of the 1-in., and the 3-in. circle exactly 9 times, Fig. 2.

FIG. 2

Areas

Diam -->1   2   3   4

Areas of triangles pictured below in Fig. 3 (each half a square) are actually ½, 2, 4½ and 8, but still in the proportion 1, 4, 9, 16, as before.

FIG. 3

Areas

Base-->1   2   3   4

We begin to see that this relationship holds for any shape; for example, that shown below, in Fig. 4.

FIG. 4

Areas

Base-->1   2   3   4

Now this same rule holds for the surfaces of solid bodies as well as for the shapes drawn on a flat surface. This is most obvious for a cube. It has six sides, each a square. So the surface of any cube is 6 times the square of one edge, Fig. 5.

FIG. 5

Surfaces

Side

A sphere or ball evidently has somewhat less surface than a cube whose edge equals the sphere's diameter; yet the same proportion holds. Surface of a 4-in. ball is exactly 16 times that of a 1-in. ball, Fig. 6.

FIG. 6

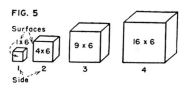

Surfaces

Diameter

Now let's work a few problems illustrating effect of dimension changes on surface. First: What is effect on quantity of blueprint paper required if drawing scale is reduced from ½ in. per ft to ⅜ in. New dimension is ¾ of the first, so new area is ¾x¾ = 9/16, or 56%; saving is 44%.

Another problem: A shape cut from plywood weighs 12 lb. What would it weigh if a certain dimension were increased from 32 in. to 48 in. with no change in shape? Note that 48 ÷ 32 = 1.5. Then 1.5x1.5 = 2.25, and 2.25x12 = 27-lb weight.

How much lumber (area) and outside paint can be saved by cutting a building plan 12% in all dimensions? The final dimension is 88% of original; so the final surface is 0.88x0.88 = 0.7744, or 77.4%. Saving is 22.6%.

Problem: A certain part is a brass hex rod, ¾ in. across flats. How much is brass cost increased if rod is thickened to ⅞ in. across flats? Ratio of dimensions is 7/8 ÷ 6/8 = 7/6. Then 7/6x7/6 = 49/36 = 1.36, and cost increases 36%, assuming a fixed cost per lb for the hex rod.

If any solid shape is enlarged without changing the proportions, volume and weight go up in proportion to cube of any dimension. This holds for actual cubes. Volumes are 1, 8, 27 and 64 for edges of 1, 2, 3 and 4. Weights must be proportional to volumes. Fig. 7.

FIG. 7

Weight

Base

It is equally true for spheres or any other shapes, as the irregular pyramids shown below in Fig. 8.

FIG. 8

Weight

Base edge

Problem: How much will weight of a steel cone be reduced by cutting off the point to a depth equal to ¼ length of cone? Both cutoff and original cones have same shape. Since little cone has ¼ the length, its relative weight will be ¼x¼x¼ = 1/64 = 0.0156. So weight loss is 1.56%, Fig. 9.

FIG. 9

Cut here

1"

4"

Problem: Conical coal pile is to be built up from fixed central spout until pile base is a circle with 80-ft dia. To find how much coal will be stored here, engineer makes model pile of the same coal with 10-ft dia. cone, by spilling it from a single spout and letting it find its own angle of repose (same as large pile). If coal in sample pile weighs 4560 lb, what is tonnage of big pile?

Ratio of dimensions is 8, so ratio of weights is 8x8x8 = 512. Pile holds 4560x512 ÷ 2000 = 1167 tons.

Useful Facts From Geometry: 3

10 — ELLIPSE

11 — ANNULUS

12 — SECTOR

13 — COMPOSITE SHAPE

1 — SQUARE

2 — RECTANGLE

3 — PARALLELOGRAM

4 — TRAPEZOID

5 — RHOMBUS

6 — TRIANGLE

7 — REGULAR HEXAGON

8 — REGULAR OCTAGON

9 — CIRCLE

▶ HERE ARE handy rules for areas of common geometrical shapes:

1 Area of a *square* is the square of its side: Area $= A \times A$.

2 Area of a *rectangle* is the product of its two dimensions: Area $= A \times B$.

3 Area of a *parallelogram* (4-sided figure with opposite sides parallel and equal) is the product of any side by the perpendicular distance between that side and its parallel side: Area $= B \times H$.

4 Area of a *trapezoid* (4-sided figure with 2 sides parallel) equals product of distance between parallel sides by average length of parallel sides: Area $= H \times (B_1 + B_2) \div 2$.

5 Area of a *rhombus* (4-sided figure with all sides equal) is half the product of its diagonals: Area $= \frac{1}{2} \times d_1 \times d_2$.

6 Area of a *triangle* is half the product of base and height: Area $= \frac{1}{2} \times A \times H$, or $\frac{1}{2} \times B \times H$. Note: *Any* side may be taken as the base. Height is the perpendicular distance from apex to base (base-line extended where necessary).

7 Area of a *regular hexagon* is 86.6% of the square of the distance between flats:
Area $= 0.866 \times D \times D$.

8 Area of a *regular octagon* is 82.8% of square of distance between flats: Area $= 0.822 \times D \times D$.

9 Area of a *circle* is 78.54% of the square within which it fits. Area $= 0.7854 \times D \times D$. Note: For most engineering purposes 0.785 is close enough, and for many rough, offhand estimates 0.78 is okay.

10 An *ellipse* fills the same fraction (78.54%) of its inclosing rectangle as a circle fills of its inclosing square:
So area $= 0.7854 \times A \times B$.

11 Area of an *annulus* is difference between areas of inner and outer circles. Also, if average diameter (average of inner and outer diameter) is *D*, and width of band is *W*, area of the annulus is 3.1416 $\times$ average diameter $\times$ width: Area $= 3.1416 \times D \times W$. Use 3.14 for most practical problems.

12 *Area of a sector* is area of its circle multiplied by a fraction whose numerator is the angle of the sector in degrees, and whose denominator is 360 deg. This reduces to: Area $= 0.00873 \times$ angle of segment $\times$ side $\times$ side.

13-14 Many of the shapes in engineering may be broken up into the simple geometric forms. Thus *composite shape* of Fig. 13 can be broken up, Fig. 14, into triangles, rectangle and sector. Its area is:

area triangle AHC
plus area triangle CIE
plus area sector CHI
minus area rectangle BFGD

15 Where all other methods fail, an area can be measured by dividing it into small squares and counting them. First count full squares. Then estimate fractional squares. For precise results, use small squares.

16 Sometimes it is easier to subdivide the area into narrow strips of constant width. Add average lengths of all strips, and multiply this sum by width of strip to get the area. Lengths of strips are measured along their center lines. This is not theoretically exact, but generally gives a reasonably close answer. The narrower the strips, the less this error will be.

14 — ELEMENTS OF COMPOSITE SHAPE

15 — IRREGULAR FIGURE

16 — IRREGULAR FIGURE

Sheet-Metal Layout

$10 \div 25 = 0.4$
$0.4 \times 360° = 144°$

FIRST, draw desired cone and measure indicated radii. Use these to draw arcs on paper or metal sheet. Allow for lap

Base circle

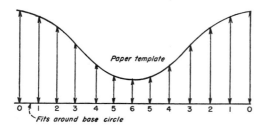

Paper template

Fits around base circle

DIVISIONS of template base line correspond to 12 divisions of pipe circumference. At each division measure distance to cut

▶ ALL PROBLEMS in sheet-metal layout are based on principles of descriptive geometry, but you can go a long way with a combination of mechanical horse sense, some knowledge of drawing and a bit of practice. For a starter try the following two easy examples of layouts often found in actual maintenance and construction.

Better get out your drawing board and follow through, step by step.

CUTOFF CONE (FUSTRUM)

First lay out a flat shape for this cutoff cone, which is a common sheet-metal shape. As a practice example you draw it small scale. If the problem is actual, you should lay out the flat full scale on a wrapping-paper template or directly on the metal sheet.

To draw the full-scale circle you can easily improvise a beam compass as long as needed. Just drive a nail (for a center) near one end of a long stick. Then drill holes (for pencil) 10 in. and 25 in. from nail.

The only real trick is to know how much of the arc to use. Merely take the same fraction of the whole circle (360 deg) as radius 10 is of radius 25. That's 144 deg. Clearly mark end of 144 deg arcs for matching purposes. Then add what metal you need for overlap.

PIPE CUT 45 DEG

Next take common case of pipe or round duct sliced off at 45 deg—either to enter a flat surface or to pair with another 45 deg to make a 90-deg turn.

End view of pipe is the circle shown. Divide this into even parts. Twelve 30-deg divisions, as pictured, are convenient and usually close enough. Project these left and then on to the base circle. Base circle is drawn at any convenient place.

Next measure pipe circumference with string or tape and lay it out full scale as the horizontal base line of the wrapping-paper template. Divide this into 12 equal divisions, and number them as shown to correspond with the divisions of the pipe circumference. On side view of pipe measure distance from base circle to cut line at each division and transfer to verticals on template. Draw smooth curve.

Wrap template around until base circles match. Then transfer template curve to metal, and cut.

1 Motor circuit drawings should be studied first before shooting trouble. Working knowledge of both wiring and schematic drawings shown above, is half the job

2 A circuit ground is quickly located using the simple series lamp test

Seven Steps For Testing

▶ You can talk all you want to about circuit testing, but putting to practical use proves the pudding. Let us say we have a 120-v dc shunt motor with a manual starter and the motor won't run. Fig. 1A shows the wiring diagram of the equipment as it appears in the plant. Fig. 1B shows the schematic or simplified layout.

Trouble Shooting. In shooting trouble take the following steps:

1. See that the motor armature turns freely. Try turning the motor by hand because drag or overload can cause trouble. The high load demand causes fuses, overload trips and such devices to open the circuit.

2. See that the line switch is closed. You might consider this too elementary but, fortunately, the most common causes of trouble are usually simple.

3. Place a test lamp across the switch terminals L_1 and L_2 to see that power is available from the supply. If it isn't, check the voltage at the next switch ahead of the supply. Continue testing back to the source until the supply interruption is located.

4. When the test lamp shows voltage at the motor switch, check the motor fuses.

Fuses may be blown by trying to start the motor too rapidly or by overloading the motor; by a ground or short; by a short circuit in the armature, field, starter or leads and by open circuit in the field.

There are other causes of motor failure that do not blow fuses.

Finding An Open. 5. To check for an open circuit in the motor note whether there are two paths for current to flow from L_2 to L_1, Fig 1B; interrupting current through either path affects operation of the motor. When you move the starter handle from the off to the first position, one path is from L_2 through the fuse, motor lead, A_2 brushes, armature windings, A_1 brushes, starter resistance, starter handle, motor lead back to the fuse and L_1.

All parts of this circuit are of low resistance, so a test lamp may be used to check the circuit without being dimmed to any extent. Place one lead of the lamp on starter terminal L_1. The points marked L_1, A_1, and F_1 in Fig. 1A are identical points in Fig. 1B. The test lamp should light if the line switch is closed and the other test lead is placed on A_2.

The most convenient place to get at A_2 is on the motor brushes. If the lamp lights at both A_2 and A_1, you can be certain that the circuit is continuous from L_2 to A_1.

If the lamp doesn't light at the brush terminals A_1 and A_2, open the motor

switch and inspect the brushes and wiring between L_2 and the point where the lamp fails to light. The fault may be a loose wire or terminal connection, or the brushes may not be touching the commutator.

After correcting any faults, close the switch and try to run the motor. If it won't start keep on checking the armature circuit by placing one lead of the test lamp on L_1 and the other on the resistor button for the run position. As long as the lamp lights with each check, the circuit between L_2 and the test-lamp lead is complete.

By trying each button of the starter resistance you may find that the lamp won't light beyond a certain button. This means that the armature circuit is open in one of the resistor coils or grids. Spot the break by tracing the lead from the button that won't light the test lamp, back to the resistor.

Testing the Field. Having checked and corrected any breaks in the armature circuit, check the path through the motor field. One reliable check for open circuit in the field is moving the starter handle from the first position to the off position with brushes removed. If there is no arc the field circuit is open at some point.

3 Armature faults, whether an open or shorted coil, are easily found

4 A defective coil in the field usually is replaced rather than repaired

Motor Circuits

Some motors are wired so that the field current is not interrupted by the starter handle. Instead current is maintained on the field as long as the line switch is closed, so that the heat will prevent condensation in the motor windings. With this arrangement opening the line switch will cause an arc if the field wiring has no open circuit.

The field circuit may be checked with a test lamp in the same manner as the armature circuit. However, because of the higher resistance of the field circuit, the lamp won't light as brightly as in other tests. Try connecting the lamp across L_1 and F_1, Fig. 1B. If the lamp lights, it proves the open is in the starter and at the same time proves the motor fields are not open.

If the open is in the motor fields, each field coil will have to be checked separately by untaping the leads connecting the individual field coils. Refer to Fig. 4 and note the method of checking. The field-coil leads are untaped at points A, B, C and D. The line switch is closed and the starter handle placed on the first position. Apply the test lamp leads to AB, BC, CD and DL,. If the field coil is open at BC the lamp will light when connected across it, but will not light across the other three coils. It is advisable to replace the open coil with a new one, rather than attempt to repair the break.

Locating Grounds. 6. Before checking for grounds, make the following test. With the line switch open, ground

one of the test-lamp leads. Connect the other lead first to L_1 and then to L_2. Assume that the test lamp lights when connected to L_2. This shows that L_1 is grounded and a circuit established from L_2 through the lamp, the grounded lead and back to L_1 through the ground in that line.

The motor and starter may be checked for ground by connecting the test lamp as in Fig. 2. Leave the line switch open and if one side of the supply is not grounded, it should be temporarily grounded through another lamp. If a ground exists in either the motor or starter the test lamp will light, except under one condition. This condition is where the ground exists in the starter handle or the wiring from that point to the switch. However, if the starter handle is moved to one of the starting buttons, this part of the circuit will also be included in the test.

If a ground is indicated by the test lamp the next step is to determine whether it is in the starter or motor. Disconnect the motor leads from the switch and starter terminals. Then by the same test as shown in Fig. 2, touch the test lamp lead to the starter terminals and then to one of the motor leads. Lamp lights on the grounded circuit.

If the ground is in the starter, the most probable fault location is in the resistor at some point where insulation has failed. The quickest way to find such a ground is by close inspection. If necessary, the location may be tracked

down by disconnecting a suspected part and repeating the ground test. When the ground appears to be in the motor, lift the brushes and check the armature by touching the test lamp lead to any commutator segment. To test the field for ground, touch the test lead to any field wire.

Testing the Armature. A ground in either the armature or the field is invariably a result of insulation failure. A grounded field coil may be found by opening the wires between field coils and checking each coil separately. Usually it is advisable to replace, rather than repair, a grounded armature or field coil.

7. Short circuited armature coils are easily located by means of a low-reading voltmeter (millivoltmeter) and a couple of dry cells. To test for a short in an armature refer to Fig. 3. The diagram shows that armature coils are connected in series, and that the ends of each armature coil are connected to the commutator bars. A low voltage from two dry cells is applied to the commutator at points 180 deg. apart. This may be done through two motor brushes provided all other brushes are lifted and any connecting wires to the brush wiring are disconnected. If the voltage is checked between adjacent bars on the commutator, it will be noted that when the millivoltmeter is across bars C and D which lead to an open armature coil, the meter will give a higher reading than when across any other adjacent bars.

On the lower half of the armature the readings between adjacent bars are about the same, except across EF. When the millivoltmeter is across the adjacent bars E-F which connect to a shorted coil, it reads zero. Between the points A-C-B-D the voltage across adjacent bars reads zero because that path is open circuited.

A shorted field coil is checked in a similar manner. Low voltage dc is applied to the fields in series similar to Fig. 4 and a voltmeter with a suitable scale is connected across the bared leads of each coil. The voltage drop as measured by the voltmeter will be less across a shorted coil than across the others. All coils that are satisfactory will give readings nearly alike.

An open-circuited coil in a field having a normally high resistance, may be located quickly using a voltmeter rather than a test lamp. There will be no voltage across the coils which are good. The open coil will give a voltage indication, the value of which depends on the resistance of the field coils.

Circuit

1 The insulation resistance tester used to check the condition of insulation in this dc generator is ideal for finding circuit grounds

▶ ELECTRICAL CIRCUIT TESTING is generally a simple job. The hokus-pokus and trade-secrets connected with trouble-shooting are stripped of their mystery when a few basic principles are understood. Boiled down to simple facts, circuit testing used by the operating engineer involves only two checks: (1) measuring voltage drop and (2) tracing through the circuit. Electrical trouble-shooting is merely a simple application of these two tests.

Voltage Drop. Measuring voltage drop is nothing more than connecting a voltmeter across two points and taking a reading. This may be further simplified in some cases by touching the leads of a test lamp across two points. If the

lamp lights with normal brilliance, you have, in effect, determined that the voltage drop across the two leads is equal to the voltage rating of the bulb.

In checking 220-v circuits, connect two 110-v bulbs of equal wattage in series. When the lamps give normal light, the voltage drop across this test set is 220 volts.

Continuous Circuit. Testing through a circuit shows whether there is a complete electrical path of low resistance between two points. The hookup in *A*, Fig. 3, would show a complete circuit between points *a* and *b*, since there is a continuous path through the motor armature and its field. The circuit in *B* shows an open when tested between

points *a* and *b* because of the blown fuse. This is called an "open circuit." It is only one example of the many troubles that interrupt a circuit.

The terms "short" and "ground" sound a familiar note to even those whose only association with electricity has been removing a light bulb or replacing a fuse. These terms are almost household words and as such have been used rather loosely. As most operating engineers know, a "short" is entirely different from a "ground". The tests for checking an open, short or ground and the location of the trouble are applications of the circuit test.

Open Circuits. Opens are usually caused by blown fuses, open switches,

Power	Open	Insulation-resistance tester

Test for Open Circuit   Test for Short Circuit   Test for Grounded Circuit

Short   Power   To load   Ground

2 The insulation resistance tester reads infinity on a line that has an open circuit. It reads zero if the line is shortcircuited, and when connected between ground and the line shows zero if the line is grounded

Testing Made Easy

3 The bell and battery shows a complete circuit at A and open at B

loose terminal connections or defective control devices. Rarely the cause is an actual break in a wire. One of the simplest ways to check for a continuous circuit is with an insulation-resistance tester. This instrument is handy when no source of electricity is available for testing. It is also useful on high voltage equipment where trouble shooting on live circuits would be dangerous.

The insulation tester has its own source of power—a hand-cranked generator. It develops either 500 or 1000 volts on open circuit. Fig. 1 shows a test for grounds as performed on a dc generator. Remember that this instrument can give the operator an unpleasant shock, although it can do him no real harm. Readings are indicated in ohms and megohms (1,000,000 ohms). Connect the instrument as shown in Fig. 2 to the line to be tested. If the line is open it will read infinity, which is maximum resistance. A low resistance break often causes trouble which the instrument may fail to indicate as such because it is not a low-resistance reading meter. Values under 100,000 ohms cannot be read accurately. A bell testing set or lowreading voltmeter is most effective in locating breaks of this kind.

Short Circuits. To check a line for short-circuit conditions, connect the instrument between both lines as shown in Fig. 2. Continuity between the two lines is maintained by the short. It will read zero in such a case. When making this check it is necessary to disconnect the line from any equipment to which it is ordinarily connected. For example, a light bulb, a motor or a transformer connected across the line will falsely indicate short circuit conditions on the instrument.

Grounds. Fig. 2 also shows how to test for a grounded line. With one lead

connected to a good ground (such as a water pipe or motor frame) and the other lead connected to the line to be tested, continuity will be established through any ground that may exist in the line, so the instrument will read zero.

If in doubt as to whether the lead is connected to a good ground, connect the other lead to another point thought to be a good ground. If the instrument reads zero, either ground connection may be used for testing purposes.

When using the insulation testing instrument for testing as in Fig. 2, make certain that power is off the line, and that other equipment or connections to the circuit are disconnected. This second precaution eliminates the possibility of misleading effects from other parts of the circuit.

Testing With Lamps. A test light can be used to check for grounds. First, open the line switch and ground one lamp lead. Then check each of the two power wires. If one is grounded, the lamp will light when the ungrounded wire is touched. The line that does not light the lamp is the grounded circuit.

Where no ground exists on the power source, but one is suspected on the load side, the test lamp will reveal it. With the line switch open, ground one side of the power line temporarily. Then check across the open switch with the test lamp to see if the circuit is complete.

The three checks described above were all tests of continuity. There are many variations in testing depending on the means available. Voltmeters, ohmmeters, test-lamps, bell sets and insulation resistance testers all can be used. Of these the test lamp is the simplest and may also be used effectively in checking a circuit like the one in Fig. 3.

With the switch closed, the motor will not run because of the blown fuse between b and d. Most fuses for industrial application are not as obviously blown as the one in the diagram. Usually they appear much the same whether good or "shot". Sometimes you can tell which fuse is blown by feeling each and noting which one is warmer. Do not be misled by the high temperature of fuses carrying full load.

Fig 4 shows a more certain way. Close the switch and connect a test lamp across the fuses at one end, as at a and b; then connect the lamp across the other end, c and d. If the test lamp lights when connected across a-b but does not light across c-d, it identifies a and b as the supply side of the fuses and also proves that one or both fuses are blown.

If the lamp lights both times it proves the fuses are good. Connecting the test lamp across b and c tests the condition of fuse ac. If the lamp lights, as shown, the fuse is good; if it does not light the fuse is blown.

If the test lamp is connected across a and d, the condition of fuse cd may be tested in like manner. In Fig. 4 it is clear that the lamp can't light because of the blown fuse.

4 A positive method of testing for a blown fuse is shown above. Close the line switch and check the power source, then the back side of the fuses. Cross test each fuse separately to find the one that is burned open

Figuring Force, Work and Power

FORCE

① Force of gravity

10 lb · Weighs 40 lb

40 lb

② Friction

③ Jib crane

A

1 ton

WORK

④ Weight lifted

B

Raised 4 ft

40 lb A

⑤ Friction work

50 lb A   B

250 lb

Moves 4 ft

⑥ Lift plus friction

Moves 10 ft

50 lb 100 lb

Lifts 3½ ft

POWER

⑦ Elevator

Drum

800 ft per min

2,000 ft per min

1,200 lb

Counterweight 6,500 lb

8,000 lb

Car

⑧

Block and fall

2,000 lb

1,000 ft per min

► OPERATING ENGINEERS deal every day with force, work and power, but may not always be completely clear on the exact meaning of the terms. So here are some notes I hope will be helpful.

DEFINITIONS

Force is a push or pull. Engineers generally measure force in *pounds* or *tons*.

Work is done when a force acts on a moving body. The amount of work, in *foot pounds*, is the *product of the pounds push or pull by the number of feet the object moves in the direction of the force.*

Power is the rate of doing work, generally measured in foot pounds per minute, foot pounds per second, horsepower or kilowatts.

To get either of the first two measures of power divide the foot pounds of work by the minutes or seconds required to do the work.

One horsepower is the doing of work at the rate of 550 foot pounds per second, or 33,000 foot pounds per minute. One kilowatt is 1.33 horsepower.

Force Explained. In (1) the force of gravity acts straight down 40 lb on the weight and is balanced by the 40-lb

upward pull of the supporting cord. When a body is standing still, or moving in a straight line at constant speed, the forces on the body always balance.

In (2) the 10-lb force applied to push the block to the right is balanced by an equal friction force pushing on the block to the left. In this case the block is not lifted, so its 40 lb weight doesn't enter in.

Many forces act on the elements of a structure, such as the jib crane (3). At *A*, for example, the load pulls straight down one ton, the tie rod pulls more than one ton up left on a slant and the beam thrusts out horizontally to the right, also with a force over one ton. These forces balance at *A*.

Work Picture. A common case of work is lifting a weight straight up. When the 40-lb weight (4) is lifted four feet work done is 4 x 40 = 160 ft lb.

But note that work is not always *weight* times distance. In (5), for example, you don't consider the 250-lb weight, but only the 50-lb push and the 4-ft movement in the direction of the push. Work here is 50 x 4 = 200 ft lb.

Sliding a block up an inclined plane (6) combines lift work and friction work. Total work is 50 x 10 = 500

ft lb. The lifting fraction is 3½ x 100 = 350 ft lb.

Figuring Power. In (7) the net lift on the car is 8000 − 6500 = 1500 lb, so the power absorbed in lifting is 1500 x 800 = 1.200,000 ft lb per min or 1,200,000 ÷ 33,000 = 36.4 hp. Actually more motor power than this would be needed to allow for friction losses.

In (8) the rope pull would be 1000 lb if there were no friction. Here the actual work input is 1200 x 2000 = 2,400,000 ft lb per min and the *useful* work 2000 x 1000 = 2,000,000 ft lb per min.

We have figured power as force times distance divided by time. We can get the same result by changing the order of these operations to distance divided by time multiplied by force.

To put this another way, one horsepower is one pound acting at a speed of 550 ft per second or 33,000 ft per min.

Then we have these rules: *Horsepower* is pounds force times speed in feet per second divided by 550, or *horsepower* is pounds force times speed in feet per minute divided by 33,000.

So lifting a 400 lb weight 900 ft per minute requires 400 x 900 ÷ 33,000 = about 11 hp (plus power for friction).

METRIC SYSTEM

The metric system was legalized by the United States government in 1866, but progress in adopting it has been slow. Today, with imports of foreign machinery skyrocketing, the plant operator often finds himself repairing and replacing machine components made to the metric system. While any machinery handbook should have complete conversion tables, here we present a few that should be useful in an emergency.

Remember these three principal units: *meter*, the unit of length; *liter*, the unit of capacity; *gram*, the unit of weight.

Q What is ISO and DIN?

A ISO is the International Standard Organization composed of representatives of the various national standard institutes. Its purpose is to coordinate engineering standards, specifications and recommended procedures to ensure interchangeability of parts and to improve technical communication. As the United States makes the transition to the metric system, it is logical and therefore probable that the future engineering standards will closely follow ISO recommendations.

DIN (Deutsche Industrie Normen) are the West German Industrial Standard Specifications. Since they were the most complete and detailed of all the metric standard systems, other countries have been largely following them. ISO standards are, therefore, closely patterned after DIN standards. Since World War II, the DIN standards have been gradually modified, following ISO recommendations. Although differences between the systems still exist (like nominal diameters of some small threads), they are insignificant for all practical engineering purposes.

CONVERSION

FRACTIONAL INCHES — DECIMAL INCHES — MILLIMETERS

Fractional Inches	Decimal Inches	mm.	Fractional Inches	Decimal Inches	mm.
1/64	.015625	0.397	33/64	.515625	13.097
1/32	.03125	0.794	17/32	.53125	13.494
3/64	.046875	1.191	35/64	.546875	13.891
1/16	.0625	1.588	9/16	.5625	14.288
5/64	.078125	1.984	37/64	.578125	14.684
3/32	.09375	2.381	19/32	.59375	15.081
7/64	.109375	2.778	39/64	.609375	15.478
1/8	.125	3.175	5/8	.625	15.875
9/64	.140625	3.572	41/64	.640625	16.272
5/32	.15625	3.969	21/32	.65625	16.669
11/64	.171875	4.366	43/64	.671875	17.066
3/16	.1875	4.762	11/16	.6875	17.462
13/64	.203125	5.159	45/64	.703125	17.859
7/32	.21875	5.556	23/32	.71875	18.256
15/64	.234375	5.953	47/64	.734375	18.653
1/4	.25	6.350	3/4	.75	19.050
17/64	.265625	6.747	49/64	.765625	19.447
9/32	.28125	7.144	25/32	.78125	19.844
19/64	.296875	7.541	51/64	.796875	20.241
5/16	.3125	7.938	13/16	.8125	20.638
21/64	.328125	8.334	53/64	.828125	21.034
11/32	.34375	8.731	27/32	.84375	21.431
23/64	.359375	9.128	55/64	.859375	21.828
3/8	.375	9.525	7/8	.875	22.225
25/64	.390625	9.922	57/64	.890625	22.622
13/32	.40625	10.319	29/32	.90625	23.019
27/64	.421875	10.716	59/64	.921875	23.416
7/16	.4375	11.112	15/16	.9375	23.812
29/64	.453125	11.509	61/64	.953125	24.209
15/32	.46875	11.906	31/32	.96875	24.606
31/64	.484375	12.303	63/64	.984375	25.003
1/2	.5	12.700	1	1	25.400

DECIMAL EQUIVALENTS OF MILLIMETERS

mm.	Inches	mm.	Inches	mm.	Inches
..01	.00039	.20	.00787	10	.39370
.02	.00079	.30	.01181	20	.78740
.03	.00118	.40	.01575	30	1.18110
.04	.00157	.50	.01969	40	1.57480
.05	.00197	.60	.02362	50	1.96850
.06	.00236	.70	.02756	60	2.36220
.07	.00276	.80	.03150	70	2.75590
.08	.00315	.90	.03543	80	3.14960
.09	.00354	1.00	.03937	90	3.54330
.10	.00394			100	3.93700

$$
\begin{array}{llllllll}
1\ km & (kilometer) & = & 1000 & m & (meter) & = & 0.621\ miles \\
1\ m & (meter) & = & 10 & dcm & (decimeter) & = & 1.094\ yards \\
1\ dcm & (decimeter) & = & 10 & cm & (centimeter) & = & 3.937\ inches \\
1\ cm & (centimeter) & = & 10 & mm & (millimeter) & = & .394\ inches \\
1\ mm & (millimeter) & = & .039\ inches
\end{array}
$$

MOST COMMON DIN NUMBERS

DIN	DESCRIPTION	DIN	DESCRIPTION	DIN	DESCRIPTION
1	Taper Pin	916	Socket Set Screw Cup Point	1084	Spanner Nut (Lock Nut)
7	Dowel Pin (Unground)	931	Hex Head Bolt Part Thread (Coarse Thread)	6325	Precision Ground Dowel Pin
84	Cheese Head Machine Screw	933	Hex Head Screw Full Thread (Coarse Thread)	6912	Low Head Socket Head Cap Screw (Socket with guide)
85	Pan Head Machine Screw	934	Hex Nut		
125	Flat Washer	935	Castle Hex Nut	7980	Lock Washer (Smaller o.d. for Socket Head cap screw)
127	Lock Washer	936	Hexagon Jam Nut		
137	Spring Washer	939	Stud	7984	Low Head Socket Head Cap Screw
551	Slotted Set Screw	960	Hex Head Cap Screw Part Thread (Fine Thread)	7985	Phillips Head Machine Screw Fillister Head Machine Screw
912	Socket Head Cap Screw	961	Hex Head Bolt Full Thread (Fine Thread)	7986	Phillips Head Pan Head Machine Screw
913	Socket Set Screw Flat Point	963	Flat Head Machine Screw (Slotted)		
914	Socket Set Screw Cone Point	985	Stop Nuts with Nylon Lock Ring	7987	Phillips Head Flat Head Machine Screw
915	Socket Set Screw Dog Point	1481	Roll Pins (Spring Pin)	7991	Flat Head Socket Cap Screw

ISO METRIC - U.S. UNIFIED
SCREW THREAD
COMPARISON

Unified Scr Thd	Decimal Equiv.	Nominal Diameters ISO & Unified Converted mm.	Unified Scr Thd	Decimal Equiv.	Nominal Diameters ISO & Unified Converted mm.
–	.0394	1.	7/16	.4375	11.113
–	.0043	1.1	–	.4724	12.
–	.0047	1.2	1/2	.5	12.7
–	.0055	1.4	–	.5512	14.
#0	.060	1.524	9/16	.5625	14.288
–	.063	1.6	5/8	.625	15.875
–	.0709	1.8	–	.6299	16.
#1	.073	1.854	11/16	.6875	17.463
–	.078	2.	–	.7087	18.
#2	.086	2.184	3/4	.75	19.05
–	.0866	2.2	–	.7874	20.
–	.0984	2.5	13/16	.8125	20.638
#3	.099	2.515	–	.8662	22.
#4	.112	2.845	7/8	.875	22.225
–	.1181	3.	15/16	.9375	23.813
#5	.125	3.175	–	.9449	24.
–	.1378	3.5	1.	1.	25.4
#6	.138	3.505	1-1/16	1.0625	26.988
–	.1575	4.	–	1.063	27.
#8	.164	4.166	1-1/8	1.125	28.575
–	.177	4.5	–	1.811	30.
#10	.190	4.826	1-3/16	1.875	30.163
–	.1969	5.	1-1/4	1.25	31.75
#12	.216	5.486	–	1.2992	33.
–	.2362	6.	1-5/16	1.3125	33.338
1/4	.250	6.35	1-3/8	1.375	34.925
–	.2756	7.	–	1.4173	36.
5/16	.3125	7.938	1-7/16	1.4375	36.513
–	.315	8.	1-1/2	1.5	38.1
3/8	.375	9.525	–	1.5354	39.
–	.3937	10.			

ISO - METRIC
COARSE THREADS
DRILL SIZES

Nominal Diameters mm		Pitch mm/Thread	Recommended Drill Size mm
1st Choice	2nd Choice		
1		0.25	0.75
	1.1	0.25	0.85
1.2		0.25	0.95
	1.4	0.3	1.1
1.6		0.35	1.25
	1.8	0.35	1.45
2		0.4	1.6
	2.2	0.45	1.75
2.5		0.45	2.05
3		0.5	2.5
	3.5	0.6	2.9
4		0.7	3.3
	4.5	0.75	3.75
5		0.8	4.2
6		1	5
8		1.25	6.75
10		1.5	8.5
12		1.75	10.25
	14	2	12
16		2	14
	18	2.5	15.5
20		2.5	17.5
	22	2.5	19.5
24		3	21
	27	3	24 or 1-5/16"
30		3.5	26.5 or 1-3/64"
	33	3.5	29.5 or 1-5/32"
36		4	32 or 1-1/4"
	39	4	35 or 1-3/8"
42		4.5	37.5 or 1-31/32"
	45	4.5	40.5 or 1-19/32"
48		5	43 or 1-11/16"

STRENGTH PROPERTIES (Commercial Steel)

CLASS	New Designation	4.6	4.8	5.6	5.8	6.6	6.8	6.9	8.8	10.9	12.9	14.9
	Formerly	4 D	4 S	5 D	5 S	6 D	6 S	6 G	8 G	10 K	12 K	–
Brinell	min.	110		140		170			225	280	330	390
	max.	170		215		245			300	365	425	–
Rockwell · HRB	min.	63		78		88			–	–	–	–
	max.	88		97		102			–	–	–	–
Rockwell · HRC	min.	–		–		–			18	27	34	40
	max.								31	38	44	49
Yield Point psi		45,000		56,000		76,000		76,000	91,000	128,000	153,000	180,000
Tensile Strength psi	min.	56,000		70,000		85,000		85,338	113,784	142,230	170,000	200,000
	max.	78,000		100,000		113,000		99,561	128,000	170,676	200,000	230,000

inch 1 2 3 4 5 6

mm 1 2 3 4 5 cm 6 7 8 9 10 11 12 13 14 15

50 mm.   100 mm.

Index